Signals and Communication Technology

This series is devoted to fundamentals and applications of modern methods of signal processing and cutting-edge communication technologies. The main topics are information and signal theory, acoustical signal processing, image processing and multimedia systems, mobile and wireless communications, and computer and communication networks. Volumes in the series address researchers in academia and industrial R&D departments. The series is application-oriented. The level of presentation of each individual volume, however, depends on the subject and can range from practical to scientific.

Indexing: All books in "Signals and Communication Technology" are indexed by Scopus and zbMATH

For general information about this book series, comments or suggestions, please contact Mary James at mary.james@springer.com or Ramesh Nath Premnath at ramesh.premnath@springer.com.

Nishu Gupta • Jose David Diaz Roman
Editors

Decision Support Systems for Industrial and Scientific Applications

Springer

I encourage readers to engage with this work, confident it will enrich their understanding and motivate them to apply these concepts in their fields.

Intelligent Monitoring
Full Professor
Depto. de Electrónica y Circuitos
Universidad Simón Bolívar
Caracas, Venezuela

Gianfranco Passariello, Ph.D.

Preface

The main theme of the book is about advances and updates in intelligent systems to support decision making. The book about recent advances in intelligent systems for decision support explores how smart and advanced computer systems are helping people make better decisions. Imagine having a virtual assistant that not only gathers information but also analyzes it, identifies patterns, and provides helpful insights to guide decision-making. These intelligent systems use technologies like Artificial Intelligence, Machine Learning, and Deep Learning to understand and process data. The book discusses real-world applications where these systems have been used to improve decision support in various fields, such as healthcare, finance, or business. In essence, it's about how technology is becoming increasingly sophisticated in assisting us in making smarter choices by handling complex information and offering valuable recommendations.

Written by international experts, this book explores the latest progress in intelligent systems designed to assist decision-making, revealing how advanced computers contribute to informed judgments. These intelligent systems leverage cutting-edge technologies such as artificial intelligence, machine learning, and deep learning to analyze and interpret data. The book illustrates successful applications of these systems to enhance decision support in various industries, including business, finance, and healthcare. Essentially, the book focuses on the evolving role of technology in improving decision-making. As technology advances, it becomes increasingly adept at processing intricate data and offering insightful guidance to users. Key features of this book include an elaboration on recent and emerging developments in various specializations for addressing curing health problems and their solutions; smart transportation systems, traffic management for smart cities, energy management, deep learning and machine learning techniques for smart health and smart cities, and concepts that incorporate Internet of Everything (IoE, including IoT and ICT). This book attempts to cover some useful IoE applications and architectures that cater to improved sustainability requirements.

The book serves as both a textbook and a reference, making it suitable for all levels of technical education and research, as well as for beginners seeking an overview of the applications of Decision Support System technologies. These

About the Editors

Nishu Gupta is a Senior Member, IEEE, and presently working as a Research Scientist at the VTT Technical Research Centre of Finland Ltd, located in Oulu, Finland. Before this position, he was a Postdoctoral Fellow in the Smart Wireless Systems research group, Department of Electronic Systems, Faculty of Information Technology and Electrical Engineering at Norwegian University of Science and Technology (NTNU) in Gjøvik, Norway. He has worked as Research Associate in the Smart Living Lab, at the University of Fribourg, Switzerland. Prior to this position, he was a Postdoctoral Fellow in the Department of Information Technology, Faculty of Computing and Information Technology, King Abdulaziz University in Rabigh, Saudi Arabia. He is also a visiting researcher at the University of Oviedo, Spain, under the research group on Systems for Multimedia and the Internet of Things. He is serving as a Member, Zero Trust Architecture working group of MeitY-C-DAC-STQC project under "e-Governance Standards and Guidelines," Ministry of Electronics and Information Technology (MeitY), Government of India. He holds a Ph.D. in Electronics and Communication Engineering Department from MNNIT Allahabad, India, with specialization in Vehicular Adhoc Networks. His research interests include IoT; 5G/6G technologies; Dynamic Spectrum Sensing, Intelligent Transportation, Edge computing; etc.

Jose David Diaz Roman obtained his title as an electronic engineer from the Polytechnic Institute of the National Armed Forces in Maracay, Venezuela, in 1997. He furthered his academic pursuits and earned a Doctorate in Engineering from the Simón Bolívar University in Caracas, Venezuela.

From 1998 to 2016, José was an integral part of the Electromedicine-Biomedical Engineering Program at the Francisco de Miranda National Experimental University in Coro, Venezuela. Since 2016, he has served as a full-time professor in the Biomedical Engineering Program. Additionally, he is actively involved in the academic cores of the Master's programs in Electrical Engineering and the Doctorate in Advanced Engineering Sciences at the Department of Electrical and Computer Engineering of the Autonomous University of Ciudad Juárez (UACJ), Chihuahua, Mexico. He currently also serves as the coordinator of the Doctorate in Advanced Engineering Sciences program at UACJ.

José Díaz Román holds a membership in the National System of Researchers in Mexico and has a recognized publication record. He has contributed to prestigious journals and books, presenting papers and book chapters. His research interests encompass digital signal and image processing, the modeling of artificial intelligence systems (Machine Learning and Deep Learning) for classification, segmentation, and prediction tasks, as well as electronic and biomedical instrumentation, and wireless sensor networks. He has supervised several master's theses and undergraduate engineering projects, as well as mentoring doctoral candidates in these areas of interest.

Part I
Advanced Techniques in Decision Support Systems for Smart and Sustainable Applications

Technologies for Vehicle Classification in Intelligent Transportation Systems: Advances, Challenges, and Perspectives

Ernesto Sifuentes, Gabriel Bravo, Francisco Enríquez-Aguilera, and Abraham Mejía-Aguilar

Abbreviations

ADAS	Advanced Driver Assistance Systems
AI	Artificial Intelligence
ANN	Artificial Neural Networks
CBAM	Convolutional Block Attention Module
CNN	Convolutional Neural Networks
CPPM	Customized Pyramid Pooling Module
DAS	Distributed Acoustic Sensors
DCNN	Deep Convolutional Neural Networks
DL	Deep Learning
FDLSR	Fisher Discriminative Least Squares Regression
FHWA	Federal Highway Administration
FPN	Feature Pyramid Network
FWT	Fractional Wavelet Transform
GBM	Gradient Boosting Machines
GVWR	Gross Vehicle Weight Rating
IoT	Internet of Things
ITS	Intelligent Traffic Systems
KNN	k-Nearest Neighbors
LIDAR	Laser Infrared Detection and Ranging
LLMs	Large Language Models
LSTM	Long Short-Term Memory
MFCC	Mel Frequency Cepstral Coefficients

E. Sifuentes (✉) · G. Bravo (✉) · F. Enríquez-Aguilera
Institute of Engineering and Technology, Universidad Autónoma de Ciudad Juárez (UACJ), Ciudad Juárez, México
e-mail: esifuent@uacj.mx; gbravo@uacj.mx; fenrique@uacj.mx

A. Mejía-Aguilar
Eurac Research, Center for Sensing Solutions, Bolzano, Italy
e-mail: abraham.mejia@eurac.edu

N. Gupta, J. D. Diaz Roman (eds.), *Decision Support Systems for Industrial and Scientific Applications*, Signals and Communication Technology,
https://doi.org/10.1007/978-3-031-89700-9_1

ML	Machine Learning
MLP	Multilayer Perceptron
RCNN	Region-based Convolutional Neural Networks
RNN	Recurrent Neural Networks
ROIs	Regions of Interest
SVM	Support Vector Machines
UAV	Unmanned Aerial Vehicle
V2X	Vehicle-to-Everything
VANETs	Vehicular Ad Hoc Networks
WIM	Weigh in Motion
YOLO	You Only Look Once

1 Introduction

The growing number of vehicles often strains urban infrastructure, leading to traffic congestion and reduced road safety. To address this, the implementation of intelligent traffic systems (ITS) is essential. A well-designed ITS can enhance urban mobility, making it safer and more efficient [18]. ITS comprises sensors, communication networks, and data analytics that facilitate real-time data collection and informed traffic management (see Fig. 1). Vehicle classification plays a vital role in ITS, enabling the identification of vehicles by size, type, and behavior to improve real-time traffic monitoring. This is crucial for smart city development and Advanced Driver Assistance Systems (ADAS) [26]. Recent advancements in sensor and machine learning technologies offer opportunities to enhance vehicle classification systems [7, 11, 13, 29, 36]. However, selecting the most appropriate vehicle classification method can be complex due to varying technologies and criteria.

A vehicle classification system (see Fig. 2) includes several components, each with a specific role. It begins with sensors—like inductive loops, cameras, radar, magnetic devices, and weigh-in-motion (WIM)—that detect vehicle characteristics such as presence, size, speed, and weight. These sensors send data to a data acquisition system, which converts the outputs into digital format. The data is then preprocessed through noise reduction, normalization, and synchronization to ensure quality. Key features like the number of axles, vehicle size, and weight are isolated, and a classification algorithm categorizes the vehicles into classes such as light-duty, medium-duty, or heavy-duty. Classification algorithms include decision trees, support vector machines (SVM), neural networks, k-nearest neighbors (KNN), and deep learning methods. This structured approach helps with ITS traffic management, road safety, and real-time vehicle identification.

This chapter reviews recent technologies in vehicle classification within ITS, focusing on sensors, artificial intelligence (AI), machine learning (ML), and advanced processing algorithms. We examine their effectiveness, limitations, and applications based on literature from Gholamhosseinian and Seitz [7], Shokravi et al. [29], Won [36], Alghamdi et al. [1], Kadłubek et al. [11], and Khalil et al. [13].

Fig. 1 An ITS ecosystem includes sensors, communication networks, and advanced data analytics techniques, all working together to enable real-time data collection, informed decision-making, and improved traffic management. (Created with Microsoft Designer)

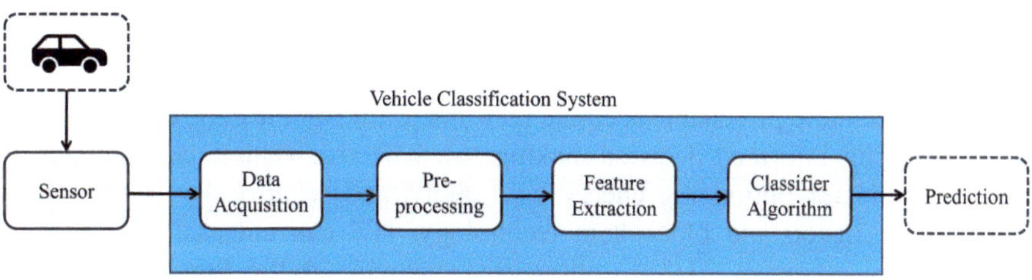

Fig. 2 Architecture of a vehicle classification system

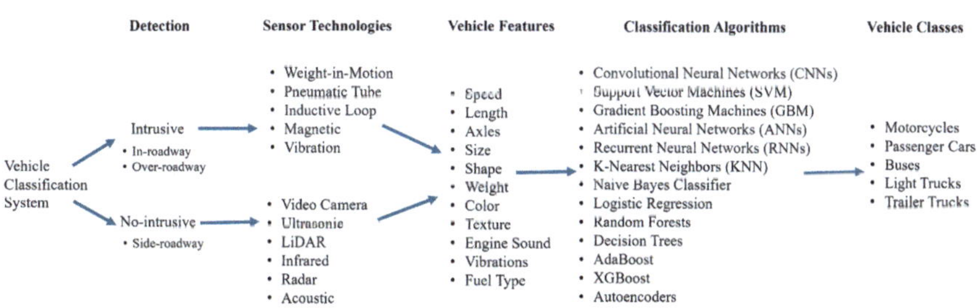

Fig. 3 Methods and cutting-edge technologies for classifying vehicles within ITS

Our search is based on the taxonomy shown in Fig. 3, which is a modified version derived from previous review papers by Gholamhosseinian and Seitz [7], Shokravi et al. [29], and Won [36]. The study categorizes vehicle classification into intrusive and non-intrusive methods, further subdividing these based on sensor technology.

We highlight innovative AI and ML techniques for feature extraction and discuss challenges like sensor fusion and real-time processing. The chapter concludes with future trends and the necessity for innovation to enhance transportation system efficiency and safety.

2 Vehicle Classification in ITS

In the 1980s, the Federal Highway Administration (FHWA) established a standardized vehicle classification system to improve traffic data collection, benefiting pavement designers and safety experts. This system categorizes vehicles by size, weight, speed, axle count, and intended use, ensuring compatibility with electronic sensors like road tubes and inductive loops [8].

Table 1 outlines the FHWA vehicle classification, which includes light-duty vehicles (motorcycles, passenger cars, and SUVs), medium-duty vehicles (buses and two- to three-axle trucks for local deliveries), and heavy-duty vehicles (four or more axles, including various types of trucks for freight transport). This classification helps with traffic surveillance, infrastructure development, and regulation compliance.

Recent advancements using deep learning, computer vision, and sensor-based systems have improved vehicle classification for traffic management [15, 28, 36, 38]. These technologies enable accurate real-time traffic monitoring, enhancing ITS and facilitating informed decision-making [6]. Effective vehicle classification in ITS is crucial for optimizing energy use, safety, and security. By incorporating multi-sensor fusion and blockchain technology, ITS can enhance classification accuracy and efficiency [31]. Furthermore, analyzing traffic flow in connected vehicle environments provides insights that improve traffic operations [16, 17].

3 Sensor Technologies for Vehicle Classification

Pneumatic tube detectors, introduced in 1920, are still used today for short-term vehicle data collection. Magnetic loop detectors measure vehicle length, while piezoelectric sensors assess weight and axle configuration in Weight-In-Motion (WIM) systems. Radar sensors classify vehicles by dimensions, and infrared sensors detect reflected light to match profiles against a database. Acoustic sensors identify vehicle types via unique sound signatures.

As noted by Shokravi et al. [29], Won [36], and Gholamhosseinian and Seitz [7], vehicle classification systems are mainly divided into two categories:

- *Intrusive detectors* (see Table 2) are sensors installed on or beneath the roadway that collect data from passing vehicles, such as vehicle length, axle count, and unique signal features. Types include pneumatic road tubes, inductive loops, and

Table 1 Vehicle classification according to Federal Highway Administration (FHWA)

Group	Class	Vehicle type	Axles	Samples
Light-duty	1. Motorcycles	Two- or three-wheeled motorized vehicles	2	Motorcycles, scooter
	2. Passenger cars	Personal use cars	2, 3 or 4	Sedans, coupes, station wagon
	3. Other two-axle, four-tire units	Light trucks and vans with gross vehicle weight rating (GVWR) up to 14,000	2, 3 or 4	Pickup trucks, minivans, SUVs
Medium-duty	4. Buses	Vehicles for more than 15 passengers, with two axles and six tires or more	2 or 3	School buses, airport shuttles
	5. Two-axle, six-tire trucks	Vehicles from 14,001 to 26,000 lbs. GVWR	2	Delivery trucks, U-haul trucks
	6. Three-axle trucks	Vehicles from 26,001 to 33,000 lbs. GVWR	3	Large box trucks, dump trucks
Heavy-duty	7. Four or more axle single-unit trucks	Trucks over 33,000 lbs. GVWR	4 or More	Heavy-duty trucks, large utility trucks
	8. Four or fewer axle single-trailer trucks	Single-trailer trucks	3 or 4	Small semi-trailers, flatbed trucks
	9. Five-axle single-trailer trucks	Standard large freight trucks with single trailer	5	Typical 18-wheelers, large semi-trailers
	10. Six or more axle single-trailer trucks	For very heavy loads	6 or More	Heavy-duty semi-trailers, large tankers
	11. Five or fewer axle multi-trailer trucks	Multi-trailer trucks	4 or 5	Short-haul multi-trailers, doubles
	12. Six-axle multi-trailer trucks	Multi-trailer combination trucks with six axles	6	Double trailers, bulk commodity transporters
	13. Seven or more axle multi-trailer trucks	Largest multi-trailer trucks	7 or More	Triple trailers, oversized load carriers

https://www.fhwa.dot.gov/publications/research/infrastructure/pavements/ltpp/13091/002.cfm

piezoelectric sensors [25]. While accurate, these systems have high installation and maintenance costs due to the need for saw cutting and lane closures. They are also susceptible to damage from traffic and weather, and limited data on their effectiveness is available.

- *Non-intrusive detectors* (see Table 3) include side-roadway and over-roadway sensors. Side-roadway sensors are positioned along the roadside and include magnetometers and accelerometers, along with advanced technologies like LIDAR and Wi-Fi. Installation is more straightforward and cheaper, but

Table 2 Intrusive sensor technologies for vehicle classification

Sensor	Detection	Advantages	Disadvantages	Use in ITS
Inductive loop	Embedded wire loop, detects inductance changes	Reliable for presence and axle counting	Disruptive installation and maintenance	Traffic flow monitoring, axle counting
Magnetic	Changes in earth's magnetic field	Reliable for presence and axle counting	Limited classification accuracy	Pre-screening, parking detection
Weight-in-motion	Deformations (strain gauges, piezoelectric)	Useful for weight data, road usage	High installation and maintenance costs	Weight enforcement, traffic management
Pneumatic Tube	Air-filled tubes	Reliable for presence and axle counting	High installation and maintenance costs	Traffic flow monitoring, axle counting
Piezoelectric	Pressure variations	Suitable for axle counting, primary classification	Requires precise installation, limited classification	Traffic flow monitoring, weight estimation
Capacitive	Changes in capacitance	Low-cost presence detection potential	Limited data for vehicle classification	Pre-screening, lane change detection
Vibration	Detects vibrations	Useful for structural health monitoring	Disruptive installation and maintenance	Bridge monitoring, traffic volume estimation

precise sensor alignment is essential. Challenges include classifying overlapping vehicles and mitigating noise interference.

Vehicle classification can utilize methods like vision systems and aerial image detection. However, video cameras are often costly and raise privacy concerns by capturing identifiable features such as faces and license plates. Their performance may also be affected by glare and rain. Solutions like weather-resistant cameras and adaptive algorithms have been proposed [3, 19, 22].

4 Developments in Vehicle Classification within ITS

4.1 Recent Review Studies

The field of vehicle classification has advanced significantly due to the integration of advanced algorithms and sensor technologies. Notably, computer vision techniques and deep learning algorithms enable real-time, accurate classification of vehicles

Table 3 Non-intrusive sensor technologies for vehicle classification

Sensor	Detection	Advantages	Disadvantages	Use in ITS
Camera	Visible light for image capture	Affordable, widely available, high precision	Affected by lighting, it requires image processing	Traffic monitoring, lane violation detection
Radar	Radio wave reflections	Detects presence, speed, direction; long range	Limited detailed classification, interference-prone	Traffic flow, collision avoidance
LiDAR	Laser pulses for 3D mapping	Highly accurate, works well in low light	Expensive, high processing requirements	High-precision classification, autonomous navigation
Infrared	Heat signature detection	Effective at night	Sensitive to fog/rain, limited range	Low-light traffic monitoring
Ultrasonic	High-frequency sound waves	Inexpensive, suitable for basic presence detection	Short-range interference from wind/noise	Traffic light optimization, parking detection
Acoustic	Sound wave analysis	Can estimate speed, primary classification	Prone to noise interference, complex processing	Traffic flow, speed estimation
Sensor fusion	Combines multiple sensors	Very high accuracy handles varied conditions	Requires complex algorithms, high computation power	Advanced traffic management, security

based on type, size, and other attributes. Convolutional neural networks (CNNs) have shown high accuracy in vehicle detection, enhancing ITS and improving traffic management. Moreover, intelligent sensors and IoT devices in vehicles facilitate effective data collection. These sensors gather information such as speed, acceleration, and dimensions, which machine learning algorithms use for accurate vehicle classification.

The following paragraphs provide a descriptive analysis of survey papers found in the literature:

- Won [36] reviewed traffic monitoring systems, focusing on vehicle classification techniques and their challenges. The study categorizes systems into in-roadway-based, side-roadway-based, and over-roadway-based types, based on sensor types. Key challenges include high costs for in-road systems, sensor placement issues, privacy concerns with cameras, and difficulties in classifying overlapping vehicles. The potential of MEMS, machine learning, wireless communication, and V2X technology to address these problems is emphasized. The study stresses the need for standardization, the development of self-learning systems, and hybrid sensor systems to enhance classification accuracy, which is crucial for traffic monitoring and transportation planning.
- Shokravi et al. [29] explored how Vehicular Ad Hoc Networks (VANETs) can address the shortcomings of current vehicle classification methods by providing

real-time global vehicle mobility data. They highlighted challenges like the local focus of existing methods, lack of physical vehicle data, and reliance on unreliable GPS. The study calls for cross-disciplinary collaboration and further research on collaborative systems using vehicular sensors and VANET technology to enhance real-time vehicle detection and classification.

- Gholamhosseinian and Seitz [7] explored various vehicle classification methods, including intrusive sensors (loop detectors, magnetometers), non-intrusive sensors (infrared, accelerometers), off-road sensors (satellite, UAVs), and hybrid methods (weigh-in-motion, VANETs). They identified challenges such as the absence of a common evaluation dataset, limited performance metrics, sensitivity to environmental factors, vehicle occlusion, and communication reliability in VANETs. The study suggests focusing on hybrid solutions that combine multiple sensors and utilize VANET technology for real-time global vehicle classification while addressing secure communication and integration issues. Current methods lack the ability to provide real-time mobility and physical information, but VANETs present a promising opportunity.
- Tippannavar et al. [34] provided an overview of real-time vehicle identification and traffic management techniques, highlighting challenges such as low-cost camera inefficiencies in low-light conditions, a 9% error rate in vehicle size distinction, and signal mixing in ultrasonic methods. The study underscores the use of image processing, machine learning, IoT, cloud systems, and distributed acoustic sensors (DAS) while emphasizing the need for real-world testing and the exploration of night vision cameras. It points out the importance of automatic vehicle counting for urban traffic signals, the effectiveness of the YOLOv5 algorithm for vehicle recognition, and the potential of DAS for overground mobility recognition. The paper concludes that image processing is more cost-effective and widely applicable in real-time environments, offering improved safety, precision, and speed compared to other approaches.
- Berwo et al. [2] provided a comprehensive overview of deep-learning techniques for vehicle detection and classification, focusing on methods, datasets, loss and activation functions, and optimization algorithms. The paper emphasizes deep convolutional neural networks (CNNs) and categorizes approaches for object detection while addressing challenges like weather conditions, large datasets, viewpoint variations, and occlusion. It discusses various techniques, including DCNNs, RCNNs, and DNNs. It suggests future research directions such as multi-task joint optimization, handling scale variations, and the development of end-to-end optimized architectures. Insights emphasize balancing lightweight model speed with accuracy to advance vehicle detection and classification frameworks.
- Alghamdi et al. [1] conducted a study on a vehicle classification system for intelligent transport systems (ITS) using deep learning techniques, particularly CNN. The process includes data collection, preprocessing, training/testing phases, and synthetic image generation to enhance model diversity. Critical challenges addressed were image noise, data quality, intra-class variability, computational constraints, and model generalization. Future directions involve improved data

collection, hardware optimization, model refinement, real-world deployment, and interdisciplinary collaboration. Integration of sensors like LiDAR, infrared, and ultrasonic is suggested for better performance. The system achieved a validation accuracy of 90.85%, demonstrating the effectiveness of deep learning. Conclusions highlight the importance of data ownership, performance evaluation, and the system's potential to enhance vehicle classification, thus improving transportation efficiency and security.
- Khalil et al. [13] conducted a survey on deep learning (DL) techniques in ITS, focusing on applications like traffic flow prediction, vehicle detection, road condition monitoring, traffic sign recognition, and autonomous vehicles. They also explored the potential of large language models (LLMs) in this field. The study identified challenges such as data quality, the need for labeled data, DL model interpretability, real-time performance, data privacy, and financial considerations. The researchers emphasized the need to address data scarcity, enhance interpretability, improve real-time performance, ensure privacy, and advance hybrid models and explainable AI. Key findings indicated that DL can significantly improve transportation efficiency, safety, and mobility through precise predictions. However, challenges such as robustness and interpretability remain. The researchers noted that emerging technologies like LLMs could further revolutionize ITS, alongside existing technologies like CNNs, RNNs, and Federated Learning, using sensors such as wireless networks, GPS, and IoT for real-time data acquisition.

Table 4 summarizes the progress, challenges, and future perspectives for vehicle classification in ITS based on previous literature reviews.

Emerging technologies like MEMS, machine learning, wireless communication, VANETs, and V2X technology are crucial for vehicle tracking and classification. However, challenges remain, including diverse vehicle types, classification ambiguities, varying environmental conditions, occlusions, high costs, and complex sensor placement. To address these challenges, there is a need for robust vehicle tracking systems, mobile real-time solutions, and self-learning systems. Various technologies are being used, including sensors, control systems, and computer vision. Deep learning models like CNNs, RNNs, and federated learning are gaining popularity, and sensors such as LiDAR, infrared, and ultrasonic technologies are being employed. Integrating these advanced technologies holds promise for improving automatic vehicle classification systems.

4.2 Recent Research Studies

Recent research in the field of ITS has made significant advancements in vehicle detection and classification using deep learning algorithms. Several key papers introduce innovative methods to address challenges related to accuracy, real-time processing, and varying environmental conditions. Utilizing CNN, YOLOv3,

Table 4 Progress, challenges, and future perspectives in vehicle classification within ITS

Source	Advancements	Challenges	Future perspectives
Won [36]	MEMS, machine learning, wireless communication, V2X technology	High costs, sensor placement, privacy concerns, complex overlaps	Universal standards, self-learning systems, hybrid sensors
Shokravi et al. [29]	VANETs, intelligent vehicles, OBUs, RSUs, V2V, V2I communication	Local terrain, limited data, unreliable GPS	Collaborative systems using vehicle sensors and VANET
Gholamhosseinian and Seitz [7]	Intrusive and non-intrusive sensors, hybrid methods (WIM, VANETs)	Lack of datasets, performance metrics, environmental sensitivity, occlusion	Hybrid VANET solutions for real-time classification
Tippannavar and SD [34]	Image processing, machine learning, IoT, cloud systems, DAS	Low-light redundancy, vehicle size error rates, signal mixing	Testing in real-world conditions with night vision cameras
Berwo et al. [2]	Deep learning (DCNNs, RCNNs, DNNs)	Weather impact, large datasets, viewpoint occlusion	Multi-task optimization, scale variation handling, end-to-end architectures
Alghamdi et al. [1]	Deep learning (CNNs), sensors (LiDAR, infrared, ultrasonic)	Image noise, limited data, intra-class variability, resources	Enhanced data, hardware optimization, interdisciplinary collaboration
Khalil et al. [13]	CNNs, RNNs, federated learning, capsule networks, LLMs, sensors	Data quality, labeling needs, real-time performance, privacy	Data scarcity solutions, explainable AI, weather resilience, multi-modal integration

YOLOv4_AF, and hybrid models, these studies focus on vehicle classification, license plate detection, and traffic monitoring. These advancements enhance intelligent transportation systems and expand applications in object detection, tracking, and traffic statistics. The following paragraphs analyze the recent findings:

- Alghamdi et al. [1] developed a vehicle classification system using CNN, achieving 90% accuracy even in low-quality image environments. They emphasized the need for data ownership protection through watermarking and suggested future improvements by training on higher-resolution datasets, increasing sample sizes, and addressing hardware and software limitations.
- Rajput et al. [24] presented an automatic vehicle identification and classification model using the YOLOv3 algorithm for toll management systems, achieving 94.1% accuracy. The model demonstrated strong precision and recall for vehicle identification at toll plazas. Future work includes identifying additional vehicle classes and developing vehicle counting capabilities, utilizing AI techniques such as deep learning and custom datasets.

- Zhao et al. [39] introduced YOLOv4_AF, an optimized version of the YOLOv4 model aimed at enhancing object detection and classification, especially for vehicles. The model achieved accuracies of 83.45% on the BIT-Vehicle dataset and 77.08% on the UA-DETRAC dataset, with corresponding F1 scores of 0.816 and 0.808. Key improvements include the integration of a Convolutional Block Attention Module (CBAM) for better attention mechanisms and modifications to the Feature Pyramid Network (FPN) for improved multi-scale feature extraction. While YOLOv4_AF shows potential for applications like object tracking and traffic statistics, its main limitation is increased computational complexity due to the CBAM module. Future research may explore its capabilities beyond vehicle detection.
- Srividhya et al. [32] examined an intelligent transportation system that uses video surveillance to classify vehicles and detect license plates. The system employs critical algorithms, such as the Inner-Outer Outline Profile (IOOPL) for shadow removal and vehicle detection, and the delta learning rule for vehicle classification and log generation. The paper aims to enhance traffic monitoring, reduce congestion, and improve traffic security. Future developments will focus on night surveillance, occlusion handling, 3D modeling, and vehicle tracking. Key features include the application of video surveillance, machine learning algorithms for vehicle classification, and the use of real-time video feeds as data sources.
- Shokri et al. [30] assessed the performance of YOLO, RCNN, and SSD deep learning algorithms for vehicle detection and classification in highway camera images. They found that YOLOv7 outperformed the others, achieving a vehicle detection accuracy of 98.77% and a classification accuracy of 97.37%. The YOLO models were praised for their high accuracy and near real-time capabilities on a CPU, making them suitable for traffic flow monitoring in intelligent transportation systems. The researchers recommended that future studies focus on evaluating deep learning architectures using UAV (Unmanned Aerial Vehicle) videos to enhance accuracy and explore precise vehicle localization and tracking. The study utilized data from highway cameras in Quebec, Canada, along with additional datasets from YouTube and Kaggle.
- Hayee et al. [9] introduced a hybrid CNN model that employs Fisher Discriminative Least Squares Regression (FDLSR) for view-independent car make and model classification, achieving an impressive accuracy of 94.62. The study details the fine-tuning of deep CNN models, feature extraction, and FDLSR application to enhance feature quality for classifier training. Focusing on datasets like Stanford Cars and BoxCars21k, it also presents a new dataset of 129,000 images across 94 vehicle classes on Pakistani roads. Future experiments on this dataset and the implementation of incremental learning for feature extraction are also planned, showcasing significant potential for vehicle classification advancements.
- Kamencay et al. [12] introduced a hybrid convolutional neural network architecture that improves vehicle classification accuracy from 90% to 97% by combining data from CCTV and fiber Bragg grating sensors. This architecture

features separate branches for sensor and image data. The study also discusses smart city applications, such as automatic vehicle counting, dynamic road usage optimization, and speed. The CNN extracts both low-level features, like edges and colors, and higher-level features, such as shapes and objects.

- Rafique et al. [23] proposed a framework for ITS utilizing aerial imagery, which includes semantic segmentation, vehicle detection, classification, and tracking through advanced deep learning techniques. The study employs CNN, customized pyramid pooling, Kalman filter, and kernelized filter-based tracking, achieving accuracies of 95.78% on VAID, 95.18% on VEDAI, and 93.13% on DLR3K datasets. Key methods involve CNN-based semantic segmentation and customized pyramid pooling for vehicle detection, along with Kalman filter and kernelized correlation filter for tracking. The authors also recommend future improvements in vehicle tracking using an end-to-end deep learning approach for enhanced traffic monitoring and surveillance.
- Maiga et al. [20] developed a CNN-based model for vehicle classification using low-quality images from standard security cameras in poor lighting and weather conditions. The model achieved 95.8% accuracy, making it suitable for IoT-based ITS applications with limited computational resources. It can classify various vehicles, including bicycles, cars, trucks, minibuses, pickups, and juggernauts. The study also plans to create a vehicle detection algorithm that extracts vehicle images from low-resolution video frames captured by affordable surveillance cameras. The dataset consisted of 4800 low-quality images taken under various low-light and weather conditions.
- Chen and Hu [5] introduced a vehicle detection and classification method utilizing static appearance and motion features, achieving 98.92% accuracy. The multi-detector framework involves training on small sets of RGB and binary images, manually labeling ground truth bounding boxes and vehicle classes, eliminating false detections, and merging overlapping detections for final results. The vehicle categories assessed include complete and incomplete small and large vehicles. The method shows strong performance in accuracy, recall, precision, F1-score, and Kappa. Future work aims to optimize the approach by incorporating vehicle relationships across frames and expanding the training data to cover smaller vehicles and rear views.
- Wu et al. [37] implemented a vehicle classification and counting system that utilizes YOLO object detection technology to analyze surveillance camera footage of different road types. The system uses the YOLOv3 model and assesses accuracy through precision rate, recall rate, and F-measure. It categorizes vehicles into motorcycles, small cars, and heavy vehicles, aiming for future application in ITS. The paper also delves into the use of acoustic sensors, magnetometers, accelerometers, and other technologies, as well as the dataset comprising video footage from surveillance cameras and training images of vehicles.
- Butt et al. [3] proposed a vehicle classification system based on CNN. The system was developed using a self-constructed dataset and then fine-tuned on the VeRi dataset to ensure its applicability across different regions. The system utilizes a

pre-trained ResNet-152 model and fine-tunes it to enhance vehicle classification performance. It achieved an accuracy of 99.68% on the self-constructed dataset and 97.66% on the VeRi dataset. The paper aims to extend its work by developing a more detailed vehicle classification system to enhance its effectiveness in ITS. This will involve incorporating both manually crafted features and deep features learned automatically by CNN.

- Kim [14] introduced a method for detecting vehicles and classifying their types and colors to enhance autonomous driving safety. This deep learning-based approach achieved a 93.2% accuracy by using YOLOv4 for detection and ResNet-50 CNN for classification. The study notes limitations in low-light conditions and suggests future research on infrared and thermal imaging sensors. It emphasizes the crucial role of accurate vehicle identification for safe autonomous driving and recommends exploring advanced sensor technologies.
- Hashima et al. [10] presented a novel deep learning method for vehicle classification using seismic data. Their approach combines the Fractional Wavelet Transform (FWT) for feature extraction and LSTM networks for classification, achieving an impressive 98% accuracy. Utilizing a comprehensive dataset from Kyushu University that included buses, cars, motorcycles, and seismic noise, their study demonstrates the method's effectiveness and its potential to enhance ITS and inform road infrastructure planning and traffic management.
- Moallemi et al. [21] presented a method for classifying vehicles as light, heavy, or super-heavy using vibration data from MEMS accelerometers. The approach involves preprocessing raw data, extracting features, labeling them with WiM data, and employing unsupervised machine learning algorithms for clustering. The Mean shift algorithm achieved the highest classification accuracy of 96.87%. The study used data collected over 5 days, with 4 days for training and one for validation, focusing on low traffic volumes during nighttime. Key features included Maximum Amplitude, Standard Deviation, Mean, and Line Length. Overall, the unsupervised classification method yielded an accuracy of 96.87% for vehicles passing over a bridge.
- Sarcevic et al. [27] developed a real-time vehicle classification system using a single magnetometer that classifies vehicles into nine classes with up to 74.67% accuracy. The system employs an algorithm for vehicle detection, feature extraction, and classification through a multilayer perceptron neural network, utilizing over 30,000 samples acquired from a magnetometer sensor and corresponding camera images collected over several months. Key features for classification include detection length, value locations, number of range changes, local maxima and minima, mean values, zero crossings, and root mean square. Future directions include enhancing recognition efficiency, adding features for better truck and bus differentiation, employing feature selection algorithms, and expanding data collection to improve geographic diversity.
- Chen et al. [4] presented a method for detecting and classifying road vehicles with a roadside magnetic sensor. They extracted features from magnetic signals using Mel Frequency Cepstral Coefficients (MFCC) and classified vehicles into four types (sedan, van, truck, bus) using a 3D Vector Quantization (VQ)

algorithm. While the paper does not provide specific accuracy results, it discusses the features, classification method, and future directions, including the use of multiple sensors for monitoring multiple lanes and examining the effects of sensor-vehicle distance.

- Alghamdi et al. [1] created a hybrid model that combines deep learning and genetic algorithms to classify eight vehicle categories, achieving an accuracy of 99.7%. The model utilizes a pre-trained CNN (VGG16) for feature extraction, a Genetic Algorithm for feature optimization, and various SVM kernels for classification. Using the Stanford car dataset with 8000 images.

Table 5 summarizes key information from each paper, detailing proposed algorithms, data types, accuracy, classification features, vehicle classes, and future perspectives.

5 Advances, Limitations, and Perspectives in Vehicle Classification

Figure 4 summarizes advancements, challenges, and perspectives in vehicle detection and classification systems. Advancements include vision-based methods, GPS, WIM, hybrid systems, machine learning, VANETs, and cloud computing. Challenges involve environmental dependence, limited information, infrastructure costs, data privacy, computational needs, and vehicle penetration rates. Perspectives emphasize machine learning, sensor fusion, data standardization, privacy-preserving techniques, and emerging technologies like V2X communication, edge computing, and intelligent sensors/IoT.

Advancements

- Vision-based: Uses computer vision and high-res cameras for accurate classification but is affected by lighting and weather, with high installation costs.
- GPS-based: Provides precise location data but limited information on vehicle types and privacy concerns.
- Weight-in-Motion (WIM): Accurately measures vehicle weight, ideal for heavy-duty classification, and enforcement, though expensive to implement.
- Machine learning: Automates classification, with neural networks improving accuracy through pattern recognition.
- Deep learning (CNNs): Extracts features from images and videos, achieving high classification accuracy across vehicle types.
- VANETs: Uses vehicle-to-infrastructure communication to enhance real-time and collaborative classification.
- Cloud computing and big data: Processes large datasets for real-time classification and decision-making in dynamic traffic conditions.

Table 5 A summary of recent developments in vehicle classification within ITS

Source	Algorithm	Data	Accuracy	Classification	Perspectives
Chen and Hu [5]	Multi-detector	Various video datasets	0.9892	Small/large vehicles	Improve algorithms, add diverse vehicle types
Wu et al. [37]	YOLOv3	Surveillance footage	PR, RR, F-measure	Motorcycles, cars, trucks	Enhance stability and accuracy for smart transport
Butt et al. [3]	CNN	Self-constructed, VeRi datasets	99.68% 97.66%	Car, bus, Van, truck	Develop detailed classification
Rajput et al. [24]	YOLOv3	Custom vehicle images	94.1%	Various vehicle types	Expand class identification, improve vehicle counting
Zha et al. [39]	YOLO v4_AF	BIT-vehicle, UA-DETRAC datasets	83.45% (BIT), 77.08% (UA)	Various vehicle types	Improve tracking, explore other object types
Srividhya et al. ([32])	IOOPL, Delta	Real-time videos	92%	Various vehicle tasks	Improve night detection, 3D modeling
Tas et al. [33]	CNN	Low-res images from Konya, Turkey	92.9%	Bike, Car, Truck, etc.	Research standard security cameras for ITS
Alghamdi et al. [1]	CNN	Google vehicle images	90%	Various types	Improve model accuracy for low-resolution data
Shokri et al. [30]	YOLO, RCNN, SSD	Highway, YouTube, Kaggle images	98.77% (YOLOv7)	Car, Truck, Bus	Evaluate on UAV videos, improve classification
Alghamdi et al. [1]	Hybrid CNN, FDLSR	Stanford, BoxCars21k, Pakistani cars	94.62%	Car make and model	Incremental learning, feature extraction
Kamencay et al. [12]	Hybrid-CNN	CCTV, FBG sensor data	90–97%	Tire width, construction type	Combine image/sensor data for better classification
Rafique et al. [23]	CNN, pooling	Aerial imagery datasets	95.78%	Various vehicle types	End-to-end learning for traffic monitoring
Maiga et al. [20]	CNN	Low-quality images	95.8%	Bike, Car, Truck, etc.	Develop low-res video frame detection, expand dataset
Vikram et al. [35]	Deep learning	Remote sensing images	99.44%	Car, Bus, Truck	Generate vehicle density maps, license plate recognition
Kim [14]	Deep learning	Road images	93.2%	Sedan, Bus, Truck, SUV	YOLO for low-light, infrared/thermal imaging research
Hashima et al. [10]	FWT, LSTM	Kyushu University data	98%	Buses, Cars, Motorcycles	Utilize geophones
Moallemi et al. [21]	Unsupervised ML	Sensor nodes, weight-in-motion	96.87%	Light, Heavy, Super-heavy	Expand to other vehicle types
Sarcevic et al. [27]	MLP	Magnetometer and camera images	74.67%	9 vehicle classes	Improve recognition, add features
Chen et al. [4]	MFCC, VQ	Roadside magnetic sensor data	Not mentioned	Sedan, Van, Truck, Bus	Monitor multiple lanes, study sensor distance effects
Alghamdi et al. [1]	VGG16, GA	Stanford car dataset	99.78%	8 vehicle categories	Increase dataset, explore new pathways

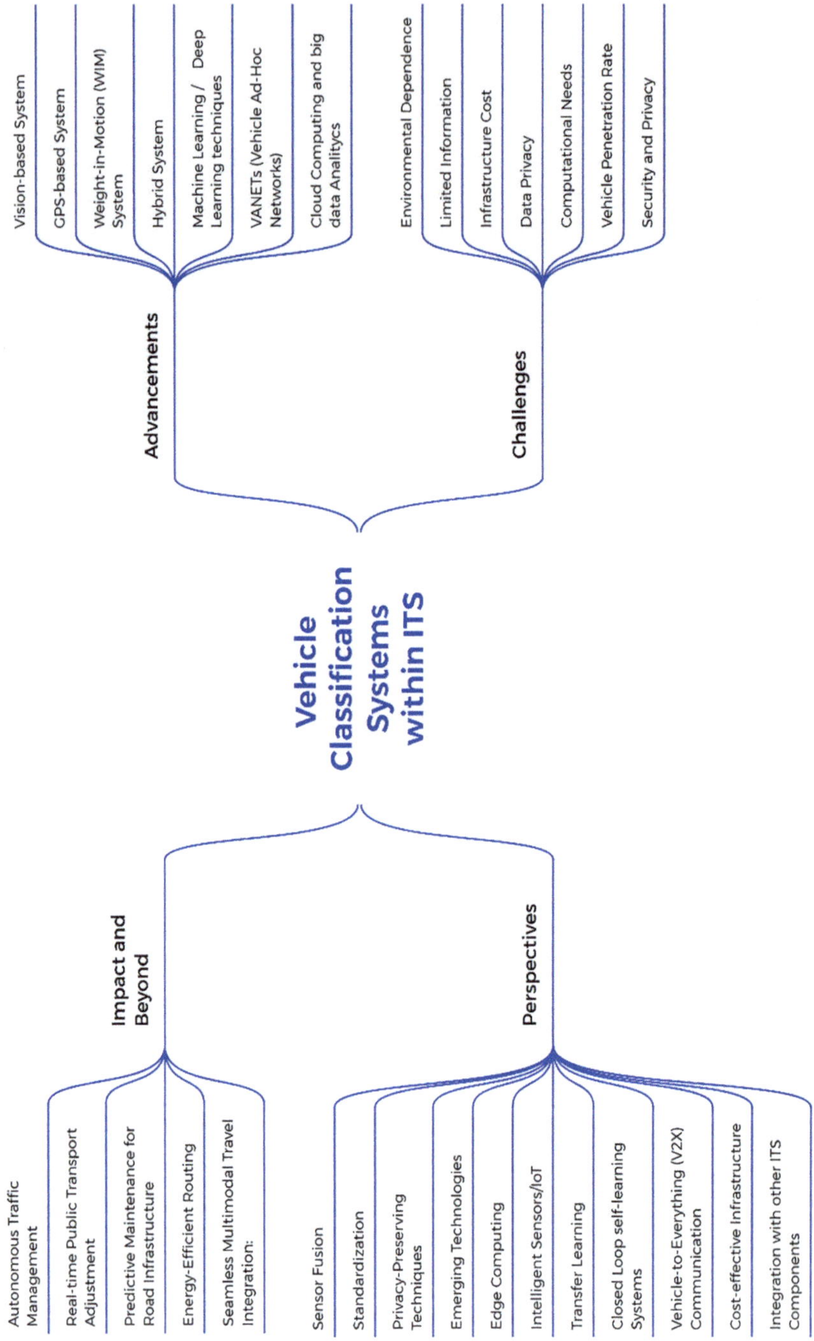

Fig. 4 Vehicle classification in ITS: challenges, limitations, and perspectives

Challenges

- Environmental dependence: Vision-based systems are affected by poor lighting and adverse weather.
- Limited information: GPS provides only location data, limiting vehicle details.
- Infrastructure cost: Intrusive systems are costly to install and maintain.
- Data privacy: GPS and camera systems raise privacy issues.
- Computational needs: Machine learning requires large datasets and high processing power.
- Vehicle adoption: VANETs need widespread use to be effective.
- Security and privacy: Robust security is needed to protect VANET data.

Perspectives

- Privacy-preserving techniques: Ensure vehicle data security with anonymization and differential privacy methods.
- Self-learning systems: Develop autonomous systems to enhance classification accuracy, reducing manual input.
- Standardization: Create standardized metrics for evaluating classification systems (accuracy, cost-effectiveness).
- V2X Communication: Use V2X for vehicle-to-infrastructure data sharing, improving real-time traffic management.
- Machine learning: Increase classification accuracy with adaptable algorithms, even under poor visibility.
- Transfer learning: Address data scarcity by transferring knowledge across datasets.
- Cost-effective infrastructure: Innovate to reduce the costs of vision-based systems for broader use.
- ITS integration: Link classification with traffic signals and routing to optimize traffic flow.
- Sensor fusion: Combine data from cameras, radar, LiDAR, and GPS to enhance classification in varied conditions.
- Edge computing: Perform data processing near the source to reduce latency and boost real-time response.

6 Conclusions

Vehicle classification is essential for ITS, as it enhances traffic management, safety, and overall transportation efficiency. By recognizing different vehicle types, systems can customize functions such as traffic rule application, axle counting for weigh-in-motion (WIM), and emissions estimation, while improving road safety through an understanding of vehicle behaviors. Advanced technologies, including machine learning, computer vision, and sensors, have made significant strides in detecting and classifying vehicles on the road.

Diverse sensor technologies are used to classify vehicles, including inductive loops, magnetic, and WIM. These sensors capture specific details like the number of axles, vehicle size, axle spacing, weight, speed, and length. Inductive loops and magnetic sensors detect a vehicle's presence and size by measuring changes in magnetic fields, while WIM sensors measure the weight of the vehicle as it moves over a sensor-equipped roadway. By analyzing patterns such as axle configurations, vehicle dimensions, and load distributions, classification algorithms use these collected details to accurately determine vehicle types.

Recent studies on vehicle classification systems using deep learning, particularly with CNN and YOLO algorithms, have shown high accuracy in detection and classification. These advancements suggest promising applications in ITS, toll management, and traffic monitoring. AI techniques are overcoming challenges like low image quality and real-time processing needs. Future work is directed toward improving model accuracy with better datasets, addressing hardware limitations, and classifying more vehicle types, highlighting the potential of deep learning to enhance traffic monitoring and reduce congestion.

Recent research on vehicle classification systems using deep learning has shown significant advancements. Algorithms like YOLOv3 and YOLOv4_AF demonstrate high accuracy in vehicle detection, with promising applications in ITS, toll management, and traffic monitoring. These studies reveal the potential of AI techniques to overcome challenges such as limited image quality and real-time processing needs. A recurring theme is the emphasis on future work to enhance model accuracy, utilize higher-resolution datasets, classify additional vehicle types, and broaden applications.

Reviews highlight the importance of developing cost-effective, privacy-conscious systems amidst challenges like high computational demands and the need for VANETs. Future directions point to advancements in machine learning, sensor fusion, and privacy-preserving methods to improve system robustness and public acceptance. Ultimately, the future of ITS relies on integrating diverse approaches for comprehensive traffic management solutions.

References

1. Alghamdi, A. S., Imran, T., Mursi, K. T., Ejaz, A., Kamran, M., & Alamri, A. (2023). A vehicle classification system for intelligent transport system using machine learning in constrained environment. *International Journal of Advanced Computer Science and Applications, 14*(7). https://doi.org/10.14569/IJACSA.2023.0140746
2. Berwo, M. A., Khan, A., Fang, Y., Fahim, H., Javaid, S., Mahmood, J., et al. (2023). Deep learning techniques for vehicle detection and classification from images/videos: A survey. *Sensors, 23*(10), 4832. https://doi.org/10.3390/s23104832
3. Butt, M. A., Khattak, A. M., Shafique, S., Hayat, B., Abid, S., Kim, K. I., & Adnan, A. (2021). Convolutional neural network based vehicle classification in adverse illuminous conditions for intelligent transportation systems. *Complexity, 2021*(1), 6644861. https://doi.org/10.1155/2021/6644861

4. Chen, X., Kong, X., Xu, M., Sandrasegaran, K., & Zheng, J. (2019). Road vehicle detection and classification using magnetic field measurement. *IEEE Access, 7*, 52622–52633. https://doi.org/10.1109/access.2019.2908006
5. Chen, Y., & Hu, W. (2021). A video-based method with strong-robustness for vehicle detection and classification based on static appearance features and motion features. *IEEE Access, 9*, 13083–13098. https://doi.org/10.1109/access.2021.3051659
6. Ennehar, B. (2023). Master-slave convolutional deep architecture for vehicle identification and type classification. *Traitement Du Signal, 40*(2), 619–627. https://doi.org/10.18280/ts.400220
7. Gholamhosseinian, A., & Seitz, J. (2021). Vehicle classification in intelligent transport systems: An overview, methods and software perspective. *IEEE Open Journal of Intelligent Transportation Systems, 2*, 173–188. https://doi.org/10.1109/OJITS.2021.3096756
8. Gupte, S., Masoud, O., Martin, R., & Papanikolopoulos, N. (2002). Detection and classification of vehicles. *IEEE Transactions on Intelligent Transportation Systems, 3*(1), 37–47. https://doi.org/10.1109/6979.994794
9. Hayee, S., Hussain, F., & Yousaf, M. H. (2023). A novel FDLSR-based technique for view-independent vehicle make and model recognition. *Sensors, 23*(18), 7920. https://doi.org/10.3390/s23187920
10. Hashima, S., Saad, M. H., Hatano, K., & Rizk, H. (2023). Vehicle classification in intelligent transportation systems using deep learning and seismic data. In *2023 IEEE international conference on intelligence and security informatics (ISI)* (pp. 1–6). IEEE. https://doi.org/10.1109/ISI58743.2023.10297252
11. Kadłubek, M., Thalassinos, E., Domagała, J., Grabowska, S., & Saniuk, S. (2022). Intelligent transportation system applications and logistics resources for logistics customer service in road freight transport enterprises. *Energies, 15*, 4668. https://doi.org/10.3390/en15134668
12. Kamencay, P., Markovic, M., Dubovan, J., Dado, M., & Benedikovic, D. (2023). Neural-network-optimized vehicle classification using clustered image and fiber-sensor datasets. *IEEE Access, 11*, 41315–41324.
13. Khalil, R. A., Safelnasr, Z., Yemane, N., Kedir, M., Shafiqurrahman, A., & Saeed, N. (2024). Advanced learning technologies for intelligent transportation systems: prospects and challenges. *IEEE Open Journal of Vehicular Technology, 5*, 397–427. https://doi.org/10.1109/OJVT.2024.3369691
14. Kim, J. (2024). Deep learning-based vehicle type and color classification to support safe autonomous driving. *Applied Sciences, 14*(4), 1600. https://doi.org/10.3390/app14041600
15. Lin, C., Jeng, S., & Lioa, H. (2021). A real-time vehicle counting, speed estimation, and classification system based on virtual detection zone and yolo. *Mathematical Problems in Engineering, 2021*, 1–10. https://doi.org/10.1155/2021/1577614
16. Liu, Y. (2024). Research on the characteristics of mixed traffic flow in an intelligent connected environment. In *International conference on smart transportation and city engineering (STCE 2023)* (Vol. 13018). https://doi.org/10.1117/12.3024823
17. Luo, Y., Luo, Z., & Liang, Z. (2023). Research on vehicle recognition based on deep learning. In *International conference on computer graphics, artificial intelligence, and data processing (ICCAID 2022)* (Vol. 126043H). https://doi.org/10.1117/12.2674558
18. Lv, Z., & Shang, W. (2023). Impacts of intelligent transportation systems on energy conservation and emission reduction of transport systems: A comprehensive review. *Green Technologies and Sustainability, 1*(1), 100002. https://doi.org/10.1016/j.grets.2022.100002
19. Ma, S., & Yang, J. J. (2023). Image-based vehicle classification by synergizing features from supervised and self-supervised learning paradigms. *Eng, 4*, 444–456. https://doi.org/10.3390/eng4010027
20. Maiga, B., Dalveren, Y., Kara, A., & Derawi, M. (2023). Convolutional neural network-based vehicle classification in low-quality imaging conditions for internet of things devices. *Sustainability, 15*(23), 16292. https://doi.org/10.3390/su152316292
21. Moallemi, A. A., Zanatta, L., Burrello, A., Salvaro, M., Longo, M., Darò, P., et al. (2023). Unsupervised vehicle classification using a structural health monitoring system. In *Structural health monitoring 2023*. https://doi.org/10.12783/shm2023/36839

22. Neupane, B., Horanont, T., & Aryal, J. (2022). Real-time vehicle classification and tracking using a transfer learning-improved deep learning network. *Sensors, 22*(10), 3813. https://doi.org/10.3390/s22103813
23. Rafique, A. A., Al-Rasheed, A., Ksibi, A., Ayadi, M., Jalal, A., Alnowaiser, K., et al. (2023). Smart traffic monitoring through pyramid pooling vehicle detection and filter-based tracking on aerial images. *IEEE Access, 11*, 2993–3007. https://doi.org/10.1109/ACCESS.2023.3234281
24. Rajput, S. K., Patni, J. C., Alshamrani, S. S., Chaudhari, V., Dumka, A., Singh, R., et al. (2022). Automatic vehicle identification and classification model using the YOLOv3 algorithm for a toll management system. *Sustainability, 14*(15), 9163. https://doi.org/10.3390/su14159163
25. Rajab, S., Kalaa, M., & Refai, H. (2016). Classification and speed estimation of vehicles via tire detection using single-element piezoelectric sensor. *Journal of Advanced Transportation, 50*(7), 1366–1385. https://doi.org/10.1002/atr.1406
26. Saharan, S., Bawa, S., & Kumar, N. (2020). Dynamic pricing techniques for intelligent transportation system in smart cities: A systematic review. *Computer Communications, 150*, 603–625. https://doi.org/10.1016/j.comcom.2019.12.003
27. Sarcevic, P., Pletl, S., & Odry, A. (2022). Real-time vehicle classification system using a single magnetometer. *Sensors, 22*(23), 9299. https://doi.org/10.3390/s22239299
28. Sharma, D., & Jaffery, Z. (2021). Categorical vehicle classification and tracking using deep neural networks. *International Journal of Advanced Computer Science and Applications, 12*(9). https://doi.org/10.14569/ijacsa.2021.0120964
29. Shokravi, H., Shokravi, H., Bakhary, N., Heidarrezaei, M., Rahimian Koloor, S. S., & Petrů, M. (2020). A review on vehicle classification and potential use of smart vehicle-assisted techniques. *Sensors, 20*(11), 3274. https://doi.org/10.3390/s20113274
30. Shokri, D., Larouche, C., & Homayouni, S. (2023). A comparative analysis of multi-label deep learning classifiers for real-time vehicle detection to support intelligent transportation systems. *Smart Cities, 6*(5), 2982–3004. https://doi.org/10.3390/smartcities6050134
31. Song, K., Pan, X., Hu, Z., Jia, Y., & Huang, J. (2023). Motion control of connected and automated vehicles in a lane-changing scenario with guaranteed tracking performance. *International Journal of Robust and Nonlinear Control, 33*(6), 3399–3419. https://doi.org/10.1002/rnc.6567
32. Srividhya, S. R., Kavitha, C., Lai, W. C., Mani, V., & Khalaf, O. I. (2022). A machine learning algorithm to automate vehicle classification and license plate detection. *Wireless Communications and Mobile Computing, 2022*(1), 9273233. https://doi.org/10.1155/2022/9273233
33. Tas, S., Sari, O., Dalveren, Y., Pazar, S., Kara, A., & Derawi, M. (2022). Deep learning-based vehicle classification for low-quality images. *Sensors, 22*(13), 4740. https://doi.org/10.3390/s22134740
34. Tippannavar, S., & SD, Y. (2023). Real-time vehicle identification for improving the traffic management system a review. *Journal of Trends in Computer Science and Smart Technology, 5*(3), 323–342. https://doi.org/10.36548/jtcsst.2023.3.007
35. Vikram, A., Akshya, J., Ahmad, S., Rubini, L. J., Kadry, S., & Kim, J. (2024). Deep learning based vehicle detection and counting system for intelligent transportation. *Computer Systems Science & Engineering, 48*(1). https://doi.org/10.32604/csse.2023.037928
36. Won, M. (2020). Intelligent traffic monitoring systems for vehicle classification: A survey. *IEEE Access, 8*, 73340–73358. https://doi.org/10.1109/access.2020.2987634
37. Wu, J. D., Chen, B. Y., Shyr, W. J., & Shih, F. Y. (2021). Vehicle classification and counting system using YOLO object detection technology. *Traitement du Signal, 38*(4), 1087–1093. https://doi.org/10.18280/ts.380419
38. Yang, B., & Lei, Y. (2015). Vehicle detection and classification for low-speed congested traffic with anisotropic magnetoresistive sensor. *IEEE Sensors Journal, 15*(2), 1132–1138. https://doi.org/10.1109/jsen.2014.2359014
39. Zhao, J., Hao, S., Dai, C., Zhang, H., Zhao, L., Ji, Z., & Ganchev, I. (2022). Improved vision-based vehicle detection and classification by optimized YOLOv4. *IEEE Access, 10*, 8590–8603. https://ieeexplore.ieee.org/document/9681804

IoT Fusion: Empowering Decision-Making with Integrated Intelligence

M. Bhuvaneswari, Bharani M., M. Suresh Kumar, and Ahmad Hoirul Basori

1 Introduction

The convergence of IoT, advanced analytics, and AI is revolutionizing decision-making across sectors. This fusion enables businesses to use enormous data sources to improve insights, operational efficiency, and strategic planning. IoT fusion is the process of merging IoT devices, advanced data analytics, and AI to create a unified, intelligent decision-making ecosystem. IoT fusion collects huge amounts of real-time data from several sources by connecting sensors, devices, and systems. When combined with AI and machine learning algorithms, this data is examined and processed to create actionable insights that help decision-makers in a number of industries.

M. Bhuvaneswari (✉)
Department of Mechatronics Engineering, Sri Krishna College of Engineering and Technology, Coimbatore, Tamilnadu, India
e-mail: bhvaneswarim@skcet.ac.in

Bharani M.
Department of AI & ML, Panimalar Engineering College, Chennai, Tamilnadu, India
e-mail: bharani.aiml@panimalar.ac.in

M. S. Kumar
Department of Computer Science and Engineering, Sri Venkateswara College of Engineering, Sriperumbudur, Tamilnadu, India
e-mail: sureshm@svce.ac.in

A. H. Basori
Faculty Of Computing and Information Technology Rabigh, King Abdulaziz University, Rabigh, Saudi Arabia
e-mail: Abasori@Kau.Edu.Sa

N. Gupta, J. D. Diaz Roman (eds.), *Decision Support Systems for Industrial and Scientific Applications*, Signals and Communication Technology,
https://doi.org/10.1007/978-3-031-89700-9_2

IoT fusion improves operational efficiency in areas such as household appliances, healthcare, smart cities, vehicle monitoring, and manufacturing by enabling real-time data monitoring and predictive analytics in real-time systems. For example, in healthcare, wearable IoT devices monitor patient vitals while AI analyzes trends to forecast health hazards. Similarly, in smart cities, IoT sensors control energy consumption, traffic patterns, and garbage to improve urban living through educated judgments.

The combination of IoT with AI assures not only connectivity, but also intelligence, allowing systems to adapt autonomously to changing contexts, predict outcomes, and offer solutions. This connectivity enables organizations and governments to make data-driven choices more quickly and accurately, promoting innovation and overall performance while building a more connected and smarter society.

The comprehensive survey on IoT fusion techniques highlights their applications across various domains [1]. This synthesis draws on multiple studies to elucidate key aspects of IoT fusion. The study on the use of AI-powered IoT fusion in managing urban infrastructure, with an emphasis on public safety and traffic optimization [2]. In this paper [3], the data privacy issues in IoT fusion are addressed, and new cryptographic algorithms are proposed to guarantee secure data transmission. This article [4] investigates scalability issues in IoT systems and suggests solutions to enable interoperability among diverse devices. Scalability and interoperability of IoT fusion systems are critical for improving communication efficiency and enabling a diverse set of applications across domains. Recent research has identified fresh ways to these challenges, with a focus on dynamic protocol translation, distributed architectures, and semantic interoperability. The proposed chapter is organized as follows. Section 2 depicts the various literature and its survey.

2 Literature Review

The integration of quantum computing into the IoT poses both enormous opportunities and concerns, particularly in terms of security and efficiency [5]. Quantum technology can transform IoT by enabling sophisticated encryption methods and boosting data processing capabilities. World Economic Forum paper delves into the integration of AI and quantum computing in IoT systems, with an emphasis on global connectivity and decision-making [6].

The standard defines a reference architecture for IoT systems and specifies best practices for integrating IoT devices and systems across industries [7]. The ethical challenges of IoT fusion, particularly data ownership, permission, and societal impact, are highlighted [8]. The performance of edge computing in IoT fusion systems, particularly for real-time applications in industrial environments, is illustrated in [9]. A detailed look at how AI-powered IoT fusion is transforming healthcare, including remote monitoring, diagnosis, and individualized treatment [10]. The blockchain-based framework for secure data fusion in IoT systems addresses issues about data integrity and privacy [11]. The survey addresses the obstacles of attaining interoperability in diverse IoT environments, as well as various standardization attempts and solutions [12]. This research investigates the use of IoT fusion in precision agriculture, focusing on data-driven approaches to optimizing resource management and increasing crop yields [13].

An in-depth examination of of how IoT fusion is being used to smart manufacturing settings to enhance decision-making and operational effectiveness [14]. The paper examines the most recent advances in AI algorithms intended expressly for IoT fusion, covering both the opportunities and problems of applying these technologies [15]. This paper examines the most important developments in IoT and AI integration, with a focus on how these trends will affect the future of IoT fusion and decision-making [16]. The standard outlines best practices for managing and fusing data in IoT systems, with a focus on assuring data quality, security, and privacy [17]. A case study investigates the use of IoT fusion in autonomous vehicles, with a focus on real-time data integration from several sensors to improve decision-making and safety [18]. The technical report from Microsoft Research examines the integration of AI, edge computing, and cloud services in IoT systems, emphasizing how these technologies work together to improve IoT fusion [19]. The paper discusses the ethical difficulties of AI in IoT fusion, concentrating on bias, fairness, and openness in decision-making processes [20]. Presenting general trends of IoT review publications in the various IoT fields [21]. The findings of these IoT bibliometric statistics will be useful to identify the evolving concerns of IoT, and illustrate what research remains to be conducted in this field in the future, by exploring the existing structures and the citation trends of IoT.

Enhanced Decision-Making by Data Fusion is described [22]. It generates large datasets, which, when combined with modern analytics, dramatically increase decision-making accuracy. For example, in smart manufacturing and healthcare, novel data integration strategies have resulted in greater efficiency and actionable insights. Intelligent Systems and AI Integration shows the combination of AI and IoT creates intelligent systems capable of adaptive learning and anomaly detection, transforming traditional data handling into proactive decision-making processes [23]. A case study in agricultural IoT demonstrates how fusion intelligence can optimize pollination, showcasing the practical benefits of this integration [24]. Real-World Applications in energy distribution, a novel IoT architecture utilizing data fusion and intelligent decision-making, has improved system reliability and efficiency, supporting intelligent energy management [25]. While the combination of IoT and AI gives substantial benefits, it may also create issues such as data privacy maintenance and the necessity for solid infrastructure remains an essential consideration for future advances.

The results imply that data fusion can greatly improve data quality. The research suggests that timely data resampling is necessary for companies to improve their prediction-making processes, while also addressing potential problems such as faulty data and inconsistencies [26]. Table 1 illustrates the literature review insights on the various approaches, problems, applications, and the researchers' future directions.

Table 1 Literature review in integration of IoT and AI

Insight	Methods	Challenges	Applications	Future direction
Dynamic IoT-based enterprise management framework (IEMF) intelligent decision-making (IDM) The framework uses fusion optimization techniques to enhance decision-making in various domains [27]	Fusion optimization techniques Various methods and models such as ML and fuzzy approaches	Data analysis accuracy and interpretation for decision-making Fusion optimization techniques for enhanced decision-making in IoT-based management	Diverse data sources Self-service data resources	Fusion optimization techniques in diverse industries Enhancing decision-making with advanced data visualization techniques
Framework for secure decision-making in power IoT using federated learning Incorporates data security, transfer learning, and expert fusion for accuracy [28]	Federated learning with homomorphic encryption Data augmentation, transfer learning, and model fusion techniques	Data security and leakage concerns Insufficient local training data addressed	Image recognition using VGG-19 and ResNet-50 models Text processing with random removal and disruption methods	Design new ML models for power IoT data characteristics Research on data types for advanced security in power IoT
Integration of ML techniques has enabled self-sufficient IoT networks Combining IoT and DL technologies improves app performance and security [29]	Improved DRL models for intelligent video surveillance applications Fusion of multiple models to optimize decision-making processes	Enhancing IoT apps with deep learning technologies Improving security, performance, and accuracy by over 97.24%	Robotics, cloud environments Multimedia applications, E-systems	Fusion system design optimization for decision-making processes Improving data fusion in multimedia applications and E-systems
IoT-enabled model for fresh agri-product supply chain decision-making Integrates planting, storage, distribution stages for revenue maximization [30]	Mixed-integer programming model for predicting the real-time decision-making in IoT-enabled supply chains process Integration of planting, storage, and distribution stages for optimal decisions	Challenges include integrating storage, distribution decisions and planting, effectively Balancing day-to-day market factors strategy and agricultural characteristics for optimal decision-making	Integrative decision made on model for IoT-enabled organic Agri-product supply chains Optimal decisions for planting, storage, and distribution to maximize revenue	Explore IoT-enabled decision-making in manufacturing Investigate integrative models for other agricultural supply chains

Data fusion techniques for smart IoT system with decision-making intelligence Approaches based on probability, AI, evidence theory for efficient data aggregation [31]	Probability, artificial intelligence, evidence theory Techniques for cognitive decision-making under uncertainty and limited resources	Challenges include interferences between devices and harsh operation conditions. Challenges also involve extracting useful data from multiple sources	Data fusion techniques for smart IoT system Intelligent decision-making applications in smart IoT environment	Enhancing data fusion techniques for improved decision-making in IoT. Addressing challenges to optimize data aggregation in smart IoT systems
IoT devices and data fusion techniques are popular Multi-sensor fusion and signal-image encoding proposed [32]	Multi-sensor approach	Complex data fusion from heterogeneous IoT sensors Intelligent identification and detection using deep learning models	IoT wearable devices	Deep learning and anomaly detection models for hand gesture data Intelligent identification and detection operations using IoT wearable devices
IoT enhances marketing strategies for comprehensive customer-centric decisions IoT offers opportunities for product development and enhancements in business [33]	Data collection and analysis, innovative service development, integration of IoT into business processes, a customer-centric approach, and the establishment of feedback mechanisms	Marketers or investors must gain knowledge to think outside the box for effective relevant services Implementing IoT for decisions-making in business, development in product, and service enhancements	Marketing and business decisions	IoT implementation in product development Enhancing business decisions through IoT usage

The researchers explore a wide range of topics under IoT Fusion, including technical issues and solutions, ethical implications, and future trends. They are intended to lay a solid foundation for any research paper or study on IoT Fusion and its applications.

3 The Advancement in Internet of Things (IoT)

The IoT is a system that connects equipment and people to the internet, allowing them to interact in real time. The IoT has grown throughout time, and its applications have expanded into many domains, such as transportation, medical, and smart cities. The theory of the Internet of Things is not new, but it has evolved dramatically over time due to advances in connectivity, sensors, computation, and cloud technologies. Figure 1 demonstrates the implementation of IoT in several sectors.

IoT has evolved since the 1980s, when early systems such as RFID tags were employed to track shipments. Mark Weiser coined the term "ubiquitous computing" in 1991, referring to the seamless integration of technology into everyday life. Kevin Ashton created the term IoT in 1999, picturing a society in which computers automatically track and manage physical objects via networked systems. In the 2000s, IoT grew in popularity as more items became internet-linked, and breakthroughs in wireless technology and sensor downsizing paved the path for smarter, more connected gadgets. By the 2010s, IoT had grown high with the emergence great giants like cloud computing and big data, which provided the infrastructure to ease the massive amounts of data-driven service created by IoT devices. In the 2020s, the introduction of 5G and edge computing pushed IoT into new realms, allowing for real-time data processing and speedier device connectivity. This progress continues to fuel innovation in domains such as self-driving cars, healthcare, and smart infrastructure, establishing IoT as a critical enabler of the digital transformation of industries and everyday life.

The rapid rise of the IoT has been propelled by a number of technological advancements, all of which contribute to improved connectivity, data processing, and automation. Key technological advancements driving IoT development are as follows:

- Sensor Miniaturization: The development of smaller, more energy-efficient sensors has enabled them to be embedded in ordinary things, enabling for seamless data collection from wide sources, including appliances, and industrial equipment.
- Wireless communication: The rise in wireless technologies such as the terms like Wi-Fi, Bluetooth, and cellular networks has dramatically reduced the cost of connecting devices across vast distances. The development of 5G has boosted the Internet of Things by providing faster, more reliable connections, which are critical for real-time service applications such as autonomous vehicles and smart cities.

Fig. 1 IoT applications

- Cloud Computing: The upgradation of cloud services has created scalable platforms for storing and predicting the massive volumes of data driven by IoT devices. Cloud infrastructure enables real-time service-driven analytics and machine learning concepts to be applied to IoT data, transforming raw data into useful insights.
- Edge Computing: In addition to cloud computing, edge computing processes closer view to the source node (the network's "edge"), lowering latency and increasing the performance of time-sensitive application service in industrial automation and healthcare.
- AI and Machine Learning: Integrating AI with IoT provides more intelligent decision-making and automation. Machine learning models examine trends in IoT data to help with predictive maintenance, autonomous systems, and intelligent home gadgets.

IoT combination refers to the integration of data and functionality from numerous IoT systems, creating a unified platform that enhances the overall capability of connected devices. This concept combines diverse IoT systems, such as sensors, networks, and software platforms, into a cohesive framework. The goal is to hold the collective power of these systems for enhancing decision-making, automation, and innovation across industries. The importance of IoT Fusion lies in its ability to overcome the limitations of isolated IoT systems. Individual IoT networks often operate in silos, collecting and processing data relevant to specific functions. By merging these systems, IoT Fusion enables the cross-utilization of data from various sources, offering a more comprehensive view of processes and environments. This, in turn, supports more intelligent, data-driven decisions and fosters innovation in areas in smart cities, healthcare, and industrial automation.

4 Challenges in IoT Data Integration

Integrating multiple IoT systems is crucial for achieving advanced decision-making capabilities. When data from disparate sources—like in sensors monitoring air quality, energy consumption, and traffic patterns—is combined, a more accurate and holistic concern of the environment is obtained. This comprehensive data pool improves more informed decisions, such as optimizing resource allocation, predicting equipment failures, or enhancing public safety.

The IoT Fusion facilitates real-time data analysis and predictive modeling. It allows for more responsive systems capable of adapting to changing conditions, thereby increasing efficiency, reducing operational costs, and improving user experiences.

5 IoT Fusion: A Holistic Approach

IoT Fusion is a holistic and linked method to maximizing the potential of the IoT. It entails the seamless combination of IoT devices, data analytics, AI, and communication technologies to build an intelligent ecosystem that functions in tandem. Unlike traditional IoT systems, which function in silos, IoT Fusion stresses the holistic integration of diverse elements sensors, networks, data, and intelligence resulting in better decision-making, automation, and system adaptability.

This strategy ensures that data acquired from various devices and sources is not just aggregated, but also processed and analyzed to deliver actionable insights. For example, in the sector of smart manufacturing, IoT Fusion allows for real-time monitoring of machinery via sensors, while AI-driven analytics forecast possible breakdowns, optimize maintenance schedules, and increase production efficiency. This integrated system decreases downtime, lowers expenses, and increases productivity.

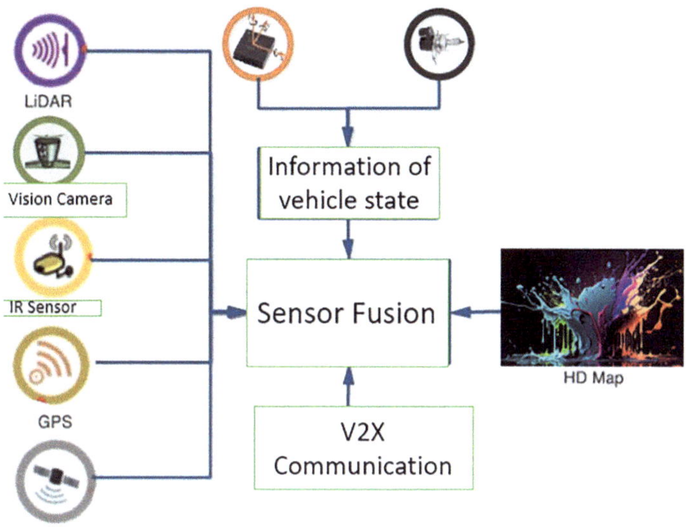

Fig. 2 Sensor fusion in vehicle communication

In smart cities, IoT Fusion integrates data from diverse subsystems like in traffic management, energy usage, and safety for environment. The combination of data from these several sectors provides an elongate view of the urban environment, allowing municipal planners and administrators to make more predictive decisions further, hence increasing overall efficiency and sustainability.

IoT Fusion also allows organizations to improve client experiences by offering more personalized offerings. Retailers, for example, might use real-time customer data analysis to adapt recommendations and streamline processes depending on purchase habits.

IoT Fusion takes a holistic approach, moving beyond individual applications to enable collaboration between technologies and systems, resulting in smarter surroundings, better decision-making, and disruptive innovations across industries. Sensor blending is the act of combining data from several sensors to get more precise, dependable, and complete results than could be achieved from individual sensors alone. Sensor fusion uses data from a more number of sources, including accelerometers, gyroscopes, magnetometers, cameras, and radar, to capitalize on each sensor type's strengths while correcting for their respective deficiencies. Figure 2 shows the Sensor Fusion in vehicle communication.

Sensor fusion is the application of algorithms and data processing techniques to combine data streams in real time, resulting in a unified, coherent depiction of the environment or system being monitored. In autonomous vehicles, for example, sensor blending combines data from lidar, radar, cameras, and GPS to create a thorough awareness of the vehicle's environment, hence increasing navigation and safety. Figure 3 shows the benefits of sensor fusion.

Sensor fusion is widely utilized in industries such as robotics, aircraft, automotive, and healthcare, where complex surroundings and dynamic conditions

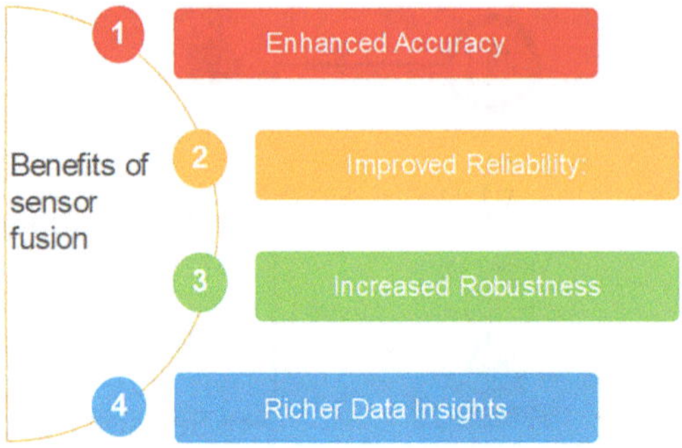

Fig. 3 Benefits of sensor fusion

necessitate the integration of several data sources for optimal performance and safety.

6 Leveraging Unified Data Streams

In the field of IoT Fusion, harnessing unified data sources is critical for empowering decision-making through integrated intelligence. This method entails combining data from several IoT devices and sensors into a single, coherent data stream that gives a complete picture of the system or environment. Unified data streams enable the real-time gathering and predicting data from various sources, including temperature sensors, motion detectors, and video cameras. By combining different data sources, companies can acquire a comprehensive picture of complex systems, resulting in more informed and timely decisions. For example, in smart cities, combining data from traffic sensors, weather stations, and transportation systems can enhance traffic flow, increase safety, and improve overall urban administration.

The power of unified data streams stems from their capacity to support advanced data analysis and machine learning techniques. These systems can handle and evaluate enormous amounts of integrated data, revealing patterns and trends that are not apparent from isolated data sources. This allows for predictive maintenance, effective resource management, and individualized user experiences.

Furthermore, utilizing unified data streams in IoT Fusion improves decision-making by offering a richer, more accurate database, resulting in smarter, more responsive systems and better operational outcomes across various sectors.

7 Enabling Predictive Analytics and Insights

Enabling predictive analytics and insights is a transformative capability within modern data-driven environments. Predictive analytics holds historical data, statistical algorithms, and machine learning techniques to forecast future patterns, behaviors, and outcomes. By analyzing past and present data, predictive models identify patterns and correlations, enabling organizations to hold future events and make predictive decisions.

For example, in manufacturing, predictive analytics can forecast equipment failures by analyzing data from sensors monitoring machine performance. This allows for preemptive maintenance, reducing downtime and operational costs. In healthcare, predictive data models can analyze patient data to identify potential health risks, enabling early interventions and personalized treatment service.

The procedure entails combining several data sources, using advanced algorithms, and constantly updating models with fresh data to improve predictions. This dynamic method ensures that findings are both relevant and actionable. Predictive analytics enables firms to change from reactive to proactive tactics, optimizing operations, improving customer experiences, and reducing risks. Businesses and organizations can use predictive insights to improve decision-making, streamline operations, and achieve strategic goals more confidently and efficiently.

8 Improving Operational Efficiency

IoT Fusion improves operational efficiency by combining many data sources and using advanced analytics to streamline procedures and maximize performance. In an IoT Fusion architecture, several sensors and devices collected from wide variety of data from a variety of sources, including production lines, supply chains, and facilities management systems. This data is then consolidated and evaluated to provide a complete picture of activities.

On integrating intelligence, IoT Fusion enables several key improvements in operational efficiency as follows:

- Real-Time Monitoring: Continuous data gathering and processing enables real-time monitoring of processes. In manufacturing, for instance, sensors can observe equipment performance, detect anomalies, and provide fast feedback, resulting in quicker modifications and reduced downtime.
- Predictive Maintenance: Advancement in analytics can predict the faults in equipment and maintenance requirements before they occur. This proactive strategy reduces disruptions and increases the lifespan of machines.
- Resource Optimization: IoT Fusion provides efficient resource management by evaluating consumption patterns and optimizing the use of energy, materials, and labor. This leads to huge cost savings and less waste.
- Enhanced Decision-Making: Integrated data streams provide a full perspective of operations, permitting managers to make more informed decisions based on comprehensive insights rather than fragmented data.

Overall, IoT Fusion enables enterprises to improve in efficiency in operations by facilitating smarter, decision-making, lowering costs, and increasing productivity. IoT technology helps to improve operational efficiency across multiple industries

[34]. It covers how IoT Fusion integrates data from various sources to enhance real-time monitoring, predictive support, resource enhancement, and decision-making processes. The paper focuses on case studies and applications, emphasizing the benefits and limitations of deploying IoT Fusion for operational efficiency.

9 The Future of IoT Fusion: Unlocking Endless Possibilities

Advancements in IoT Fusion will transform industries, as discussed [35]. Key advancements include improved data integration for real-time insights, sophisticated analytics and AI for more precise forecasts, edge computing for lower latency, and tighter security measures. It also emphasizes the growth of IoT Fusion into new industries like healthcare and smart cities, which promise smarter systems and good decision-making.

The future of IoT Fusion promises to open up disruptive possibilities, owing to ongoing technological and data analytics advancements. As IoT Fusion advances, it will further empower decision-making through integrated intelligence, providing unparalleled capabilities across several industries. Figure 4 shows the future of IoT Fusion is set to unlock limitless possibilities by merging cutting-edge technologies and analytics, improving decision-making capabilities, and driving innovation across several industries.

IoT Fusion improves smart city applications by bringing together data from many sensors [36]. It emphasizes advantages such as improved city services and resource management while also addressing issues such as data privacy and integration problems. IoT Fusion improves industrial automation, focusing on increased efficiency and predictive maintenance [37]. It also discusses challenges related to data integration and security. The influence of IoT Fusion on healthcare, including benefits like increased patient monitoring and individualized therapy, as well as obstacles such as data protection and interoperability, is represented in [38].

Fig. 4 Future of IoT fusion

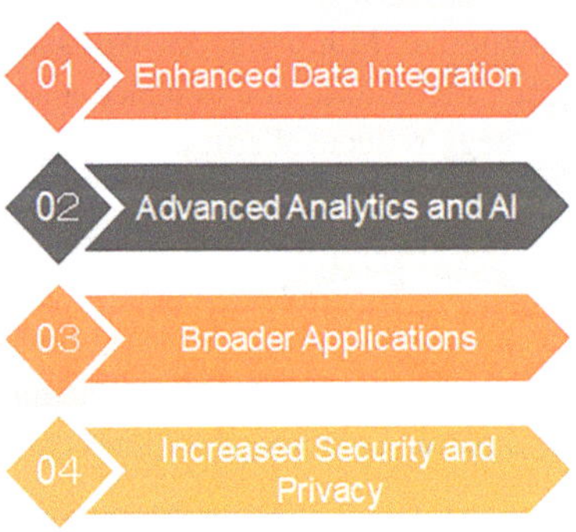

10 Conclusion and Future Scope

IoT Fusion provides a big step forward in the integration and study of data from various IoT devices and systems. IoT Fusion improves decision-making capabilities across multiple domains by seamlessly integrating heterogeneous data streams.

The combination of data from various sources allows for a more complete knowledge of complex systems and settings. This holistic perspective enables firms to make better educated, data-driven decisions. In industries like manufacturing, healthcare, smart cities, and agriculture, IoT Fusion leads to increased operational efficiency, predictive measures, and efficient resource management.

The future of IoT Fusion seems promising, with advances in data integration, artificial intelligence, and edge computing expanding its potential. However, data security, integration complexity, and scalability are all issues that must be solved before it can completely achieve its promise. IoT Fusion is a breakthrough technology that opens up new opportunities for smarter, more responsive systems. By continuing to innovate and solve existing hurdles, IoT Fusion will play a more important role in driving efficiency, creativity, and better decision-making in a more linked society.

References

1. Smith, J., & Liu, Y. (2023). A comprehensive survey on IoT fusion techniques and their applications. *IEEE Internet of Things Journal, 10*(4), 241–259.
2. Patel, A., & Zhang, H. (2024). Enhancing decision-making with AI-driven IoT fusion: A case study on smart cities. *Journal of Urban Computing, 18*(1), 13–29.
3. Rahman, M., & Khan, S. (2022). Data privacy in IoT fusion: Challenges and solutions. *ACM Transactions on Internet Technology, 22*(3), 17–34.
4. White, E., & Green, T. (2024). Scalability and interoperability in IoT fusion systems. *International Journal of Distributed Sensor Networks, 20*(2), 89–102.
5. Miller, P., & Fernandez, J. (2023). Quantum computing in IoT: Potential and challenges for future IoT fusion. *IEEE Access, 11*, 1345–1360.
6. World Economic Forum. (2023). *The future of IoT: Integrating AI and quantum computing for global connectivity*. WEF Industry Report.
7. ISO/IEC 30141:2022. (n.d.). *Internet of Things (IoT) – Reference architecture*. International Organization for Standardization.
8. Brown, D., & Shah, M. (2024). Ethical considerations in IoT fusion: Balancing innovation and responsibility. *Journal of Ethics in Technology, 12*(1), 45–60.
9. Wang, Q., & Chen, X. (2024). Edge computing for real-time IoT fusion: A performance evaluation. *IEEE Transactions on Industrial Informatics, 20*(3), 451–463.
10. Gupta, R., & Rao, S. (2023). AI-enhanced IoT for healthcare: A review of current trends and future prospects. *IEEE Journal of Biomedical and Health Informatics, 27*(4), 983–998.
11. Kumar, V., & Bansal, P. (2022). Secure data fusion in IoT: A Blockchain-based approach. *Journal of Parallel and Distributed Computing, 163*, 122–135.
12. Ahmed, S., & Taylor, G. (2023). Interoperability in heterogeneous IoT environments: Challenges and solutions. *ACM Computing Surveys, 55*(6), 1–28.
13. Chaudhry, A., & Malik, M. (2023). IoT fusion in precision agriculture: A data-driven approach. *Computers and Electronics in Agriculture, 205*, 107467.

14. Lee, S., & Park, H. (2024). Smart manufacturing with IoT fusion: Enhancing efficiency through integrated systems. *Journal of Manufacturing Systems, 68*, 77–89.
15. Nguyen, T., & Ho, D. (2023). Next-generation AI algorithms for IoT fusion: Opportunities and challenges. *Neural Computing and Applications, 35*(9), 11345–11358.
16. Gartner. (2023). *Top trends in IoT and AI for 2024*. Gartner Research Report.
17. ISO/IEC 27555:2023. (n.d.). *Guidelines for IoT data management and fusion*. International Organization for Standardization.
18. Zhou, J., & Sun, X. (2022). Real-time IoT fusion for autonomous systems: A case study in autonomous vehicles. *IEEE Transactions on Vehicular Technology, 71*(12), 12023–12034.
19. Microsoft Research. (2023). *The future of IoT: Integrating AI, edge, and cloud computing*. Microsoft Technical Report.
20. Smith, A., & Kaur, P. (2024). Ethical AI in IoT fusion: Addressing bias and fairness. *Journal of AI and Ethics, 3*(2), 67–82.
21. Sadeghi-Niaraki, A. (2023). Internet of thing (IoT) review of review: Bibliometric overview since its foundation. *Future Generation Computer Systems, 143*, 361–377. https://doi.org/10.1016/j.future.2023.01.016
22. Deepa, S., Radhika, M., Diwakar, M., Rufina, P., Nithya, A., & Mohanapriya, M. (2023). Data fusion and analysis techniques for enhanced decision-making in IoT environments. In *2023 7th international conference on electronics, communication and aerospace technology (ICECA)* (pp. 1513–1518). IEEE.
23. Kumar, K., & Kumar, V. (2023, November). Integration of artificial intelligence and machine learning for internet of things. In *2023 international conference on sustainable communication networks and application (ICSCNA)* (pp. 491–497). IEEE.
24. Kalavakonda, R. R., Huan, J., Dehghanzadeh, P., Jaiswal, A., Mandal, S., & Bhunia, S. (2024). Fusion intelligence: Confluence of natural and artificial intelligence for enhanced problem-solving efficiency. arXiv preprint arXiv:2405.09763.
25. Bao, S., Xu, K., Zheng, L., Shan, Y., Liu, Y., & Li, L. (2023). IoT architecture for electricity distribution considering data fusion and intelligent decision making. In *2023 IEEE 5th international conference on power, intelligent computing and systems (ICPICS)* (pp. 109–113). IEEE.
26. Ram, M. S., Sreeram, A., Poongundran, M., Singh, P., Prajapati, Y. N., & Myrzahmetova, S. (2022). Data fusion opportunities in IoT and its impact on decision-making process of organisations. In *2022 6th international conference on intelligent computing and control systems (ICICCS)* (pp. 459–464). IEEE.
27. Ahmed, S. S., Jawad, M., Ja'afar, A., Abd, S. K., Mohammed, A., & Majeed, A. H. (2023). Intelligent decision making in IoT-based enterprise management through fusion optimization with deep learning models. *Fusion: Practice & Applications, 11*(2).
28. Li, W., Zhang, N., Liu, Z., Ma, S., Ke, H., Wang, J., & Chen, T. (2023). A trusted decision fusion approach for the power internet of things with federated learning. *Frontiers in Energy Research, 11*, 1061779.
29. Hussein, A., Ahmed, S., Abed, S. K., & Thamer, N. (2023). Enhancing IoT-based intelligent video surveillance through multi-sensor fusion and deep reinforcement learning. *Fusion: Practice & Applications, 11*(2), 21–34. https://doi.org/10.54216/fpa.110202
30. Han, J., Li, L., Sun, Z., Feng, X., Lin, N., & Ruan, J. (2023). An integrative decision-making model for the internet of things-enabled supply chains of fresh agri-product. *International Journal of Production Research, 61*(13), 4358–4373.
31. Ullah, I., & Youn, H. Y. (2020). Intelligent data fusion for smart IoT environment: A survey. *Wireless Personal Communications, 114*(1), 409–430.
32. Sharma, P. K., Dennison, M., & Raglin, A. (2021). Iot solutions with multi-sensor fusion and signal-image encoding for secure data transfer and decision making. arXiv preprint arXiv:2106.01497.
33. Bajpai, S., & Pillai, S. (2022). Decision making with IoT-paving an integrated approach. In *2022 international conference on decision aid sciences and applications (DASA)* (pp. 1305–1310). IEEE.

34. Singh, S., Kumar, V., & Kumar, N. (2022). Internet of things (IoT) for operational efficiency: A review. *International Journal of Computer Applications*.
35. Patel, J., Liu, A., & Sharma, R. (2023). The future of IoT fusion: Unlocking new possibilities. *Journal of Emerging Technologies*.
36. Zhao, Y., & Chen, L. (2022). IoT fusion and its impact on smart cities: A review. *Sensors, 22*(15), 5567. https://doi.org/10.3390/s22155567
37. Huang, J., & Li, S. (2024). Enhancing industrial automation with IoT fusion: Opportunities and challenges. *Journal of Industrial Information Integration, 85*, 102413. https://doi.org/10.1016/j.jii.2024.102413
38. Nguyen, T., & Garcia, R. (2023). The role of IoT fusion in modern healthcare: A systematic review. *Journal of Healthcare Engineering, 2023*, 473928. https://doi.org/10.1155/2023/473928

Decision Support Systems for Smart Manufacturing Using Digital Twins

Labaran Isiaku, Damla Karagozlu, and Abdelmalek Hamada

1 Introduction

Envision a factory where machines communicate seamlessly, predict maintenance needs before failures occur, and optimize production lines in real-time. This vision of a hyper-connected, intelligent production environment is not a far aspiration but rather a tangible reality enabled by smart manufacturing. As the next phase of industrial evolution, smart manufacturing strengthens new technologies to create more accurate, adaptable, and interconnected development environments [33]. It integrates advanced manufacturing techniques with digital transformation tools such as the Internet of Things (IoT), artificial intelligence (AI), big data analytics, and cloud computing [24]. Collectively, these technologies empower manufacturers to boost their processes, develop product quality, and swiftly respond to market dynamics. By fostering a more agile and responsive manufacturing ecosystem, smart manufacturing aims to optimize resource utilization, reduce operational costs, and drive innovation across the production lifecycle [44].

Smart manufacturing is built on the conception of cyber-physical systems, which are computer-based algorithms that monitor and control physical processes while being tightly integrated with the Internet and its users [32]. This integration enables real-time data collection and analysis, leading to improved decision-making, predictive maintenance, and operational efficiency. Given the increasing importance of digital transformation across industries, the adoption of smart manufacturing practices is crucial to maintain a competitive edge in the global market [31].

L. Isiaku · D. Karagozlu (✉) · A. Hamada
Department of Management Information Systems, Cyprus International University, Nicosia, Turkey
e-mail: lisiaku@ciu.edu.tr; dkaragozlu@ciu.edu.tr; ahamada@ciu.edu.tr

N. Gupta, J. D. Diaz Roman (eds.), *Decision Support Systems for Industrial and Scientific Applications*, Signals and Communication Technology,
https://doi.org/10.1007/978-3-031-89700-9_3

Among the various technologies underpinning smart manufacturing, digital twins and decision-support systems stand out for their transformative potential. A digital twin is a virtual replica of a physical asset, process, or system that is created to simulate, analyze, and optimize real-world performance [29]. By mirroring the physical world in a digital environment, digital twins enable manufacturers to visualize processes, monitor systems in real-time, and predict future outcomes based on current data [12]. This ability to simulate and test scenarios helps to identify potential issues before they arise, reduce downtime, and enhance overall productivity [12, 42].

Digital twins collect data from sensors and IoT devices embedded in physical assets and feed them into sophisticated models that replicate the behavior and performance of these assets [38]. This continuous flow of data allows for dynamic updates to the digital twin, ensuring that it remains an accurate representation of its physical counterpart. Through advanced analytics and machine learning algorithms, digital twins provide deep insights into operational efficiencies, equipment health, and potential failure points, enabling proactive maintenance and informed decision-making.

When integrated with digital twins, decision support systems (DSSs) amplify these capabilities by providing a structured framework for analyzing data and generating actionable insights. DSS leverages AI and machine learning to process vast amounts of data, identify patterns, and recommend optimal courses of action [15]. This integration facilitates data-driven decision-making, allowing manufacturers to swiftly address challenges, optimize resource allocation, and enhance production processes.

The combined use of digital twins and decision support systems in smart manufacturing creates powerful synergy that drives operational excellence [35]. By providing real-time visibility to manufacturing processes and enabling predictive and prescriptive analytics, these technologies empower manufacturers to achieve higher levels of efficiency, reliability, and product quality. As the manufacturing landscape continues to evolve, the adoption of digital twins and decision support systems will be pivotal in realizing the full potential of smart manufacturing. Therefore, the insights gained from this research will provide valuable frameworks and best practices for implementing these technologies, driving innovation and excellence in smart manufacturing.

2 Concept and Architecture of Digital Twins

The concept of digital twins has emerged as a transformative approach in the realm of technology, enabling the seamless integration of the physical and digital worlds. A digital twin is a virtual representation or model of a physical entity, which can be anything from a product, a process, or even an entire system [40]. This virtual counterpart is designed to mimic the state and behavior of its physical counterpart in real-time, allowing for enhanced monitoring, optimization, and decision-making. The origins of the digital twin concept can be traced back to the 1960s, when NASA utilized a virtual model of the Apollo 13 spacecraft to aid in the rescue mission

[17]. Since then, the concept has evolved, and the advent of Industry 4.0 has further propelled its widespread adoption across various industries, from manufacturing to healthcare [1].

According to the literature, a digital twin is characterized by three key elements: a representation of the physical system, a bidirectional data exchange between the physical and virtual models, and a connection throughout the entire lifecycle of the system [21, 46]. This interconnectivity enables the digital twin to learn from the physical system's historical performance and virtual processes, allowing it to continuously evolve and improve [22].

The architecture of a digital twin typically comprises several interconnected components, each playing a crucial role in the overall system [39]. These components include sensors and data acquisition systems, which gather real-time data from the physical system; a virtual model, which simulates the physical system's behavior; and a communication interface, which facilitates the bidirectional flow of data between the physical and virtual realms [4, 29, 45, 46].

The virtual model, often referred to as the "digital twin," is a key component in this architecture [29]. This model is designed to replicate the physical system's characteristics, such as its geometry, material properties, and dynamic behavior. By leveraging advanced modeling techniques, such as finite element analysis, computational fluid dynamics, and machine learning, the digital twin can accurately predict the system's performance, identify potential issues, and simulate various scenarios [45].

The communication interface, which enables the exchange of data between the physical and virtual systems, is another critical element of the digital twin architecture [46]. This interface typically includes sensors, communication protocols, and data storage and processing capabilities, allowing for the seamless transfer of information in both directions [4].

The real-time data collected from the physical system is fed into the digital entity, which then uses this information to continuously update and refine its virtual model. Conversely, the digital twin can provide insights and recommendations back to the physical system, enabling its optimization and adaptation to changing conditions. Figure 1 illustrates the concept of digital twin technology.

The implementation of digital twins has been transformative across various industries, leading to significant improvements in product design, manufacturing processes, asset management, and predictive maintenance. By providing a comprehensive, data-driven understanding of physical systems, digital twins empower organizations to make more informed decisions, reduce costs, and enhance overall operational efficiency [9, 26].

2.1 *Characteristics of Digital Twin System*

The digital twin comprises three main elements: (1) a physical entity, (2) its virtual counterpart, and (3) the interconnected systems that facilitate information exchange between the physical and virtual entities, as illustrated in Fig. 2.

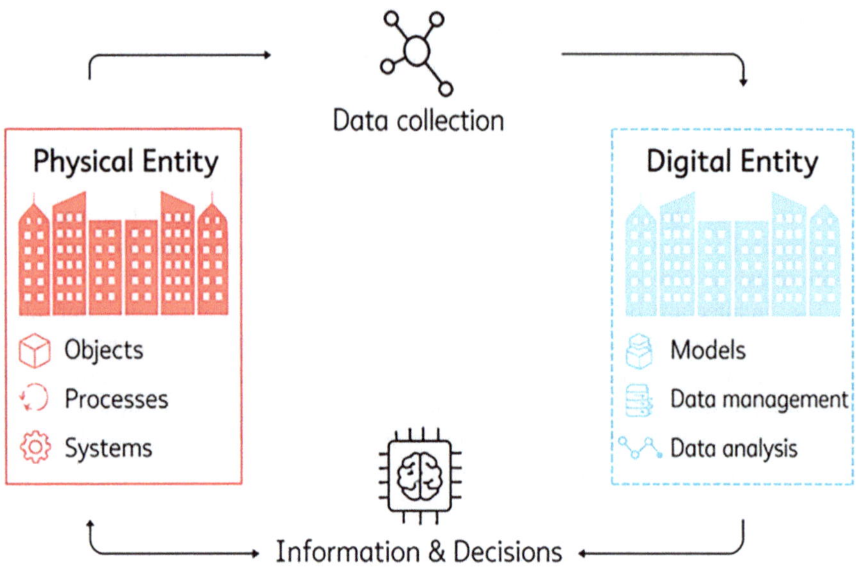

Fig. 1 Schematic of physical and digital twin integration. (Adapted from HSLU)

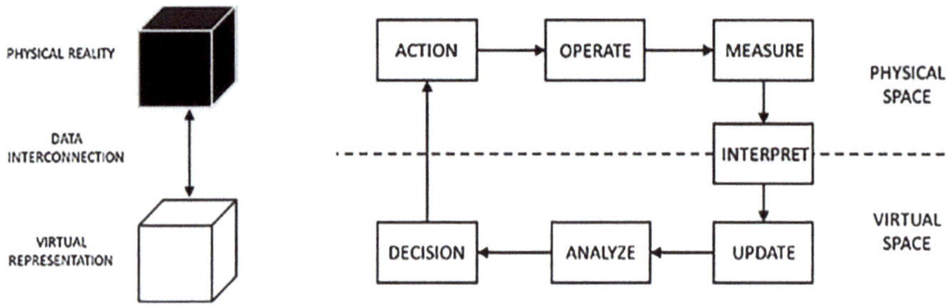

Fig. 2 Components and high-level processes of a digital twin. (Adapted from VanDerHorn and Mahadevan [46])

The following subsections will provide a more comprehensive examination of these three aspects.

2.2 Physical Reality

In the literature of digital twins, there is a wide array of terms used to explain the physical feature of interest, with the phrasing often differing based on the domain. To address this, the term "physical reality" is offered, as the most comprehensive term available to illustrate what is being modeled [46]. This term encompasses the entire system, including what is common and not elements. Physical reality is often divided into three components: the physical system, the physical environment, and physical processes.

2.3 Virtual Representation

In relation to digital twin, the phrase "virtual representation" refers to a digital model or simulation that replicates the physical system and its environment [36]. This virtual representation was created using collected data from sensors and other references to mirror the real-world system's behavior, characteristics, and interactions [13]. It allows for real-time monitoring, analysis, and manipulation of the physical system through simulations, providing valuable insights for decision-making, predictive maintenance, and optimization without the need for direct interaction with the physical system itself [20, 36].

2.4 Relationship Between Physical and Virtual

An integral aspect of the digital twin concept is the interconnection between physical reality and its virtual representation, enabling bi-directional exchange of data and information: from the physical to the virtual realm, and vice versa [46].

The following Table 1 provides a summary of the aforementioned explanations of the various elements of a digital twin.

3 Applications of Digital Twins in Smart Manufacturing

The implementation of digital twins in manufacturing industry has considerably revolutionized the industry, providing numerous advantages such as increased efficiency, diminished downtime, and enhanced product quality [20, 42]. In the past, manufacturing often involved a disconnect between the physical and digital realms, resulting in inefficiencies and reactive maintenance strategies. However, the introduction of digital twins has eliminated this gap, facilitating the smooth exchange of information and promoting proactive approaches to manufacturing processes [30]. As a result, the adoption of digital twins has enabled manufacturers to achieve higher efficiency and cost-effectiveness in their production processes. In addition, digital twins have also allowed manufacturers to simulate and test various production scenarios, enabling them to optimize their operations and improve overall process performance [2, 26]. This has led to significant reductions in costs

Table 1 An overview description of digital twin components

Summary of the various elements of digital twin components	
Components	Description
Physical reality	
Physical system	The segment of the physical reality that has been selected for modeling and comprises all the relevant connected and interconnected entities that function as a cohesive unit
Physical environment	The elements that have an impact on a physical system and are situated beyond its boundaries
Physical processes	The physical system's functions that cause it to transition between different states
Virtual representation	
Virtual system	The specific details and computer simulations of the physical system of interest, at a particular level of abstraction
Virtual environment	The virtual replica of the physical environment for the physical system being studied
Virtual processes	The digital illustrations of physical processes, used to simulate their impact on physical systems, serve as the primary means of generating insights to achieve desired outcomes
Information interconnection	
Physical-to-virtual connection	The methods used to gather, analyze, and convey information from a physical system, and apply it to update the virtual representation's state
Virtual-to-physical connection	The process of determining and executing an action based on a virtual representation, with the aim of bringing about alterations in the states of a physical system

and improved efficiency for manufacturers who have implemented digital twins in their production processes [12, 23].

Digital twins are a dynamic digital representation of physical assets, processes, and systems that manufacturers can use to simulate, predict, and optimize performance [26]. In smart manufacturing, which emphasizes the use of advanced technologies such as the Internet of Things (IoT), artificial intelligence (AI), and big data analytics, digital twins are essential for achieving higher levels of automation, agility, and accurate decision-making. Digital twins enable smart manufacturing to achieve higher levels of automation, agility, and accurate decision-making by providing real-time data and insights into the performance of physical assets and processes, allowing for proactive maintenance and optimization [23, 26]. This chapter delves into the applications of digital twins in smart manufacturing, highlighting their impact on predictive maintenance and asset management, process optimization and real-time monitoring, and quality control and defect detection.

3.1 Predictive Maintenance and Asset Management

Predictive maintenance and asset management are essential applications for digital twins in the manufacturing industry [10]. Historically, maintenance strategies have been limited to either reactive or preventive measures, which often resulted in unanticipated downtimes or unnecessary maintenance activities. However, digital twins have transformed this approach by allowing for predictive maintenance, which leverages real-time data and sophisticated analytics to anticipate potential equipment failures and optimize maintenance schedules.

3.2 Process Optimization and Real-Time Monitoring

Optimizing processes and closely monitoring them in real-time are crucial for attaining top-notch performance in manufacturing. Digital twins are of utmost importance in these areas, as they offer a highly detailed, real-time perspective of manufacturing processes and empower data-driven choices. In this context, the use of digital twins can lead to optimized production processes and improved outcomes [16]. By leveraging the capabilities of digital twins, businesses can simulate and analyze various scenarios to identify the most efficient production methods and achieve better results. In addition to improving production efficiency, digital twin technology enables manufacturers to forecast the quality of their final products, allowing them to make better-informed decisions regarding modifications to materials and processes [48]. By using digital twins, manufacturers can ensure consistency in mass production, making them an essential tool in the industrial manufacturing the process. Figure 3 illustrates a digital twin mapping strategy, which shows the process of updating and optimizing the model [42].

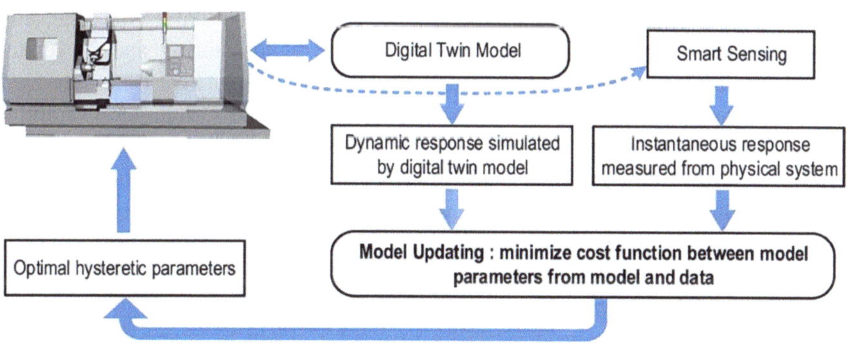

Fig. 3 The mapping scheme for digital twin model for updating and optimization process. (Adapted from Soori et al. [42])

3.3 *Quality Control and Defect Detection*

Quality control and defect detection are critical components of the manufacturing process, and their effectiveness is enhanced by the use of digital twins. With the ability to simulate product performance in real-time, digital twins can identify defects and quality issues before they occur, ensuring that products meet the necessary standards and reducing the risk of costly recalls or customer complaints [14].

4 Decision Support Systems for Digital Twins

The evaluation of digital twins has been substantially impacted by the integration of Decision Support Systems (DSS), which augment the functionalities of digital twins by delivering actionable insights and promoting well-informed decision-making. Decision Support Systems for digital twins utilize data, models, and user interfaces to bolster organizational decision-making, enabling more precise, efficient, and prompt reactions to arising challenges and opportunities in manufacturing and other industries [22]. Decision Support Systems for Digital Twins are becoming increasingly popular in manufacturing and other industries due to their ability to analyze vast amounts of data and provide insights that inform strategic decision-making [35, 48]. This leads to more precise, efficient, and prompt reactions to arising challenges and opportunities, enabling organizations to stay ahead of the competition.

4.1 *Integration of Decision Support Systems with Digital Twins*

The integration of Decision Support Systems (DSS) with Digital Twins represents a substantial breakthrough in the realm of smart manufacturing and its related fields [28]. This integration enables the effortless merging of real-time data from physical systems with sophisticated computational models, thereby promoting more resilient and responsive decision-making processes. The ability to seamlessly combine real-time data from physical systems with advanced computational models not only enhances decision-making processes but also facilitates the creation of more adaptive and efficient strategies [27]. This integration of real-time data and computational models allows for more informed decision-making and the development of strategies that are responsive to changing conditions, leading to improved performance and outcomes. The ability to quickly adapt to changing conditions and make informed decisions is crucial in today's fast-paced and dynamic business environment, making the integration of real-time data and computational models an indispensable tool for organizations looking to stay ahead of the competition. There are several advantages that can be gained by combining decision support systems with digital twins.

4.1.1 Enhancing Real-Time Monitoring and Control

The use of DSS in conjunction with digital twins offers a number of advantages, particularly in the realm of real-time monitoring and control. Through the continuous receipt of data from sensors and other IoT devices embedded within a physical system, the digital twin is able to provide an up-to-date, virtual representation of the system's current state [39]. This information can then be analyzed by the DSS to identify anomalies, predict potential issues, and suggest corrective actions. By quickly addressing any deviations from expected performance, the real-time interaction between the digital twin and DSS helps to minimize downtime and enhance operational efficiency [25]. Additionally, the real-time interaction between the digital twin and DSS allows for proactive identification of potential issues, which can further improve operational efficiency by minimizing the need for unscheduled downtime [42].

4.1.2 Process Optimization

The implementation of DSS in conjunction with digital twins enhances process optimization by allowing for the simulation and analysis of a variety of situations. DSS can make use of the digital twin to explore different production methods, pinpoint bottlenecks, and examine the consequences of process modifications on overall performance [48]. Through simulating these scenarios in a virtual environment, manufacturers can evaluate and fine-tune their processes before making any changes to the physical system, resulting in reduced risks and heightened efficiency. By implementing these simulations in real-world scenarios, companies can ensure that their new processes are effective and efficient before implementing them, resulting in a reduction in errors and delays.

4.1.3 Decision-Making Support

Decision Support Systems (DSS) play a vital role in decision-making by furnishing data-driven insights and recommendations [5]. By merging with digital twins, DSS can harness real-time data and simulations to yield actionable insights, thereby aiding decision-makers in selecting the most optimal course of action. This can include modifying production schedules, allotting resources, or adopting new technologies [7]. The DSS guarantees that decisions are founded on thorough, current information. The DSS also ensures that these decisions are communicated and implemented in a timely and effective manner [43]. To achieve this, the DSS establishes clear communication channels with relevant stakeholders and monitors progress to ensure that the decisions are implemented as intended.

4.1.4 Enhanced Collaboration

The use of DSS and digital twins encourages cooperation among various stakeholders. By presenting a unified, digital depiction of the physical system, the digital twin empowers engineers, managers, and other stakeholders to observe and analyze the system's performance in real-time [20]. The DSS then supports collaborative decision-making by providing a forum for sharing ideas, considering alternatives, and determining the optimal approach. The DSS is an effective tool for enhancing the decision-making process by allowing users to exchange diverse views, evaluate various options, and reach a consensus on the most suitable course of action [8].

4.2 A Case Study Implementation of Digital Twin in Automotive Industry

In the automotive industry, a digital twin is essential for simulating and testing various aspects of vehicles, including their design, development, manufacturing, and maintenance [3]. Creating a digital twin of a vehicle allows engineers and designers to analyze its performance and behavior in different scenarios, identify potential issues, and optimize its design and features before commencing production [3, 18]. This approach significantly reduces time and cost while enhancing the quality and safety of the car production process. Furthermore, digital twins play a crucial role in predictive maintenance, allowing real-time monitoring of a vehicle's condition to detect potential problems before escalation [41]. This practice helps minimize downtime, prevent breakdowns, and extend the lifespan of the vehicle. Additionally, digital twins can speed up car manufacturing by providing real-time data on part production, reducing data processing time, and improving communication among automotive development teams [42]. According to Sooriet al [42], by creating an intelligent link between the automaker and driver, this technology aids Maintenance Repair Operations (MROs). For instance, a digital twin of an engine can be used to monitor its performance and detect potential issues, such as wear or damage to components, allowing for proactive maintenance and reducing the risk of breakdowns. In addition to monitoring performance and detecting potential issues, a digital twin of an engine can also be used to simulate different operating conditions and optimize engine performance, resulting in increased efficiency and reduced fuel consumption. Figure 4 demonstrates the integration of digital twin technology within MROs.

Digital twins are incredibly beneficial during the system design phase for a range of applications, including car parts and mechanical arms. They greatly improve the

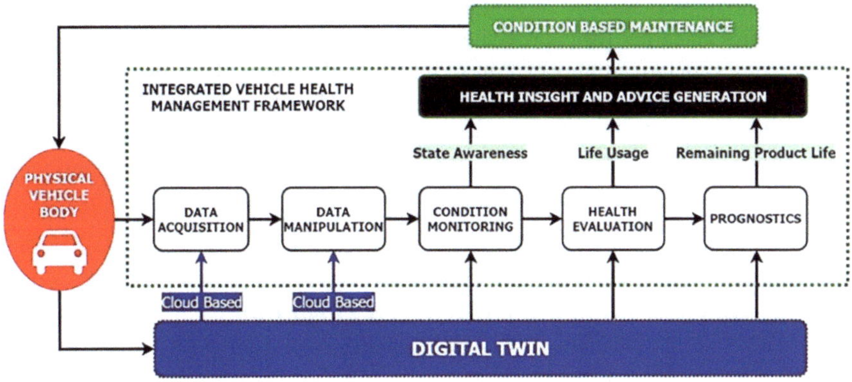

Fig. 4 Digital twin technology integrated within MRO. (Adapted from Soori et al. [42]

dependability of vehicle design and development. By speeding up simulations and carrying them out simultaneously, digital twins can replicate thousands of hours of driving in a realistic environment with accurate predictions of gravity, weight, and collisions [42] For instance, a digital twin of an electric car can simulate various conditions, such as different driving scenarios, battery usage, and charging patterns [49]. This capability allows manufacturers to optimize the design of electric cars for optimal efficiency and performance while also identifying potential problems before they occur. Furthermore, the implementation of digital twins in the automotive industry can lead to a reduction in costs and an increase in efficiency, ultimately benefiting both manufacturers and consumers [3]. In conclusion, digital twins have the potential to transform the automotive industry by streamlining processes in design, development, production, and maintenance.

5 Benefits and Impact

5.1 Enhanced Operational Efficiency and Productivity

Digital twins significantly improve operational efficiency and productivity in smart manufacturing by providing real-time insights and data-driven decision support [12, 28]. These virtual representations of physical assets and processes allow manufacturers to simulate, predict, and optimize their operations without interrupting actual production lines. By integrating digital twins with advanced analytics and AI, manufacturers can anticipate potential bottlenecks, streamline workflows, and reduce cycle times. The continuous feedback loop between the physical and virtual systems enables proactive adjustments, ensuring optimal performance [29]. This real-time optimization not only enhances resource utilization but also decreases waste, energy consumption, and operational costs. Additionally, digital twins facilitate the implementation of lean manufacturing principles, as they help identify inefficiencies and areas for improvement. With the ability to model various scenarios and outcomes, manufacturers can make informed decisions about process changes, production scheduling, and resource allocation, leading to increased throughput and profitability. Overall, digital twins provide a solid foundation for smart manufacturing, driving enhanced operational efficiency and productivity through continuous monitoring, analysis, and optimization [47].

5.2 Reduced Maintenance Costs and Downtime

Incorporating digital twins into smart manufacturing processes results in a significant decrease in maintenance expenses and downtime [42, 47]. By creating a comprehensive digital replica of physical assets, manufacturers can monitor equipment health in real-time, predict failures, and schedule maintenance activities more effectively. With the aid of predictive maintenance, digital twins transition their focus from reactive to proactive maintenance strategies, identifying potential issues before they lead to costly breakdowns. This approach reduces unexpected downtime, ensures uninterrupted production, and minimizes the financial impact of equipment failure. Moreover, digital twins facilitate condition-based maintenance, allowing interventions based on the actual state of the equipment rather than on a predetermined schedule. This not only prolongs the lifespan of machinery, but also optimizes maintenance resources and inventory management. By simulating various maintenance scenarios and outcomes, digital twins enable manufacturers to plan and execute maintenance activities with minimal disruption to the production [11]. The ability to foresee and address maintenance needs in advance results in improved asset reliability, higher operational availability, and substantial cost savings, thereby enhancing the overall efficiency of the manufacturing process.

5.3 Improved Product Quality and Consistency

Digital twins are essential for improving product quality and uniformity in smart manufacturing [6]. These virtual representations of the entire production process, ranging from raw materials to finished goods, allow manufacturers to monitor and control the quality parameters at every stage. This enables real-time detection and correction of variations, ensuring that products consistently meet strict quality criteria. Digital twins facilitate comprehensive quality control by merging data from various sources, such as sensors, machines, and inspection systems, to identify potential defects and optimize production parameters [19]. By simulating different production scenarios, manufacturers can assess the impact of process changes on product quality, resulting in more informed decision making and continuous improvement. Additionally, digital twins support root cause analysis by tracing quality issues back to their source, enabling manufacturers to implement corrective actions and prevent recurrence [37]. The predictive capabilities of digital twins also aid in anticipating quality issues and taking preemptive measures, thus reducing waste and rework [20]. In conclusion, incorporating digital twins in smart manufacturing guarantees higher product quality and consistency, enhancing customer satisfaction and competitive advantage.

6 Challenges and Future Directions

6.1 Technical and Integration Challenges

The introduction of digital twins in smart manufacturing presents several technical and integrative challenges. One of the primary obstacles is the seamless integration of digital twins with existing manufacturing systems and infrastructure [34]. This necessitates ensuring compatibility between various hardware and software components that often come from different vendors with proprietary technologies. Furthermore, real-time data collection for effective digital twins requires advanced sensor technology and dependable communication networks such as 5G [30]. Another challenge is the development of sophisticated algorithms and models that can accurately replicate the intricate behavior of physical systems. These models must be continuously updated and refined to account for the dynamic nature of manufacturing environments. Moreover, the high computational power required for real-time simulations and analyses can be a limiting factor, particularly for small- and medium-sized enterprises (SMEs) with limited resources. To overcome these technical and integrative challenges, ongoing research and development and collaborative efforts among industry stakeholders to establish common standards and protocols are necessary.

6.2 Data Security and Privacy Concerns

Data security and privacy are pressing concerns when implementing digital twins for smart manufacturing because of the extensive data collection and sharing involved in creating and maintaining them [42]. This increases the risk of cyber-attacks and unauthorized access, making it crucial for manufacturers to safeguard sensitive information such as intellectual property and proprietary processes. To achieve this, robust cybersecurity measures including encryption, secure communication protocols, and regular security audits must be implemented. Moreover, the use of cloud-based platforms for data storage and processing introduces additional risks because data may be vulnerable to breaches during transmission or stored on remote servers. Privacy concerns also arise when personal data of employees or customers are collected during the manufacturing process. Compliance with data protection regulations such as the General Data Protection Regulation (GDPR) is essential to avoid legal consequences and maintain trust with stakeholders. Developing and maintaining secure and privacy-conscious digital twin solutions necessitate a proactive approach to cybersecurity, continuous monitoring of vulnerabilities, and adherence to best practices in data protection.

6.3 Future Trends and Research Opportunities in Digital Twins and Smart Manufacturing

The future of digital twins in smart manufacturing appears to be bright, with several emerging trends and research opportunities on the horizon that are expected to improve their capabilities. One such trend is the integration of artificial intelligence (AI) and machine learning (ML) with digital twin technology, which can enable more sophisticated data analysis, predictive maintenance, and optimization of manufacturing processes [19]. Another promising development is the advancement of edge computing, which allows data processing to occur closer to the source of data collection, reduces latency, and improves real-time decision-making. Additionally, the development of more advanced and cost-effective sensors will enhance the accuracy and granularity of the data collected for digital twins. Research has also focused on improving the interoperability of digital twins across different platforms and industries, which will facilitate broader adoption and collaboration. Furthermore, the concept of hybrid digital twins, which combines physical and virtual simulations, is gaining traction and offers new possibilities for enhancing manufacturing efficiency and innovation. As these trends and research efforts continue, digital twins are expected to become increasingly integral to smart manufacturing, driving significant improvements in productivity, quality, sustainability and informed decision-making.

7 Conclusion

The incorporation of digital twins alongside decision support systems represents a groundbreaking development in the realm of smart manufacturing. This cohesive approach enables live monitoring, proactive maintenance, and data-informed enhancement of the production procedures. By fashioning a digital duplicate of a tangible setup, manufacturers can experiment with various situations, foresee potential problems, and apply corrective measures with exceptional accuracy. This results in elevated operational productivity, minimized disruptions, and superior product performance.

The implementation of digital twins in smart manufacturing not only simplifies production processes, but also encourages innovation, enabling rapid adjustments to market fluctuations and technological enhancements. Given the growing intricacy and interconnectedness of manufacturing systems, digital twins and decision support systems will play a crucial role in managing these complexities. This chapter has delved into the concept and structure of digital twins, paying particular attention to their constituent parts, data collection and integration, and real-time data analysis and processing. The information presented highlights the significant potential of this technology for transforming manufacturing operations on a large scale.

The future of smart manufacturing will be shaped by the ongoing development and use of digital twins in conjunction with advanced decision support systems. These technologies can lead to the creation of more resilient, efficient, and adaptable production environments. This study aims to improve our understanding of these technologies and facilitate their wider adoption and integration across various manufacturing domains.

Conflict of Interest The authors state that they have no actual conflict of interest.

References

1. Augustine, P. (2020). The industry use cases for the digital twin idea. In *Advances in computers* (pp. 79–105). Elsevier.
2. Bécue, A., et al. (2020). A new concept of digital twin supporting optimization and resilience of factories of the future. *Applied Sciences, 10*(13), 4482.
3. Bhatti, G., Mohan, H., & Singh, R. R. (2021). Towards the future of smart electric vehicles: Digital twin technology. *Renewable and Sustainable Energy Reviews, 141*, 110801.
4. Boschert, S., & Rosen, R. (2016). Digital twin – The simulation aspect. In *Mechatronic futures: Challenges and solutions for mechatronic systems and their designers* (pp. 59–74).
5. Bousdekis, A., et al. (2021). A review of data-driven decision-making methods for industry 4.0 maintenance applications. *Electronics, 10*(7), 828.
6. Chen, Y., et al. (2020). Digital twins in pharmaceutical and biopharmaceutical manufacturing: A literature review. *PRO, 8*(9), 1088.
7. Coelho, F., Relvas, S., & Barbosa-Póvoa, A. P. (2021). Simulation-based decision support tool for in-house logistics: The basis for a digital twin. *Computers & Industrial Engineering, 153*, 107094.
8. Courtney, J. F. (2001). Decision making and knowledge management in inquiring organizations: Toward a new decision-making paradigm for DSS. *Decision Support Systems, 31*(1), 17–38.
9. Das, S. (2023). Digital twin technology: Enhancing efficiency and decision-making in industry 4.0. Available at SSRN 4459204 [Preprint].
10. van Dinter, R., Tekinerdogan, B., & Catal, C. (2022). Predictive maintenance using digital twins: A systematic literature review. *Information and Software Technology, 151*, 107008.
11. Errandonea, I., Beltrán, S., & Arrizabalaga, S. (2020). Digital twin for maintenance: A literature review. *Computers in Industry, 123*, 103316.
12. Friederich, J., et al. (2022). A framework for data-driven digital twins of smart manufacturing systems. *Computers in Industry, 136*, 103586.

13. Frontoni, E., et al. (2018). Cyber physical systems for industry 4.0: Towards real time virtual reality in smart manufacturing. In *Augmented reality, virtual reality, and computer graphics: 5th international conference, proceedings, part II 5* (pp. 422–434). Springer.
14. Grieves, M. (2023). Digital twin certified: Employing virtual testing of digital twins in manufacturing to ensure quality products. *Machines, 11*(8), 808.
15. Gupta, S., et al. (2022). Artificial intelligence for decision support systems in the field of operations research: Review and future scope of research. *Annals of Operations Research, 308*(1), 215–274.
16. Hasidi, O., et al. (2022). Digital twins-based smart monitoring and optimisation of mineral processing industry. In *International conference on smart applications and data analysis* (pp. 411–424). Springer.
17. Hazrathosseini, A., & Afrapoli, A. M. (2023). The advent of digital twins in surface mining: Its time has finally arrived. *Resources Policy, 80*, 103155.
18. Hu, W., et al. (2021). Digital twin: A state-of-the-art review of its enabling technologies, applications and challenges. *Journal of Intelligent Manufacturing and Special Equipment, 2*(1), 1–34.
19. Huang, Z., et al. (2021). A survey on AI-driven digital twins in industry 4.0: Smart manufacturing and advanced robotics. *Sensors, 21*(19), 6340.
20. Javaid, M., Haleem, A., & Suman, R. (2023). Digital twin applications toward industry 4.0: A review. *Cognitive Robotics, 3*, 71–92.
21. Jiang, H., et al. (2021a). How to model and implement connections between physical and virtual models for digital twin application. *Journal of Manufacturing Systems, 58*, 36–51.
22. Jiang, Y., et al. (2021b). Industrial applications of digital twins. *Philosophical Transactions of the Royal Society A, 379*(2207), 20200360.
23. Kamble, S. S., et al. (2022). Digital twin for sustainable manufacturing supply chains: Current trends, future perspectives, and an implementation framework. *Technological Forecasting and Social Change, 176*, 121448.
24. Karagozlu, D., Ajamu, J., & Mbombo, A. B. (2020). Adaptation and effects of cloud computing on small businesses. *BRAIN. Broad Research in Artificial Intelligence and Neuroscience, 11*(4), 149–167.
25. Kunath, M., & Winkler, H. (2018). Integrating the digital twin of the manufacturing system into a decision support system for improving the order management process. *Procedia CIRP, 72*, 225–231.
26. Leng, J., et al. (2021). Digital twins-based smart manufacturing system design in industry 4.0: A review. *Journal of Manufacturing Systems, 60*, 119–137.
27. Ma, M., et al. (2018). Data and decision intelligence for human-in-the-loop cyber-physical systems: Reference model, recent progresses and challenges. *Journal of Signal Processing Systems, 90*, 1167–1178.
28. Ma, S., et al. (2022). Digital twin and big data-driven sustainable smart manufacturing based on information management systems for energy-intensive industries. *Applied Energy, 326*, 119986.
29. Madni, A. M., Madni, C. C., & Lucero, S. D. (2019). Leveraging digital twin technology in model-based systems engineering. *Systems, 7*(1), 7.
30. Mihai, S., et al. (2022). Digital twins: A survey on enabling technologies, challenges, trends and future prospects. *IEEE Communications Surveys & Tutorials, 24*(4), 2255–2291.
31. Mittal, S., et al. (2018). A critical review of smart manufacturing & industry 4.0 maturity models: Implications for small and medium-sized enterprises (SMEs). *Journal of Manufacturing Systems, 49*, 194–214.
32. Mittal, S., et al. (2019). Smart manufacturing: Characteristics, technologies and enabling factors. *Proceedings of the Institution of Mechanical Engineers, Part B: Journal of Engineering Manufacture, 233*(5), 1342–1361.
33. Morgan, J., et al. (2021). Industry 4.0 smart reconfigurable manufacturing machines. *Journal of Manufacturing Systems, 59*, 481–506.

34. Perno, M., Hvam, L., & Haug, A. (2022). Implementation of digital twins in the process industry: A systematic literature review of enablers and barriers. *Computers in Industry, 134*, 103558.
35. Qi, Q., & Tao, F. (2018). Digital twin and big data towards smart manufacturing and industry 4.0: 360 degree comparison. *IEEE Access, 6*, 3585–3593.
36. Rasheed, A., San, O., & Kvamsdal, T. (2020). Digital twin: Values, challenges and enablers from a modeling perspective. *IEEE Access, 8*, 21980–22012.
37. Rosen, K. M., & Pattipati, K. R. (2023). Operating digital twins within an enterprise process. In *The digital twin* (pp. 599–659). Springer.
38. Segovia, M., & Garcia-Alfaro, J. (2022). Design, modeling and implementation of digital twins. *Sensors, 22*(14), 5396.
39. Semeraro, C., et al. (2021). Digital twin paradigm: A systematic literature review. *Computers in Industry, 130*, 103469.
40. Singh, M., et al. (2021). Digital twin: Origin to future. *Applied System Innovation, 4*(2), 36.
41. Singh, M., et al. (2022). Applications of digital twin across industries: A review. *Applied Sciences, 12*(11), 5727.
42. Soori, M., Arezoo, B., & Dastres, R. (2023). Digital twin for smart manufacturing, a review. *Sustainable Manufacturing and Service Economics, 2*, 100017.
43. Sutton, R. T., et al. (2020). An overview of clinical decision support systems: Benefits, risks, and strategies for success. *NPJ Digital Medicine, 3*(1), 17.
44. Tao, F., et al. (2018). Data-driven smart manufacturing. *Journal of Manufacturing Systems, 48*, 157–169.
45. Tao, F., et al. (2022). Digital twin modeling. *Journal of Manufacturing Systems, 64*, 372–389.
46. VanDerHorn, E., & Mahadevan, S. (2021). Digital twin: Generalization, characterization and implementation. *Decision Support Systems, 145*, 113524.
47. Warke, V., et al. (2021). Sustainable development of smart manufacturing driven by the digital twin framework: A statistical analysis. *Sustainability, 13*(18), 10139.
48. West, S., et al. (2021). Digital twin providing new opportunities for value co-creation through supporting decision-making. *Applied Sciences, 11*(9), 3750.
49. Wu, B., et al. (2020). Battery digital twins: Perspectives on the fusion of models, data and artificial intelligence for smart battery management systems. *Energy and AI, 1*, 100016.

AI-Driven Decision Support Systems for Sustainable Environmental Management

Abdelmalek Hamada, Damla Karagozlu, and Labaran Isiaku

1 Introduction

1.1 Overview of Environmental Management

Environmental management is a systematic approach to dealing with environmental problems, focusing on sustainable improvement and the conservation of natural resources. The main objective is to produce a balance between economic improvement and environmental preservation, maintaining that the needs of the present are the same without balancing the ability of future generations to meet their own needs [1].

1.2 Importance of Decision Support Systems (DSS) in Sustainability

One of the effective benefits of DSS in sustainability is their ability to handle huge amounts of data from different sources. Environmental data, such as climate patterns, resource availability, and pollution levels, can be combined with economic and social data to give an overall view of the factors affecting sustainability. This point of view allows decision-makers to classify potential trade-offs and

A. Hamada · D. Karagozlu (✉) · L. Isiaku
Department of Management Information Systems, Cyprus International University, Nicosia, Turkey
e-mail: ahamada@ciu.edu.tr; dkaragozlu@ciu.edu.tr; lisiaku@ciu.edu.tr

N. Gupta, J. D. Diaz Roman (eds.), *Decision Support Systems for Industrial and Scientific Applications*, Signals and Communication Technology,
https://doi.org/10.1007/978-3-031-89700-9_4

synergies between different sustainability targets, such as economic development, environmental protection, and social equity [2].

DSS ease the stakeholder engagement and collaboration, which are crucial for sustainable decision-making process. These systems often have features that allow the visualization and communication data and scenarios in an accessible manner.

Decision support systems are crucial for improving sustainability as they enable data-driven, transparent, and collaborative decision-making processes. By implementing diverse types of information, modeling potential outcomes, and easing stakeholder engagement, DSS help organizations and policymakers make easier choices that promote long-term environmental, social, and economic success [2].

1.3 Role of AI in Enhancing Environmental Management

Artificial intelligence has a main role in enhancing environmental management by providing creative solutions for monitoring, measuring, and addressing environmental challenges [3].

AI strengthens environmental management through predictive modeling and scenario analysis. Machine learning models can predict the effect of various environmental policies and management practices under different scenarios [3].

2 AI Techniques for Environmental Management

2.1 Machine Learning Algorithms for Predictive Analysis

Machine learning algorithms for predictive analytics are impactful tools that allow organizations to predict future results based on previous literature and data. By strengthening these predictive possibilities, businesses and researchers can make informed decisions, optimize processes, and improve the whole performance [4].

Linear regression models the connection between a dependent variable and one or more independent variables by adding a linear equation to the observed data [5].

Another key algorithm is decision trees. Decision trees work by continuously dividing the data into subsets based on the values of input features, ultimately creating a tree-like figure where each branch represents a certain decision rule [6].

Random forests are a way of having multiple deep decision trees, which is also used in predictive analytics. A random forest consists of a group of decision trees, where each tree is trained on a random subset of the data and features. The final prediction is taken by the average of all trees [7].

2.2 *Optimization Techniques for Resource Management*

According to Abouelyazid [8], optimization techniques for resource management are the main tools that support organizations and systems in assigning their resources comfortably, maintaining maximum productivity and sustainability. By strengthening optimization methods, organizations can decrease costs, minimize waste, and boost overall operational efficiency.

Integer programming (IP) is different from linear programming because the decision variables must be integers. This technique is particularly useful when dealing with discrete resources, such as the allocation of personnel or the scheduling of tasks, where fractional solutions are not practical [9].

Dynamic programming (DP) is another main optimization method, mainly useful for issues that can be broken down into a series of connected decisions over time. It is used in industries like energy management [10].

2.3 *Remote Sensing and Data Analytics for Environmental Monitoring*

Remote sensing relates to the collection of information about an object or area from a distance, typically using satellite or aerial sensors. These sensors capture data in various electromagnetic spectrum wavelengths, involving visible light, infrared, and radar. This data is then used to create images or maps that shows information about the Earth's surface. Remote sensing is vital for monitoring large and often inaccessible areas, such as forests. It allows the detection of changes in land use, vegetation health, water bodies, and atmospheric conditions over time [11].

One of the main advantages of remote sensing is its ability to provide high resolution and frequent observations across wide areas. For instance, satellite imagery can be used to track deforestation, monitor agricultural crop health, and assess the effect of natural disasters such as floods and wildfires. Furthermore, remote sensing technology helps environmental research by offering data on climate change indicators, such as sea level rise and glacier retreat [12].

Data analytics complements remote sensing by transforming the raw data collected into actionable findings. Data analytics can be used to identify and predict changes in environmental monitoring [13, 14].

3 Applications in Environmental Management

AI-driven Decision Support Systems (DSS) are transforming sustainable environmental management by providing advanced tools for data analysis, prediction, and optimization [15–17]. In forest management, these systems can predict fire risks and biodiversity loss, enabling proactive conservation strategies.

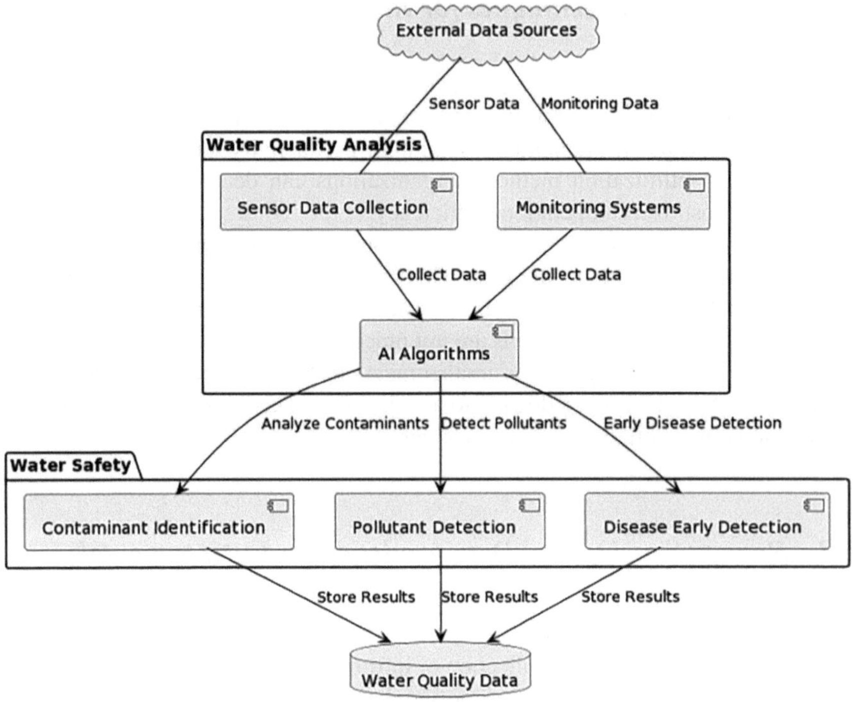

Fig. 1 Significant components and their relationships in an AI-powered water quality monitoring system. (Adapted from Nova [18])

3.1 Water Resource Management and Conservation

AI-driven Decision Support Systems (DSS) are revolutionizing water quality monitoring by providing real-time, accurate analysis of water bodies [18]. By analyzing historical data alongside real-time inputs, these systems can forecast water quality changes and recommend timely interventions to prevent environmental damage [19] (Fig. 1).

In this system the "Water Quality Analysis" module is essential for gathering and evaluating water quality data, while the "Water Safety" module focuses on ensuring the delivery of safe drinking water.

3.2 Air Quality Monitoring and Pollution Control

AI-driven Decision Support Systems (DSS) are considerably advancing air quality monitoring and pollution control by offering precise, real-time analysis and predictive capabilities [20]. AI-driven DSS can predict pollution trends, identify sources

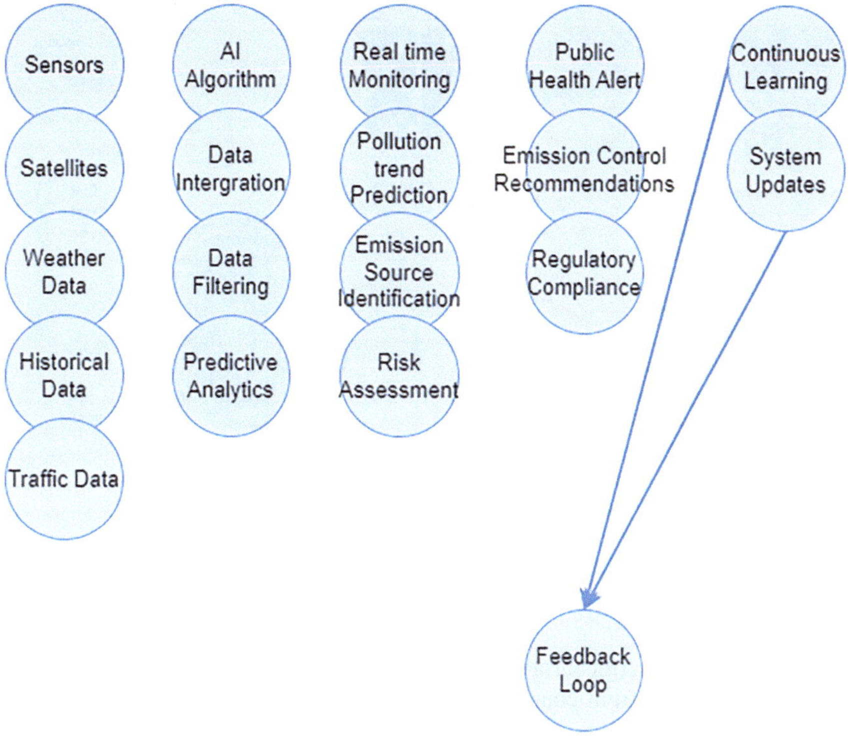

Fig. 2 AI-driven decision support systems for air quality monitoring and pollution control

of emissions, and assess the impact of various factors like weather conditions on air quality [20, 21].

Figure 2 above illustrates the structure of an AI-Driven Decision Support System (DSS) for air quality monitoring and pollution control. The system's Decision-Making Module uses this processed data to perform real-time monitoring, predict pollution trends, identify emission sources, and assess associated risks.

3.2.1 Optimized Maintenance Scheduling

Digital Twins allow manufacturers to optimize maintenance schedules based on the real-time condition of assets instead of adhering to predefined intervals [38, 39].

3.2.2 Asset Performance Management

Digital Twins provide a comprehensive perspective of asset performance, allowing manufacturers to monitor and evaluate vital performance indicators (KPIs) [40].

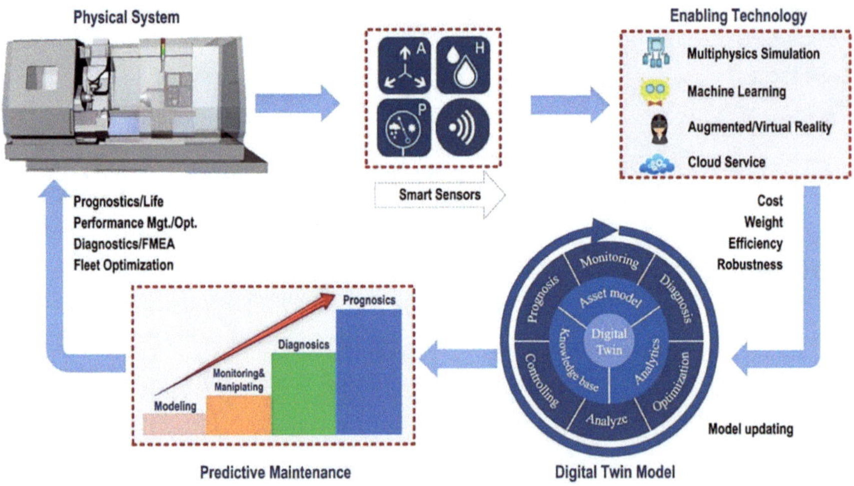

Fig. 3 A framework for problem diagnostics using Digital Twin [43]

3.2.3 Lifecycle Management

Digital Twins play a critical role in managing the lifecycle of assets, encompassing their entire lifespan from conception to retirement [41]. Through the integration of data from various stages of an asset's lifecycle, manufacturers can make well-informed decisions about upgrades, refurbishments, and replacements.

3.2.4 Failure Prediction and Prevention

Digital Twins use machine learning and predictive analytics to forecast asset failures [42]. Figure 3 depicts a framework for defect diagnosis that integrates a Digital Twin [43].

3.3 Waste Management and Recycling Optimization

AI-driven decision support systems (DSS) are transforming waste management and recycling by enhancing efficiency and sustainability [22–24]. AI algorithms also optimize collection routes and schedules, minimizing operational costs and carbon footprints [25] (Fig. 4).

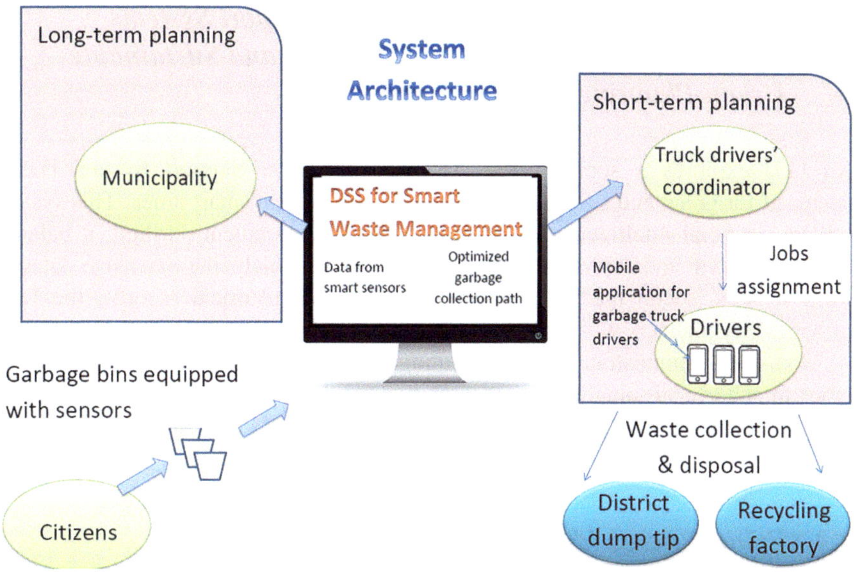

Fig. 4 The architecture of the smart waste management DSS. (Adapted from Rybnytska et al. [23])

3.4 Sustainable Agriculture and Land Use Planning

AI-driven decision support systems (DSS) are transforming sustainable agriculture and land use planning by enhancing efficiency and precision [26, 27]. In agriculture, AI DSS can predict crop yields, monitor soil health, and recommend personalized irrigation and fertilization strategies, thus promoting resource conservation and reducing environmental impact [28, 29].

4 AI-Driven Decision Supports Systems

AI-driven Decision Support Systems (DSS) leverage advanced algorithms, machine learning, and data analytics to process vast amounts of data, providing actionable insights and recommendations [15, 30]. Unlike traditional DSS, which often rely on predefined rules and static data, AI-driven systems can adapt to new information, learn from historical trends, and predict future outcomes with greater accuracy.

4.1 Case Studies of AI-Driven Decision Support Systems in Environmental Management: Green and Sustainable Urban Planning in Smart Cities

Xu [17] explores a sophisticated AI-driven Decision Support System (DSS) designed for green and sustainable urban planning within smart cities. This system utilizes artificial intelligence to optimize urban development, striking a balance between growth and environmental sustainability. By analyzing extensive datasets from sensors, social media, weather forecasts, and historical records, the DSS provides city planners with actionable insights.

This study demonstrates that integrating AI into urban planning enables smart cities to achieve sustainable growth, improve residents' quality of life, and reduce environmental impact.

4.2 Smart Energy Management

Lévy [31] examines an advanced AI-driven Decision Support System (DSS) designed for smart energy management. By applying machine learning algorithms, the AI-driven DSS predicts energy demand, optimizes the operation of energy systems, and proposes cost-effective strategies for energy consumption. Furthermore, the DSS improves the integration of renewable energy sources by forecasting their availability and optimizing their utilization.

4.3 Enhancing End-of-Life Value Recovery from E-Wastes

Simaei and Rahimifard [24] investigate an AI-driven Decision Support System (DSS) designed to enhance the recovery of end-of-life value from electronic wastes (e-wastes).

The study demonstrates that by improving the efficiency and effectiveness of value recovery, AI-driven DSS contributes significantly to sustainable resource management, reduces environmental impact, and supports the circular economy by ensuring valuable materials are reused rather than discarded.

5 Benefits and Impact

5.1 Improved Resource Efficiency and Sustainability

AI-driven DSS can evaluate big amounts of data to optimize the allocation of resources such as water, energy, and raw materials. By predicting demand, assessing availability, and considering environmental constraints, these systems maintain that resources are used precisely, reducing waste and decreasing environmental impact [32].

AI-driven DSS helps organizations in meeting with environmental regulations by offering real-time monitoring and reporting abilities. These systems can track resource usage, emissions, and environmental impacts, maintaining that organizations adhere to regulatory standards. Not only it helps in avoiding legal sanctions but also supports a culture of accountability and responsibility for the environment [33].

5.2 Enhanced Predictive Capabilities and Proactive Management

Predictive analytics given by AI-driven DSS supports identifying potential environmental risks before they grow into serious problems. By evaluating trends and patterns, these systems can suggest areas of concern, such as increasing pollution levels or potential habitat loss [12, 25].

Proactive management endorses long-term sustainability by dealing with potential environmental challenges before they become essential. By continuously monitoring and evaluating data, AI-driven systems can identify emerging trends, predict risks, and implement timely interventions that minimize environmental impact and support sustainable development [34, 35].

Proactive management facilitated by AI-driven DSS supports collaboration among stakeholders by providing a common platform for data sharing and decision-making [36, 37].

5.3 Reduced Environmental Footprint and Better Compliance

AI-driven DSS can evaluate complex datasets from different operational processes to identify inefficiencies that support in waste and emissions. By optimizing production schedules, resource use, and energy consumption, these systems support organizations decrease their environmental footprint. For example, in manufacturing, AI can adjust processes to reduce material waste, energy use, and greenhouse gas emissions, leading to a more sustainable operation overall [44].

AI-driven DSS can evaluate and interpret complex regulatory frameworks, helping organizations adjust to their practices in real-time to meet improving compliance requirements [45].

6 Challenges and Future Directions

6.1 Technical and Implementation Challenges

AI-driven DSS depend heavily on high-quality, comprehensive data to produce precise and practical insights. However, obtaining reliable environmental data can be tough due to challenges such as data fragmentation, inconsistencies, and gaps. In different cases, data from different sources may be incomplete or incompatible, making it difficult to integrate and evaluate.

Future research can focus on developing data collection methods and integrating different data sources more effectively. This contains improving new technologies for real-time data acquisition, strengthening data interoperability, and acknowledging issues related to data quality [45].

6.2 Ethical Considerations and Data Privacy Concerns

The use of AI in environmental management increases ethical and privacy concerns, certainly regarding data collection and usage. Maintaining that data is collected and used in compliance with privacy regulations and ethical standards is crucial [46].

AI systems can unintentionally amplify existing biases present in the data they are trained on. In environmental management, this could lead to inappropriate outcomes, such as excessive impacts on marginalized communities or biased resource allocation [47].

AI technologies can have wider environmental and social effects beyond their immediate applications. Furthermore, the adoption of AI-driven solutions may affect local communities and ecosystems in ways that need to be carefully analyzed [46, 48].

6.3 Future Research Opportunities and Advancements in AI for Environmental Management

One key area for future research is the advancement of developed predictive modelling methods. By taking advantage of new AI methods such as deep learning and ensemble models, researchers can develop the accuracy and granularity of environmental forecasts.

Future research can also address the ethical and equity challenges linked with AI in environmental management. This involves developing frameworks and techniques for maintaining fairness, transparency, and accountability in AI-driven decisions. Considering these ethical considerations will be vital for building trust and maintaining that AI contributes positively to sustainable advancement [47].

Research into the long-term effects of AI-driven environmental management practices is crucial for maintaining sustainability. Future research can focus on assessing the long-term environmental, social, and economic effects of AI interventions. This involves analyzing how AI-driven decisions affect future resource availability, ecosystem health, and community well-being.

7 Conclusion

AI-driven Decision Support Systems (DSS) play a pivotal role in advancing sustainable environmental management. By harnessing the power of artificial intelligence, these systems provide comprehensive tools for analyzing complex environmental data, predicting impacts, and optimizing resource use. They enhance decision-making processes in various domains, including urban planning, energy management, and waste recovery, by offering actionable insights and efficient solutions. In urban planning, AI-driven DSS help balance growth with sustainability by recommending optimal locations for green spaces and energy-efficient infrastructure. Overall, the integration of AI into environmental management not only promotes more informed and effective decisions but also supports the broader goals of sustainability and resilience, paving the way for an eco-friendlier future.

References

1. Marchese, D., et al. (2018). Resilience and sustainability: Similarities and differences in environmental management applications. *Science of the Total Environment, 613*, 1275–1283.
2. Deveci, M., et al. (2022). A decision support system for assessing and prioritizing sustainable urban transportation in metaverse. *IEEE Transactions on Fuzzy Systems, 31*(2), 475–484.
3. Nishant, R., Kennedy, M., & Corbett, J. (2020). Artificial intelligence for sustainability: Challenges, opportunities, and a research agenda. *International Journal of Information Management, 53*, 102104.
4. Bowles, M. (2019). *Machine learning with spark and python: Essential techniques for predictive analytics*. Wiley.
5. Maulud, D., & Abdulazeez, A. M. (2020). A review on linear regression comprehensive in machine learning. *Journal of Applied Science and Technology Trends, 1*(2), 140–147.
6. Charbuty, B., & Abdulazeez, A. (2021). Classification based on decision tree algorithm for machine learning. *Journal of Applied Science and Technology Trends, 2*(01), 20–28.
7. Reis, I., Baron, D., & Shahaf, S. (2018). Probabilistic random forest: A machine learning algorithm for noisy data sets. *The Astronomical Journal, 157*(1), 16.
8. Abouelyazid, M. (2021). Machine learning algorithms for dynamic resource allocation in cloud computing: Optimization techniques and real-world applications. *Journal of AI-Assisted Scientific Discovery, 1*(2), 1–58.
9. Hooker, J. N. (2024). Integer programming duality. In *Encyclopedia of optimization* (pp. 1–13). Springer.
10. Bertsekas, D. (2022). *Abstract dynamic programming*. Athena Scientific.
11. Li, J., et al. (2020). A review of remote sensing for environmental monitoring in China. *Remote Sensing, 12*(7), 1130.

12. Wang, L., et al. (2021). Environmental fate, toxicity and risk management strategies of nanoplastics in the environment: Current status and future perspectives. *Journal of Hazardous Materials, 401*, 123415.
13. Huang, Y., et al. (2018). Agricultural remote sensing big data: Management and applications. *Journal of Integrative Agriculture, 17*(9), 1915–1931.
14. Mohindru, G., Mondal, K., & Banka, H. (2020). Internet of things and data analytics: A current review. *Wiley Interdisciplinary Reviews: Data Mining and Knowledge Discovery, 10*(3), e1341.
15. Elshaikh, A., Mabrouki, J., & Mohamed, A. A. O. (2024). The future of water management: Leveraging AI for effective decision support. In *Advancements in climate and smart environment technology* (pp. 213–225). IGI Global.
16. Gupta, S., et al. (2022). Artificial intelligence for decision support systems in the field of operations research: Review and future scope of research. *Annals of Operations Research, 308*(1), 215–274.
17. Xu, C. (2024). AI-driven decision support system for green and sustainable urban planning in smart cities. *Applied Mathematics and Nonlinear Sciences, 9*(1).
18. Nova, K. (2023). AI-enabled water management systems: An analysis of system components and interdependencies for water conservation. *Eigenpub Review of Science and Technology, 7*(1), 105–124.
19. Gunasekaran, K., & Boopathi, S. (2023). Artificial intelligence in water treatments and water resource assessments. In *Artificial intelligence applications in water treatment and water resource management* (pp. 71–98). IGI Global.
20. Miles, A., Zaslavsky, A., & Browne, C. (2018). IoT-based decision support system for monitoring and mitigating atmospheric pollution in smart cities. *Journal of Decision Systems, 27*(sup1), 56–67.
21. Popescu, S. M., et al. (2024). Artificial intelligence and IoT driven technologies for environmental pollution monitoring and management. *Frontiers in Environmental Science, 12*, 1336088.
22. Lanzalonga, F., et al. (2024). The application of artificial intelligence in waste management: Understanding the potential of data-driven approaches for the circular economy paradigm. *Management Decision*.
23. Rybnytska, O., et al. (2018). Decision support for optimizing waste management. *Journal of Decision Systems, 27*(sup1), 68–78.
24. Simaei, E., & Rahimifard, S. (2024). AI-based decision support system for enhancing end-of-life value recovery from e-wastes. *International Journal of Sustainable Engineering, 17*(1), 1–17.
25. Wu, C.-J., et al. (2022). Sustainable ai: Environmental implications, challenges and opportunities. *Proceedings of Machine Learning and Systems, 4*, 795–813.
26. Adewusi, A. O., et al. (2024). AI in precision agriculture: A review of technologies for sustainable farming practices. *World Journal of Advanced Research and Reviews, 21*(1), 2276–2285.
27. Asolo, E., Gil-Ozoudeh, I., & Ejimuda, C. (2024). AI-powered decision support systems for sustainable agriculture using AI-Chatbot solution. *Journal of Digital Food, Energy & Water Systems, 5*(1).
28. Fuentes-Peñailillo, F., et al. (2024). Transformative technologies in digital agriculture: Leveraging internet of things, remote sensing, and artificial intelligence for smart crop management. *Journal of Sensor and Actuator Networks, 13*(4), 39.
29. Hassan, M., Malhotra, K., & Firdaus, M. (2022). Application of artificial intelligence in IoT security for crop yield prediction. *ResearchBerg Review of Science and Technology, 2*(1), 136–157.
30. Fathia, A. (2024). Combining AI-driven knowledge frameworks with ML for real-time financial decision support and automation.
31. Lévy, L.-N. (2024). *Advanced clustering and AI-driven decision support systems for smart energy management*. Université Paris-Saclay.

32. Halabian, H. (2019). Distributed resource allocation optimization in 5G virtualized networks. *IEEE Journal on Selected Areas in Communications, 37*(3), 627–642.
33. Lihua, W. U., et al. (2020). Improvement of regional environmental quality: Government environmental governance and public participation. *Science of the Total Environment, 717*, 137265.
34. Koul, B., Yakoob, M., & Shah, M. P. (2022). Agricultural waste management strategies for environmental sustainability. *Environmental Research, 206*, 112285.
35. Yacob, P., Wong, L. S., & Khor, S. C. (2019). An empirical investigation of green initiatives and environmental sustainability for manufacturing SMEs. *Journal of Manufacturing Technology Management, 30*(1), 2–25.
36. Reiter, D., Meyer, W., & Parrott, L. (2019). Stakeholder engagement with environmental decision support systems: The perspective of end users. *The Canadian Geographer/Le Géographe canadien, 63*(4), 631–642.
37. Reza, M., & Hassan, M. (2023). AI-driven solutions for enhanced waste management and recycling in urban areas. *International Journal of Sustainable Infrastructure for Cities and Societies, 8*(2), 1–13.
38. Friederich, J., Francis, D. P., Lazarova-Molnar, S., & Mohamed, N. (2022). A framework for data-driven digital twins of smart manufacturing systems. *Computers in Industry, 136*, 103586.
39. Negri, M., Cagno, E., Colicchia, C., & Sarkis, J. (2021). Integrating sustainability and resilience in the supply chain: A systematic literature review and a research agenda. *Business Strategy and the Environment, 30*(7), 2858–2886.
40. Mihai, S., Yaqoob, M., Hung, D. V., Davis, W., Towakel, P., Raza, M., ... & Nguyen, H. X. (2022). Digital twins: A survey on enabling technologies, challenges, trends and future prospects. *IEEE Communications Surveys & Tutorials, 24*(4), 2255–2291.
41. Hu, W., Zhang, T., Deng, X., Liu, Z., & Tan, J. (2021). Digital twin: A state-of-the-art review of its enabling technologies, applications and challenges. *Journal of Intelligent Manufacturing and Special Equipment, 2*(1), 1–34.
42. Van Dinter, R., Tekinerdogan, B., & Catal, C. (2022). Predictive maintenance using digital twins: A systematic literature review. *Information and Software Technology, 151*, 107008.
43. Soori, M., Arezoo, B., & Dastres, R. (2023). Digital twin for smart manufacturing, a review. *Sustainable Manufacturing and Service Economics, 2*, 100017.
44. Matuštík, J., & Kočí, V. (2021). What is a footprint? A conceptual analysis of environmental footprint indicators. *Journal of Cleaner Production, 285*, 124833.
45. DhivyaShree, M., Vishnu Durai, R. S., & Pavithra, R. (2024, March). An AI-driven model for decision support systems. In *International Conference on Artificial Intelligence and Smart Energy* (pp. 75–84). Cham: Springer Nature Switzerland.
46. Singh, S. K., Chen, J., Del Giudice, M., & El-Kassar, A. N. (2019). Environmental ethics, environmental performance, and competitive advantage: Role of environmental training. *Technological Forecasting and Social Change, 146*, 203–211.
47. Rolston III, H. (2020). *A new environmental ethics: the next millennium for life on earth*. Routledge.
48. Ahmad, N., Scholz, M., AlDhaen, E., Ullah, Z., & Scholz, P. (2021). Improving firm's economic and environmental performance through the sustainable and innovative environment: evidence from an emerging economy. *Frontiers in Psychology, 12*, 651394.

Harnessing Big Data and Advanced Analytics for Intelligent Traffic Management

Bhupendra Soni, Denis Jangeed, Sunil Sharma, Kavya Soni, and Polat Goktas

1 Introduction

Managing urban traffic is becoming increasingly complex due to rapid urbanization and the increasing figure of vehicles on the roads. Traditional traffic management schemes that rely on permanent signal timings and manual interventions are no longer adequate to handle modern traffic challenges. As a result, intelligent traffic management systems (ITMS) that leverage Big Data and advanced analytics are gaining momentum. These systems collect data from many sources such as GPS, mobile devices, sensor networks, and real-time monitoring systems to enhance traffic flow, develop road welfare, and improve public transportation [1, 2]. Big Data technologies enable the group and handling of huge amounts of traffic data, while innovative analytics and machine learning algorithms assist in making real-time predictions and decisions [3, 4]. From predicting traffic congestion to optimizing traffic signal timings, these technologies offer a proactive approach to traffic management. In addition to improving the productivity of public transport

B. Soni (✉)
Department of ECE, ACEIT, Jaipur, India

D. Jangeed
Department of CE, GITS, Udaipur, India

S. Sharma
Department of ECE, GITS, Udaipur, India

K. Soni
Department of CSE, Manipal University, Jaipur, India

P. Goktas
UCD School of Computer Science, University College Dublin (UCD), Dublin, Ireland
e-mail: polat.goktas@ucd.ie

N. Gupta, J. D. Diaz Roman (eds.), *Decision Support Systems for Industrial and Scientific Applications*, Signals and Communication Technology,
https://doi.org/10.1007/978-3-031-89700-9_5

systems, ITMS helps reduce fuel consumption and emissions, contributing to environmental sustainability [5]. However, implementing ITMS is not without challenges. Issues such as data privacy, interoperability between different traffic management systems, and the need for standardized protocols must be addressed to confirm the successful adoption of these technologies [6, 7]. This chapter explores how Big Data and advanced analytics are transforming traffic management and the future of smart urban mobility. This chapter delves into how Big Data and advanced analytics can be leveraged to create ITMS, focusing on key data sources, predictive models, optimization algorithms, and the role of emerging technologies in traffic management.

1.1 Flow Chart to Implement Intelligent Traffic Management

The flowchart in Fig. 1 outlines the process of implementing Big Data and Advanced Analytics for Intelligent Traffic Management. It starts with data collection from sources like GPS and sensors, followed by storage and processing using cloud computing. After data cleaning and preparation, advanced analytics generate actionable insights for real-time monitoring and incident response. Optimization algorithms enhance traffic flow by adjusting signals and routes, while public

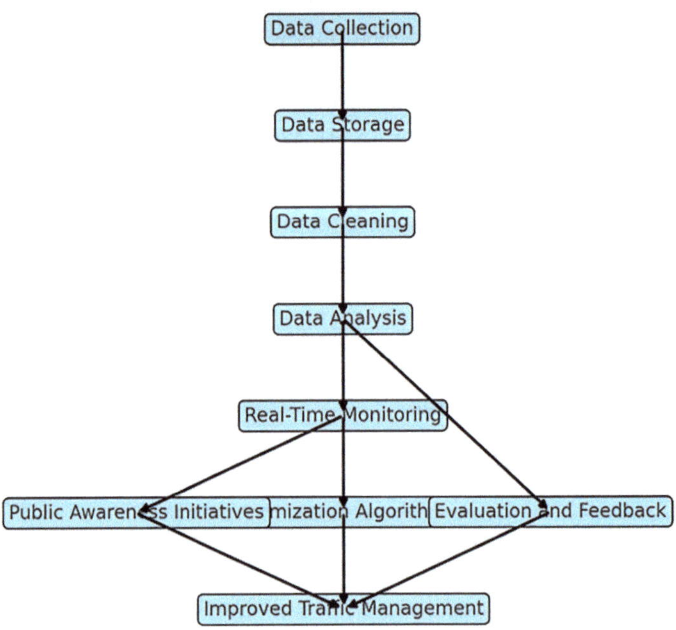

Fig. 1 Flow chart to implement intelligent traffic management

awareness initiatives keep the community informed. Continuous evaluation and feedback ensure the system adapts to changing traffic patterns, fostering smarter urban mobility.

2 Big Data Sources in Traffic Management

Efficient traffic management is crucial in today's urban environments, where traditional methods struggle to adapt. Big Data, derived from sources like GPS, mobile devices, sensors, cameras, and social media, has transformed traffic management by providing real-time insights and predictive capabilities. Figure 2 illustrates the various data sources—GPS, mobile data, sensor networks, traffic cameras, social media, weather data, and road infrastructure data—showing how they contribute to effective traffic management through interconnected data collection and analysis.

2.1 GPS and Mobile Data

GPS and mobile data have emerged as primary sources of Big Data in traffic management. Millions of vehicles and mobile devices equipped with GPS generate data that is invaluable for monitoring real-time traffic conditions. This data includes vehicle speed, travel routes, and locations, which can be aggregated to identify congestion points, optimize traffic flow, and predict travel times. As reported by *Transportation Data Analytics Journal*, the integration of GPS data with mobile apps allows cities to create more responsive and adaptive traffic systems [1]. GPS data not only helps city planners identify where congestion occurs but also helps

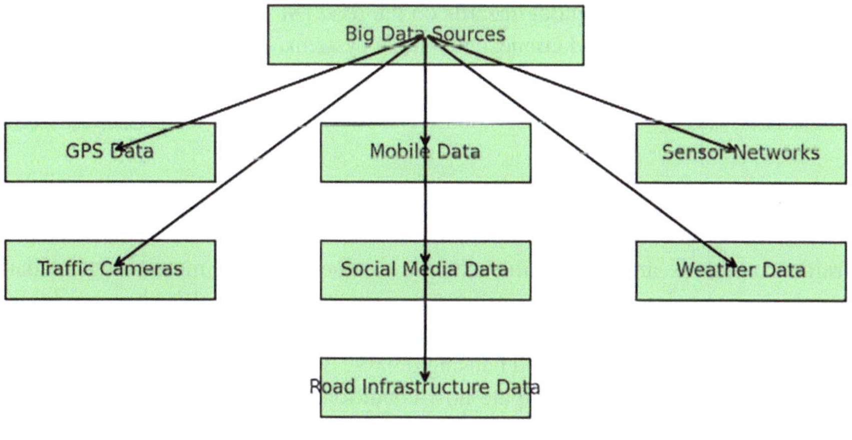

Fig. 2 Big Data sources in traffic management

them understand why it happens. Navigation apps such as Google Maps and Waze rely deeply on GPS data to provide users with real-time information about traffic conditions and alternate routes, reducing travel time and fuel consumption.

2.2 Sensor Networks

Sensor networks are another fundamental source of traffic data. These networks consist of numerous sensors, such as loop detectors, radar, infrared cameras, and inductive loops, embedded in road infrastructure. These instruments collect data on vehicle speed, traffic density, and vehicle classification, which provide a detailed understanding of traffic conditions. According to the *Journal of Urban Transportation*, sensor networks are critical for monitoring high-traffic areas like intersections and highways, enabling cities to adjust traffic indication timings in real time [2]. Additionally, sensors can detect traffic occurrences such as accidents or stalled vehicles, which can then trigger immediate responses like dispatching emergency services or alerting drivers to changing road conditions. These networks are key to creating adaptive traffic management systems that continuously study from and respond to the flow of traffic.

2.3 Real-Time Traffic Monitoring

Real-time traffic monitoring is needed for the dynamic adjustment of traffic management strategies. With the aid of sensors and traffic cameras, real-time data is self-possessed and transmitted to traffic control centers where it is processed for immediate action. Traffic cameras, for instance, provide visual data that can be analyzed using computer vision algorithms to detect congestion, analyze vehicle behavior, and identify potential hazards on the road [8]. This real-time data is also crucial for autonomous and connected vehicle systems, which rely on precise, up-to-the-minute evidence to navigate safely and efficiently.

2.4 Weather and Environmental Data

Traffic conditions are often influenced by weather, making meteorological data a crucial source for intelligent traffic management systems. Weather conditions such as rain, snow, or fog can dramatically affect vehicle behavior, reducing speeds, increasing accidents, and causing delays. By integrating weather data with traffic systems, cities can anticipate how conditions will impact traffic and adjust accordingly. For instance, if a heavy storm is predicted, traffic control systems can reroute vehicles or adjust signal timings to prevent accidents and reduce congestion.

This environmental data can also be combined with other sources, such as air quality sensors, to develop systems that optimize routes not only for congestion reduction but also for minimizing environmental impacts like emissions and pollution.

2.5 Public Transport Data

Another valuable source of traffic data comes from public transportation systems. Bus and train networks generate data on passenger flow, service delays, and capacity. Integrating this data into traffic management systems can provide a more holistic view of urban mobility. As public transportation often shares road space with private vehicles, understanding the interaction between these two can help optimize both traffic signals and bus schedules to reduce overall congestion [5].

2.6 Challenges with Big Data Sources

Despite the vast opportunities that Big Data presents, traffic management systems face challenges in handling and processing these large datasets. The diversity in data types—ranging from structured sensor data to unstructured social media posts—requires sophisticated tools for integration and analysis. Data storage solutions, such as cloud computing platforms, are crucial for managing these high volumes of data [3]. Additionally, ensuring data accuracy and cleanliness through robust data cleaning techniques is essential to avoid faulty analyses and predictions [9].

3 Advanced Analytics in Traffic Management

The emergence of advanced analytics in traffic management has revolutionized how cities handle the ever-increasing complexities of urban mobility. Traditional traffic systems, which were largely reactive and based on static models, have given way to data-driven approaches that utilize real-time information and predictive insights. Advanced analytics, which comprises machine learning (ML), artificial intelligence (AI), data visualization, and optimization techniques, enables traffic authorities to monitor, predict, and optimize traffic conditions dynamically.

3.1 Data Cleaning and Pre-processing

Before any analytics can be applied, it is essential to clean and pre-process the raw data composed from various birthplaces such as sensors, GPS devices, and social media. Traffic data, especially from real-time sources, often contains noise, inaccuracies, or missing values due to sensor failures, transmission errors, or inconsistent reporting. Advanced data cleaning techniques, such as outlier detection and imputation methods, confirm that the data is reliable and accurate, paving the way for more effective analytics [9]. Data pre-processing is another critical step, as traffic data comes in various formats and needs to be transformed into a unified structure. This involves standardizing data formats, dealing with missing values, and eliminating redundancies to create a cohesive dataset that can be analyzed more efficiently. Pre-processing ensures that predictive models and optimization algorithms function with accurate data, enhancing their overall effectiveness.

3.2 Predictive Analytics and Machine Learning

Machine learning (ML) plays a vital role in advanced traffic management systems, especially in predictive analytics, which uses historical and real-time data to forecast traffic conditions. Algorithms like regression models, decision trees, and deep learning networks predict congestion, travel times, and accident risks. For example, ML models analyze traffic speed, road occupancy, and weather conditions to enable proactive measures such as adjusting signal timings or rerouting traffic [4]. Additionally, long-term forecasts based on trends, population growth, and seasonal patterns help urban planners make data-driven decisions on infrastructure and public transport improvements [10].

3.3 Optimization Algorithms

Optimization algorithms are crucial for enhancing traffic management by solving traffic-related challenges. Techniques like linear programming, genetic algorithms, and swarm intelligence optimize signal timings, vehicle routes, and public transport schedules. Adaptive signal control, using real-time data from sensors and cameras, adjusts signal timings dynamically to reduce congestion and wait times, lowering fuel consumption and emissions [11]. Additionally, algorithms like Dijkstra's and A* optimize vehicle routing, improving travel efficiency and minimizing fuel usage, while diversion strategies help balance traffic flow during peak periods

3.4 Data Visualization

Data visualization plays a critical role in presenting complex traffic data in a way that is easy for decision-makers to understand. Advanced visualization techniques, such as heat maps, time-series graphs, and interactive dashboards, help traffic authorities monitor traffic conditions in real-time and identify patterns or anomalies. By visually representing traffic flow, congestion points, and accident-prone areas, decision-makers can quickly grasp the state of the traffic system and take appropriate actions [12]. Real-time dashboards provide an integrated view of traffic conditions across a city, showing traffic flow, sensor data, and predicted congestion points in one place. These dashboards enable traffic control centers to monitor the system continuously and react promptly to any disruptions. Moreover, historical data visualization allows urban planners to analyze long-term trends and make informed decisions about road infrastructure, public transport, and future traffic demands.

3.5 *Real-Time Traffic Monitoring*

Advanced analytics also enables real-time traffic monitoring, where data from sensors, GPS devices, and cameras is continuously collected and analyzed to provide up-to-the-minute insights into traffic conditions. Real-time monitoring allows for instant detection of traffic incidents, such as accidents, breakdowns, or construction delays, enabling authorities to respond quickly and prevent further congestion [8]. For instance, if an accident is detected on a highway, real-time monitoring systems can reroute traffic, change digital signboards, and alert emergency responders, minimizing disruptions and improving road safety.

4 Optimization and Real-Time Traffic Monitoring

Optimization and real-time traffic monitoring are two critical components of modern intelligent traffic management systems (ITMS), enabling cities to efficiently manage congestion, reduce travel times, and improve road safety. By leveraging advanced algorithms and continuous data feeds from various sources, these systems can make dynamic adjustments to traffic flow, signal timings, and vehicle routing in response to changing conditions.

4.1 *Optimization in Traffic Management*

Optimization in traffic management uses mathematical and computational models to enhance efficiency by minimizing travel time, reducing congestion, and improving traffic flow. Algorithms like linear programming, dynamic programming, and metaheuristic techniques (e.g., genetic algorithms and particle swarm optimization) analyze real-time data to dynamically adjust signal timings, reducing waiting times and preventing bottlenecks [11]. These methods prioritize emergency vehicles, public transport, or pedestrians when needed. Optimization is also applied in vehicle routing and public transport scheduling, ensuring efficient delivery routes and timely services while balancing travel time, fuel consumption, and traffic demand.

4.2 Real-Time Traffic Monitoring

Real-time traffic monitoring involves collecting and analyzing live data using sensors, cameras, GPS systems, and communication networks to manage traffic flow, detect incidents, and provide actionable insights. It enables authorities to make swift decisions to prevent congestion, reroute vehicles, and respond to accidents. Technologies like computer vision enhance monitoring by detecting issues such as accidents or stalled vehicles through video analysis [8]. Real-time data supports quick interventions, such as updating traffic signals or rerouting vehicles via navigation apps. It also plays a vital role in autonomous vehicles, providing the continuous data needed for safe navigation and effective coordination with infrastructure.

4.3 Benefits of Optimization and Real-Time Monitoring

Integrating optimization and real-time monitoring in traffic management improves efficiency, safety, and sustainability. Adaptive signal control and optimized routing reduce travel times, congestion, fuel consumption, and emissions. Real-time monitoring enhances road safety by quickly detecting accidents and adjusting traffic conditions during hazards like heavy rain or fog. These technologies also boost public transport efficiency by enabling schedule adjustments based on traffic conditions, reducing delays, and encouraging public transit use, which alleviates congestion further.

4.4 Future Directions

Looking forward, the continued development of optimization algorithms and real-time monitoring technologies will further transform traffic management. The rise of autonomous and connected vehicles will create new opportunities for real-time optimization, as vehicles, sensors, and traffic infrastructure become more interconnected. In addition, advancements in machine learning and artificial intelligence will enable more sophisticated predictive models that can anticipate traffic patterns and adjust systems proactively. Smart cities will rely heavily on optimization and real-time monitoring to ensure smooth urban mobility. As cities grow and traffic demand increases, the ability to manage traffic dynamically and efficiently will be critical to maintaining livability and sustainability in urban environments.

5 Enhancing Public Transport Systems

Public transportation is a critical element of urban mobility, providing a sustainable and efficient alternative to private vehicle use. In recent years, enhancing public transport systems has become a central focus for urban planners and policymakers worldwide as cities grapple with rising congestion, environmental concerns, and increasing demand for mobility. Through the integration of advanced data analytics, optimization algorithms, and emerging technologies, cities can significantly improve the efficiency, reliability, and accessibility of public transport systems, encouraging more people to shift away from car-centric transportation models.

5.1 Data-Driven Scheduling and Optimization

Data-driven approaches optimize public transport by dynamically adjusting schedules, routes, and vehicle allocation based on real-time demand. Traditional static timetables often caused inefficiencies, but advancements in data collection from GPS, fare systems, and passenger counters now enable predictive analytics to forecast demand and optimize capacity [5]. Algorithms analyze historical data to redesign routes, reduce delays, and address bottlenecks. Some cities implement dynamic routing, where buses adjust routes in real time based on passenger requests and traffic, improving efficiency and passenger satisfaction.

5.2 Improving Reliability and Reducing Wait Times

Reliability is crucial for encouraging public transport use, and technologies like real-time monitoring, predictive analytics, and adaptive scheduling enhance this reliability. Real-time monitoring tracks vehicles, allowing operators to adjust services for delays or disruptions, while mobile apps provide passengers with accurate arrival times and updates. Predictive maintenance algorithms detect potential vehicle or infrastructure issues before they cause breakdowns, using sensors to monitor conditions and proactively schedule maintenance. These technologies ensure smoother operations and reduce service disruptions, improving the overall passenger experience.

5.3 Enhancing Accessibility and Inclusivity

Improving public transport accessibility is crucial for inclusivity, benefiting passengers with disabilities, the elderly, and underserved communities. Technology aids this effort with features like voice guidance for visually impaired users, wheelchair accessibility info, and real-time updates on elevator availability in transport apps. Digital payment systems further simplify fare payments, removing barriers posed by traditional ticketing. Expanding transport coverage to underserved areas through data-driven route adjustments ensures access to employment, education, and healthcare. Micro-transit services, using smaller vehicles and on-demand apps, complement traditional networks by offering flexible, responsive options for low-demand or hard-to-reach areas.

5.4 Environmental Sustainability and Reducing Carbon Emissions

Enhancing public transport reduces urban transportation's environmental impact by decreasing private vehicle use, lowering emissions, and reducing energy consumption. Cities are adopting electric, hybrid, and hydrogen-powered buses, trams, and trains to further cut carbon emissions and promote sustainability [5]. Multimodal transport hubs integrating public transit with cycling and walking options encourage eco-friendly travel, creating more efficient and sustainable urban mobility

5.5 Case Study: Enhancing Public Transport Systems—Singapore's Smart Public Transport

5.5.1 Introduction

Singapore is renowned for its efficient and innovative public transport system, consistently ranked among the best globally. Facing rapid urbanization, increasing congestion, and environmental challenges, the city-state has prioritized enhancing public transport as a key solution to sustainable mobility. Through a combination of data-driven optimization, advanced technology integration, and a focus on accessibility and sustainability, Singapore has developed a highly efficient and reliable public transportation network. This case study explores how Singapore has enhanced its public transport system, offering valuable insights into improving urban mobility.

5.5.2 Key Initiatives

1. *Data-Driven Optimization*

   One of Singapore's most significant advancements has been the integration of data analytics to optimize bus and train operations. By using real-time data from GPS systems, traffic sensors, and passenger load monitors, the LTA implemented adaptive scheduling, matching services with demand. For instance, during peak hours, additional buses are deployed on high-demand routes to alleviate overcrowding, while off-peak schedules are adjusted to maintain efficiency without excessive resource use.

2. *Real-Time Information Systems*

   Singapore's public transport system has embraced real-time data sharing with passengers. The MyTransport.SG app provides real-time information on bus and train arrivals, service disruptions, and travel options. Commuters can plan their journeys based on real-time data, minimizing waiting times and improving overall convenience. Digital displays at stations and bus stops also provide live updates, further enhancing transparency and trust in the public transport system.
3. *Multimodal Integration*

   Singapore's public transport system emphasizes seamless integration between different modes of transportation, including buses, MRT (Mass Rapid Transit), LRT (Light Rail Transit), and cycling. Key transport hubs allow for smooth transfers between buses, trains, and even cycling infrastructure. The EZ-Link card system simplifies payment across all public transport modes, enhancing the convenience of multimodal travel.
4. *Sustainable Solutions*

   Singapore has invested in electric buses and a fully electrified MRT and LRT system to reduce the carbon footprint of its public transport network. The city aims to replace all diesel buses with electric ones by 2040, aligning with its broader sustainability goals. The adoption of clean energy in public transport helps reduce emissions and improve air quality.
5. *Results and Impact*

   These initiatives have transformed Singapore's public transport system into one of the world's most efficient. As of 2022, over 80% of Singapore's residents live within a 10-min walk of an MRT station, and public satisfaction rates remain high, with over 85% of commuters expressing confidence in the system. The shift toward electric buses has also contributed to the city's sustainability targets by reducing carbon emissions.

6 Future Directions

The future of intelligent traffic management (ITMS) is promising, with advancements in AI and machine learning enabling more autonomous systems that can make real-time decisions to optimize traffic flow. AI will analyze complex traffic patterns and predict congestion, while autonomous and connected vehicles will enhance coordination by sharing data with each other and traffic control systems. ITMS will also focus on sustainability, reducing emissions, optimizing energy use, and promoting public transport and cycling infrastructure. Advanced analytics will help minimize fuel consumption and lessen the environmental impact of traffic [11].

7 Technological Advancements

The rapid evolution of technology is revolutionizing public transport systems. From real-time monitoring and predictive analytics to autonomous vehicles and smart payment systems, technology is driving improvements in efficiency, safety, and user experience. Figure 3 illustrates technological advancements in intelligent traffic management. It visualizes key components such as Artificial Intelligence (AI), Internet of Things (IoT), Autonomous Vehicles, and Advanced Sensors, showing how they interact to enhance traffic flow, optimize public transport, and reduce congestion through data integration, real-time adjustments, and decision-making processes.

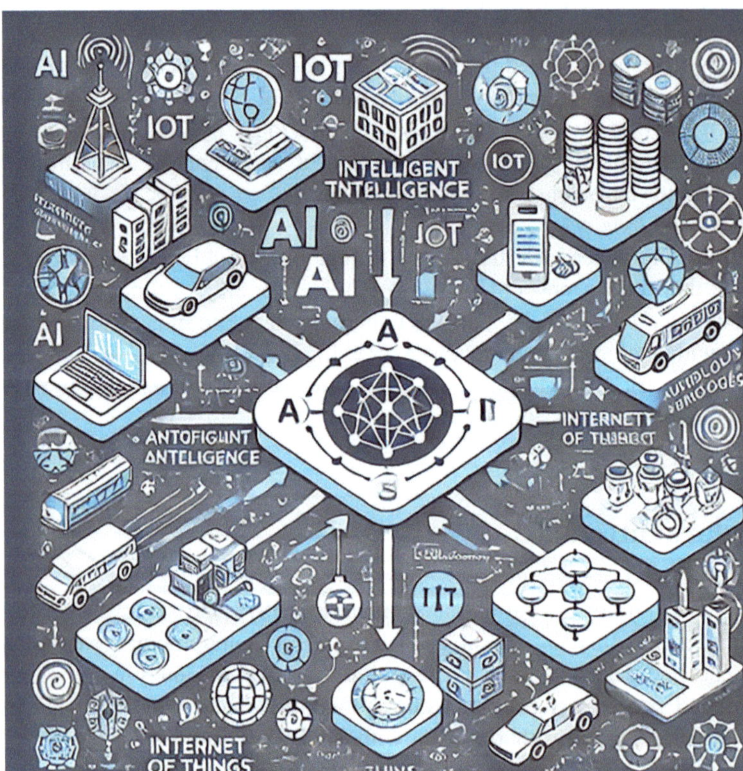

Fig. 3 Technological advancements in intelligent traffic management

(a) *Intelligent Transportation Systems (ITS)*

Intelligent transportation systems (ITS) use advanced technologies, including sensors, GPS, and data analytics, to optimize the operation of public transport systems. Real-time data from buses, trains, and traffic conditions is collected and analyzed to improve scheduling, route planning, and vehicle allocation [13]. For example, many cities use real-time data to dynamically adjust bus schedules based on demand and traffic conditions, reducing wait times and improving service reliability. Predictive analytics, powered by machine learning, is being used to anticipate demand for public transport services. By analyzing historical data on passenger movements, transport authorities can forecast peak travel times and adjust services accordingly, ensuring that sufficient capacity is available when needed.

(b) *Autonomous and Electric Vehicles*

The development of autonomous public transport vehicles has the potential to revolutionize urban mobility. Autonomous buses and shuttles, currently being tested in cities like Helsinki and Las Vegas, promise to reduce operational costs and improve safety by eliminating human error. In addition, autonomous vehicles can operate on-demand, adjusting routes and schedules dynamically to respond to real-time passenger requests. Electric vehicles are also gaining traction in public transport systems as governments seek to reduce carbon emissions. Electric buses, which produce no tailpipe emissions, are being adopted in cities across the world. For example, China has the largest fleet of electric buses, with cities like Shenzhen leading the way in transitioning to fully electric public transport.

(c) *Smart Payment Systems*

Technological advancements in payment systems have made public transport more convenient for users. Smart cards and mobile payment apps, such as Singapore's EZ-Link and London's Oyster card, allow passengers to pay fares with a tap of their card or smartphone. Contactless payment systems streamline boarding and reduce the need for physical tickets, making public transport more accessible and user-friendly. These systems also provide transport operators with valuable data on passenger movements, which can be used to optimize services.

(d) *Data-Driven Decision-Making*

Big data analytics plays an essential role in modern public transport systems. Transport authorities collect vast amounts of data from GPS devices, sensors, ticketing systems, and mobile apps, which are analyzed to improve operational efficiency. Data-driven decision-making enables transport operators to optimize routes, schedules, and vehicle maintenance. In addition, data is used to monitor passenger behavior and identify areas where service improvements are needed, ensuring that public transport networks can respond to changing demands.

7.1 Case Study: Policy and Technological Advancements in Public Transport Systems

Category	Description	Example cities	Impact
Sustainability policies	Governments are introducing policies focused on reducing emissions through electrification and cleaner energy use	London, Paris, Shenzhen	Reduced air pollution, lower greenhouse gas emissions
Low-emission zones (LEZs)	Policies limiting vehicle access based on emission standards to promote cleaner transportation	London, Paris	Encouraged use of electric and hybrid buses, improving air quality
Public-private integration	Promoting integration between public transport and private mobility services for seamless travel	San Francisco, London	Enhanced user experience with Mobility-as-a-Service (MaaS), reduced congestion
Accessibility and inclusivity	Policies aimed at making transport more accessible for disabled, elderly, and underserved communities	New York, Singapore	Broader accessibility, more inclusive public transport
Electrification	Transitioning to electric and hydrogen-powered public buses to reduce fuel consumption and emissions	Shenzhen, Oslo	Reduced emissions, lower operational costs, improved air quality
AI and Big Data analytics	Using AI for route optimization, passenger flow analysis, and schedule adjustments to improve efficiency	Singapore, Berlin	Increased operational efficiency, reduced wait times, better resource allocation
Autonomous vehicles (AV)	Piloting autonomous shuttles to enhance convenience and serve underserved areas	Las Vegas, Helsinki	Reduced labor costs, decreased traffic congestion, expanded public transport options
Real-time data and IoT	Using IoT sensors and real-time data to improve vehicle management and monitor system health	Los Angeles, Berlin	Reduced breakdowns, improved fleet management
Policy for smart infrastructure	Investment in smart infrastructure like smart traffic lights, charging stations, and connected systems	Amsterdam, Tokyo	Smoother integration of electric and autonomous vehicles, improved service delivery

(continued)

Category	Description	Example cities	Impact
Environmental and social impact	Policies aimed at increasing sustainability and improving public health through green public transport	Stockholm, Copenhagen	Reduced carbon footprint, healthier urban environments
Challenges	Infrastructure costs, cyber security risks, and data privacy concerns need to be addressed for full success	Global	Continuation of technological upgrades, security protocols

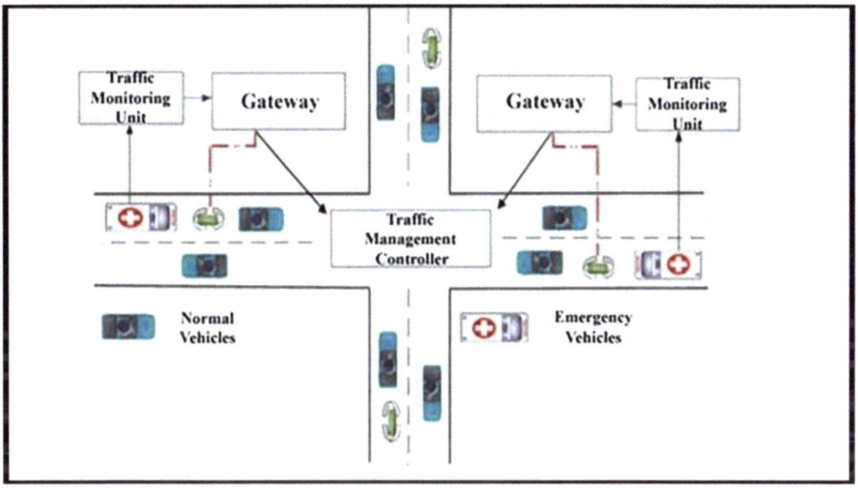

Fig. 4 Intelligent traffic management

This table provides an overview of the various policy and technological advancements in public transport systems, highlighting the associated benefits, examples, and challenges faced in their implementation.

Figure 4 shows a schematic block diagram for intelligent traffic management. It includes components such as data sources (GPS, mobile devices, and sensor networks), big data processing, advanced analytics, and outputs like optimized traffic signals and traffic predictions. The flow of data is depicted from collection through processing to the output stage, demonstrating how intelligent traffic systems work to manage urban mobility efficiently.

8 Implementation of Intelligent Traffic Management in Different Countries

(a) *United States*: A significant portion of urban areas have implemented intelligent traffic systems, especially in cities like Los Angeles and New York.
(b) *United Kingdom*: Many cities, including London, use intelligent traffic management extensively to address congestion.
(c) *Germany*: High integration of smart traffic solutions, particularly in major cities like Berlin, focusing on efficient traffic flow.
(d) *China*: Leading in the adoption of intelligent traffic management, utilizing extensive data sources and advanced technologies in cities like Shanghai.
(e) *India*: Lower implementation rate due to various challenges but increasing investments in smart city initiatives are improving the situation.

9 Comparative Analysis of Intelligent Traffic Management in the United States, the United Kingdom, Germany, China, and India

Country	City	Overview	Implementation	Results
United States	Los Angeles	Major urban area facing significant traffic congestion	Implemented an Intelligent Transportation System (ITS) utilizing GPS tracking from public transit, traffic cameras, and sensor networks for real-time analysis	20% reduction in average travel times during peak hours; data-driven infrastructure improvements
United Kingdom	London	Dense urban environment with millions of vehicles and traffic challenges	Urban Traffic Control (UTC) system integrates real-time data from sensors and GPS to monitor traffic and adjust signal timings dynamically	30% decrease in travel times in certain corridors; improved public transport efficiency
Germany	Berlin	Major European city modernizing traffic management through Big Data	Employs smart traffic lights and historical data analysis with machine learning for predictive traffic modelling and signal timing adjustments	15% improvement in traffic flow; quicker incident response times

(continued)

Country	City	Overview	Implementation	Results
China	Shanghai	Populous city utilizing intelligent traffic management for extensive road networks	Uses traffic cameras, GPS data from taxis, and social media reports to provide real-time updates and optimize signal operations through advanced analytics	20% decrease in traffic congestion during rush hours; improved bus efficiency through predictive analytics
India	Bengaluru	Rapid urbanization leading to severe traffic congestion in a growing metropolitan area	Traffic Police use data from cameras, GPS-enabled vehicles, and mobile apps to monitor conditions and analyze congestion patterns with machine learning algorithms	10% reduction in traffic congestion; enhanced incident response effectiveness

10 Conclusion

The rapid increase in urbanization and vehicle numbers has made managing urban traffic a major challenge. Traditional methods like fixed signal timings and manual interventions are no longer sufficient for modern traffic demands. This has led to the development of intelligent traffic management systems (ITMS), which utilize Big Data and advanced analytics to address these challenges. By collecting data from various sources such as GPS devices, mobile networks, and sensors, ITMS can optimize traffic flow in real-time, improving road safety and public transport efficiency. Big Data technologies manage vast amounts of traffic data, while machine learning and predictive analytics support real-time decision-making and congestion forecasting. ITMS not only improves traffic management but also reduces fuel consumption and emissions, promoting environmental sustainability. However, implementing ITMS presents challenges, including data privacy, system interoperability, and the lack of standardized protocols. Traffic data, which is sensitive in nature, requires robust security measures such as encryption and anonymization to protect privacy. Additionally, ensuring seamless integration across different systems and regions is crucial for maximizing efficiency.

Machine learning and predictive analytics are essential for improving traffic conditions. Predictive models analyze real-time and historical data to forecast traffic patterns and identify potential congestion. This helps authorities take timely action to mitigate delays. Long-term forecasting can guide city planners in making decisions about infrastructure development and public transport expansion. Optimization algorithms further enhance efficiency by adjusting signal timings, road lane assignments, and routing decisions. Despite technological advances, challenges like data privacy, standardization, and policy gaps remain. As cities collect large amounts of sensitive data, establishing governance frameworks and ensuring compliance with privacy standards is vital. As autonomous vehicles and the Internet of Things (IoT) evolve, they will significantly impact traffic management. Big Data and advanced analytics have the potential to transform urban traffic systems, making them more efficient, sustainable, and responsive to modern demands.

References

1. GPS and mobile data sources. (2023). *Transportation Data Analytics Journal, 45*(2), 102–115.
2. Sensor networks in traffic management. (2022). *Journal of Urban Transportation, 39*(4), 251–269.
3. Data storage solutions for big data. (2023). *Cloud Computing Journal, 37*(1), 78–92.
4. Machine learning for traffic prediction. (2022). *Journal of Intelligent Transportation Systems, 31*(5), 345–360.
5. Enhancing public transport systems. (2023). *Public Transportation Innovation Journal, 15*(2), 89–105.
6. Data privacy in traffic management. (2022). *Cyber Security and Data Protection Review, 20*(4), 227–242.
7. Standardization in traffic data. (2023). *Journal of Data Interoperability, 12*(1), 66–81.
8. Real-time traffic monitoring. (2022). *Traffic Management Review, 24*(4), 311–328.
9. Data cleaning techniques. (2023). *Data Management Review, 25*(3), 144–159.
10. Long-term traffic forecasting. (2023). *Urban Planning and Transportation Journal, 42*(3), 201–216.
11. Optimization algorithms in traffic management. (2023). *Operations Research and Decisions, 28*(6), 512–529.
12. Data visualization techniques. (2023). *Analytics and Visualization Review, 19*(3), 144–159.
13. Technological advancements in traffic systems. (2023). *Technology and Transportation Quarterly, 38*(2), 134–150.

Part II
Decision Support Systems for Healthcare Applications

Decision Support System in Smart Healthcare: System for Diagnosis of Skin Cancer Using Deep Learning

Veronica Angelica Villalobos Romo, Edgar Daniel Gomez Garcia, Claudia Georgina Nava Dino, José-Manuel Mejía-Muñoz, Jose David Diaz Roman, and Soledad Vianey Torres Arguelles

1 Introduction

Skin cancer has become a global public health issue, with its incidence increasing over the past decades [2, 3]. It is estimated that in 2022, there were 335,000 new cases of melanoma worldwide, and around 60,000 people died from the disease [4]. The morphological characteristics of skin lesions are crucial factors in the diagnosis and early detection of cancer [5]. The skin, the largest organ of the human body, covers between 1.5 and 2 square meters and weighs approximately 4.2 kg [3]. The term "skin cancer" encompasses several neoplasms that share malignant behavior; however, individually they show very diverse degrees of local aggressiveness, tendency to metastasize, and mortality. These differences in malignancy are one of the most distinctive characteristics of skin cancer. The least malignant skin cancer, basal cell carcinoma, develops in the skin, as does one of the most aggressive neoplasms, melanoma. There can be superficial, slow-growing carcinomas to highly destructive invasive tumors capable of metastasizing [2, 3, 5].

The American Cancer Society estimates that approximately 100,350 new cases of melanoma will be diagnosed in the United States (about 60,190 men and 40,160

V. A. V. Romo · E. D. G. Garcia · J.-M. Mejía-Muñoz · S. V. T. Arguelles (✉)
Universidad Autónoma de Ciudad Juárez, Chihuahua, Mexico
e-mail: veronica.villalobos@uacj.mx; jose.mejia@uacj.mx; vianey.torres@uacj.mx

C. G. N. Dino
Universidad Autónoma de Chihuahua, Chihuahua, Mexico
e-mail: ndino@uach.mx

J. D. D. Roman
Department of Electrical and Computer Engineering, Autonomous University of Ciudad Juárez, Ciudad Juarez, Chihuahua, Mexico
e-mail: david.roman@uacj.com

N. Gupta, J. D. Diaz Roman (eds.), *Decision Support Systems for Industrial and Scientific Applications*, Signals and Communication Technology,
https://doi.org/10.1007/978-3-031-89700-9_6

women). Melanoma is considered rare in Mexico, with an incidence of less than one case per 100,000 inhabitants, representing around 2700 cases annually [6]. However, identifying melanoma is crucial as it has been detected that those susceptible to skin cancer in Mexico include farmers, sailors, street vendors, and individuals with genetic predispositions. Most of these people can be considered low-income, leading to a lower likelihood of early diagnosis and timely treatment [6–8]. In developing countries like Mexico and Brazil, particularly in peripheral areas, there is a lack of dermatologists and dermatoscopy equipment. Mobile devices can be a useful tool in this situation, given the high number of mobile subscriptions; according to Kassianos et al. [9], there were nearly 8 billion subscriptions in 2019.

Numerous works have been proposed for melanoma detection using convolutional neural networks (CNN) on dermoscopic images. However, data imbalance in the available datasets has been a recurring problem, as the classification categories are not equally represented [10, 11]. In the medical field, a significant numerical imbalance in the number of samples of different lesion classes is common [12, 13]. In this context, machine learning has emerged as a powerful tool for medical image classification and disease detection [14–17].

For example, Dai et al. [18] developed a classifier using the HAM10000 database [19], achieving an accuracy of 70%. On the other hand, Castro et al. [20] employed the MUPEB balancing technique (Extra-Polarization and Differential Evolution Balancing) with the PAD-UFES database [21], achieving an accuracy of 83%. This result represented a 6% and 16% increase in accuracy and sensitivity, respectively, compared to previous works.

Data imbalance remains a significant challenge in developing effective classification methods. In 2019, R. Gao et al. evaluated the use of the Synthetic Minority Over-sampling Technique (SMOTE) in classifying facial skin pigmentation disorders, confirming a 7% improvement in accuracy for the two classes with the fewest data points [1, 22].

In [23], the SMOTE technique was implemented on imbalanced data from the HAM10000 skin cancer image repositories. Subsequently, a convolutional neural network-based architecture was used to classify seven classes of skin lesions. This approach has the potential to significantly improve accuracy and sensitivity in detecting melanoma and other skin lesions, thereby contributing to earlier and more effective skin cancer diagnosis [23].

1.1 Importance of Artificial Intelligence and Data Analytics in Medical Imaging

In the health sector, artificial intelligence, data analytics, and image processing are fundamental for decision-making, diagnosing, and treating various diseases, representing the most innovative advancements. Through data analysis and advanced technology, it is now possible to detect diseases, initiate immediate treatment, and increase the chances of success [24]. Modern algorithms assist doctors in utilizing everything from magnetic resonance imaging and signal acquisition equipment to laboratory test results, identifying patterns and signals that may indicate the presence of a disease. In this way, data analytics facilitates the analysis of large volumes of medical data to improve patient care and medical research [24–26]. Medical image processing and data analytics enable healthcare professionals to make more precise and personalized diagnoses. According to a study by Saba et al. (2019), the use of deep learning algorithms in medical imaging has significantly improved the ability to detect diseases such as breast cancer and melanoma [25]. Therefore, data analytics helps to manage large amounts of data to convert it into clear and useful information. This allows for more informed decision-making with less risk, as they are based on concrete data. Additionally, leveraging the potentially massive amounts of data in the medical field ensures that care is appropriate for the patient and delivered at the right time, being potentially beneficial for everyone involved in the healthcare sector [27]. Artificial intelligence and data analysis are related, although they have different goals and objectives. Artificial intelligence aims to create autonomous and intelligent systems, while data analysis focuses on extracting valuable information from data to support decision-making. Both are important in the information age and are constantly evolving, providing opportunities for development and research in various areas such as the health sector [27].

1.2 Skin Cancer and Its Diagnosis

The detection and diagnosis of skin cancer have traditionally been carried out through visual inspection combined with a systematic evaluation or examination process to identify the disease. However, this depends on the dermatologist's experience, making the process lengthy, subjective, and possibly prone to errors [28]. This is due to the complex nature of skin lesions. Additionally, diagnosing skin cancer is not just about identifying melanoma or non-melanoma. It also includes multiple skin lesions, which complicates specialized analyses, such as distinguishing a melanocytic nevus from melanoma or from basal cell carcinoma and squamous cell carcinoma, among other lesions [8]. Consequently, computer-aided diagnosis (CAD) systems become necessary for the preliminary diagnosis of different lesions [29]. The use of deep learning technology with convolutional neural networks (CNN) helps specialists identify and diagnose skin lesions, such as melanoma, more accurately and efficiently, leading to new research and developments to improve the performance of CAD in addressing many other complex clinical problems [29]. Using CAD with CNN provides a powerful tool that supports healthcare professionals in identifying skin lesions, improving diagnostic accuracy, and enabling earlier and more effective interventions. By combining medical expertise with advanced technology, more precise and efficient patient care is achieved [30]. Advances in smartphone cameras have enabled their potential use for the early detection of skin cancer. Smartphones can now recognize characteristics associated with skin cancer. To achieve this, CNNs are commonly used for the detection and classification of diseases [31]. However, only a few deep learning models can be utilized to create a mobile application, as they require high computational power and large memory, which is challenging to implement on smartphones [32]. Nonetheless, various studies have developed mobile applications on smartphones with favorable results [30–33]. Therefore, the development of a mobile application for skin cancer detection is essential, as it offers an accessible and portable tool for users to identify skin problems and seek timely medical attention. Additionally, the presence of a precise and easy-to-use diagnostic tool on a common device such as a smartphone can significantly reduce the burden on healthcare systems, allowing professionals to provide appropriate treatment for the overall well-being of patients.

2 Methodology

This section describes the methodology used to create the application that detects skin cancer using images, First, image repositories with clinically diagnosed and classified cases by healthcare professionals were sought. Then, the data were balanced with the SMOTE method, followed by the design of a fully connected dense network with a convolutional adjustable base. This network was evaluated using convolutional architectures, two of which were of low weight and the remaining two were of high weight. These architectures were sourced from the Keras library, and fine-tuning was applied. Then an application based on java language was designed to run on an Android device, and finally, the system was simulated with test users to evaluate the effectiveness of the application. Figure 1 shows a diagram describing the process carried out in this methodology.

Each section proposed in the methodology is detailed below.

2.1 Compilation of Databases

Training artificial neural networks to perform classification requires pre-labeled data, meaning each image must have an associate label indicating the class or object it contains. For this reason, the HAM10000 database [19] was optimal for

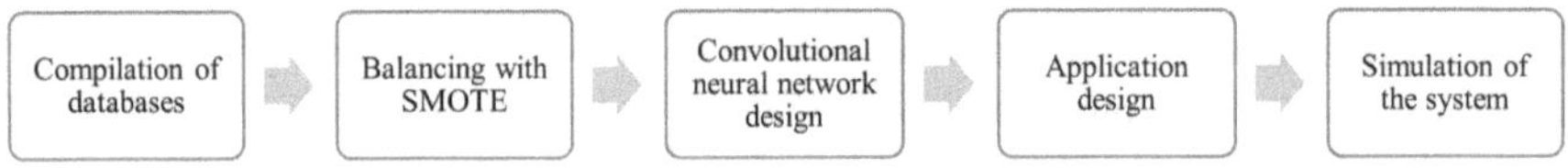

Fig. 1 Methodology process

Table 1 Distribution by class in HAM10000[a] database

Injury	Quantity
Melanocytic nevi (nv)	6705
Melanoma (mel)	1113
Benign keratosis (bkl)	1099
Basal cell carcinoma (bcc)	514
Actinic Keratoses (akiec)	327
Dermatofibroma (df)	142
Vascular skin lesions (vasc)	115

[a] HAM 10000

Table 2 Distribution by class in PAD-UFES[a] database

Injury	Quantity
Melanocytic nevi (nv)	244
Melanoma (mel)	52
Benign keratosis (bkl)	235
Basal cell carcinoma (bcc)	845
Actinic Keratoses (akiec)	730
Dermatofibroma (df)	192

[a] PAD-UFES

this project. This repository contains images of various types of skin diseases and syndromes, previously classified by healthcare professionals. It consists of 10,015 dermatoscopic images with a uniform size of 1024 × 1024 in jpg format, which are publicly available. Table 1 shows the number of images for each of the classes included in the database.

On the other hand, images from the PAD-UFES repository [21] were also used, which contains clinical images of skin lesions properly labeled. The images were acquired with mobile devices and vary in size from 256 × 256 to 1024 × 1024 in PNG format. This dataset is an effort to assist researchers in developing low-budget tools, particularly to aid in skin cancer detection. Table 2 shows the number of images per class in the PAD-UFES database.

For this work, the vascular lesion class was first eliminated from the HAM10000 database because these lesions are considered skin anomalies rather than pathologies. Subsequently, both databases were concatenated in such a way that when adding all the images from PAD-UFES to HAM1000, 1165 images of melanoma, 1334 of benign keratosis, 1359 of basal carcinoma and 1057 of actinic keratosis were obtained. Since a similar number of images per class is sought, a random undersampling of the nv class was performed. It was to use the 244 images of moles from the PAD-UFES and add them with a random sampling of 956 images from the HAM10000 to result in a final number of 1200 images of moles and thus have a balanced distribution among the above classes of lesions. As for the dermatofibroma class, the sum of images from both databases resulted in 307 images. Table 3 shows the distribution of images per class after combining both the HAM10000 and PAD-UFES databases.

Table 3 Distribution by class when combining both databases

Injury	Quantity
Melanocytic nevi (nv)	1200
Melanoma (mel)	1165
Benign keratosis (bkl)	1334
Basal cell carcinoma (bcc)	1359
Actinic Keratoses (akiec)	1057
Dermatofibroma (df)	307

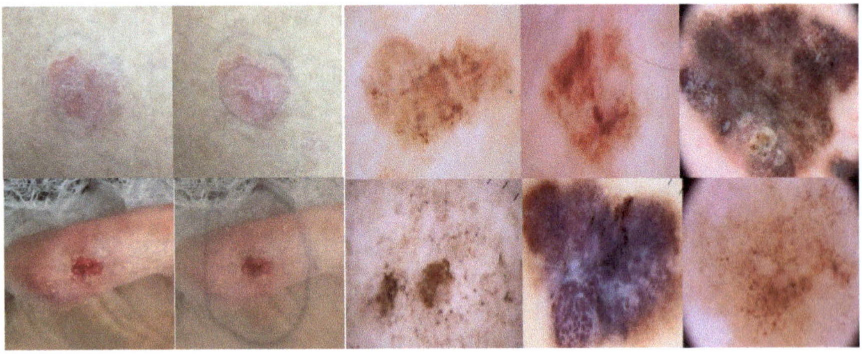

Fig. 2 Fake images generated with SMOTE of the dermatofibroma class

2.2 *Balancing with SMOTE*

The SMOTE data balancing technique was used only on the dermatofibroma class because it was the class with the smallest number of images. For this purpose, first the images of two classes were separated into different folders, wherein one of them there were the images of the minority class that needed to be balanced, i.e., "squamous cancer" and in the other one there was a reference class with the number of images to be obtained at the end; the benign keratosis class was used for this purpose. Subsequently, the images of both classes were resized to 256 × 256 pixels to standardize their dimensions, as the two databases contain images of different sizes. The processing of the data to perform the balancing by SMOTE [34] consisted of the following: the first step is to parameterize the values of each RGB (Red, Green, and Blue) combination of the pixels of each image to bring these values to a two-dimensional plane, where each point in this plane represents an image. Next, a random point from the minority class and a random point from the reference class are selected and a new point is generated at half the distance between these two points. This process was repeated until the same amount of data (images) from the minority class was obtained with respect to the reference class. Figure 2 shows examples of the results of the false images generated with SMOTE.

A total of 804 dermatofibroma images were generated and added to the total image set, resulting in the distribution shown in Fig. 3. The images were separated

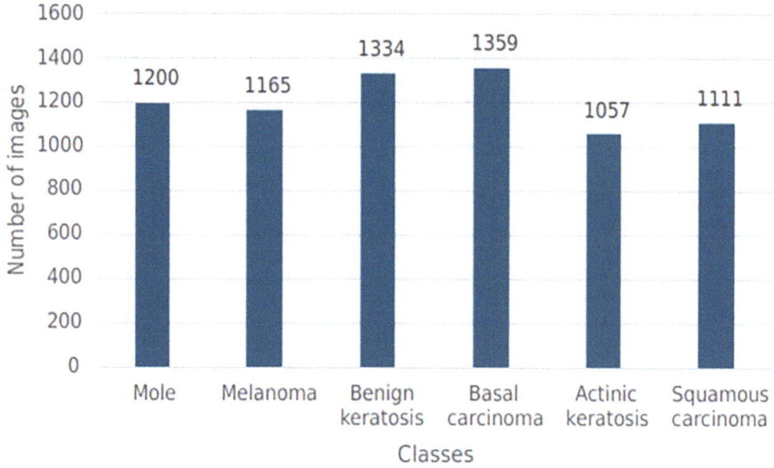

Fig. 3 Histogram of distribution by class after SMOTE

in a stratified manner, reserving 20% of all images for the test set, 64% for the trained set, and 16% for the validation set used at the end of each epoch during training. None of the images artificially generated by SMOTE were added to the test set.

2.3 Convolutional Neural Network Design

Four convolutional neural network architectures were evaluated: Xception [35], Resnet152V2 [36], EfficientnetV2 [37], and MobilenetV2 [38]. Each of these networks was loaded from the Keras library, using the pre-trained versions with weights predefined by the ImageNet challenge [39], subsequently, the last default classification layer was removed and the option to train the weights of the convolutional layers was enabled. To adapt each network to our classification task, a "GlobalAveragePooling" followed by a dense layer with six neurons and a SoftMax activation function to classify the six types of lesions in the study.

The architecture described in Fig. 4 was trained with batch sizes of 64 images for 40 epochs at a learning rate of 0.001 using an Adam optimizer and a categorical cross-entropy loss function. Additionally, a learning rate step-down function was added, along with an early stopping function to halt training if performance did not improve for 5 epochs. After training the four architectures described above and evaluating their performance using the validation set, the EfficietNetV2 convolutional neural network architecture was chosen for its superior performance at this stage. Once the layers were trained as described, the classification layer was

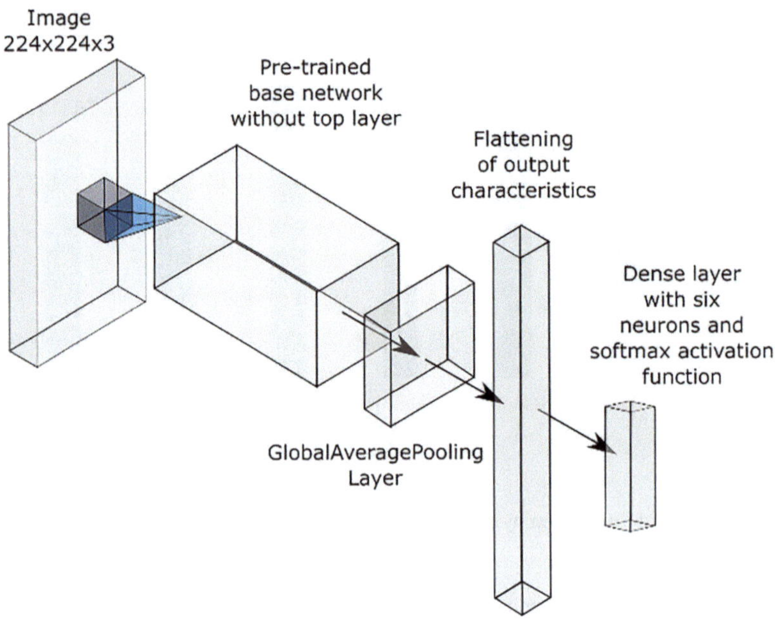

Fig. 4 Architecture for evaluating the design of convolutional networks

removed and dense layers were added to reduce the output features. These layers were sequentially counted with 1280, 1280, 640, 320, and 160 neurons, each with a GELU activation function. DropOut layer with a rate of 0.6 was added after each dense layer to prevent overfitting. Finally, a dense output layer with six neurons and a Softmax activation function was added to perform the classification. Figure 5 illustrates the described architecture.

2.4 Application Design

For the design of the application, Android Studio [40] was used, which, in order to interpret keras/tensorflow models, requires them to be in tensorflowlite (tflite) format [41]. This format is lightweight, encrypted in folders and designed to work in applications with low-performance equipment such as a cell phone, tablet, or microcomputer. Then, the following lines of code were applied to make the change to the tensorflowlite format:

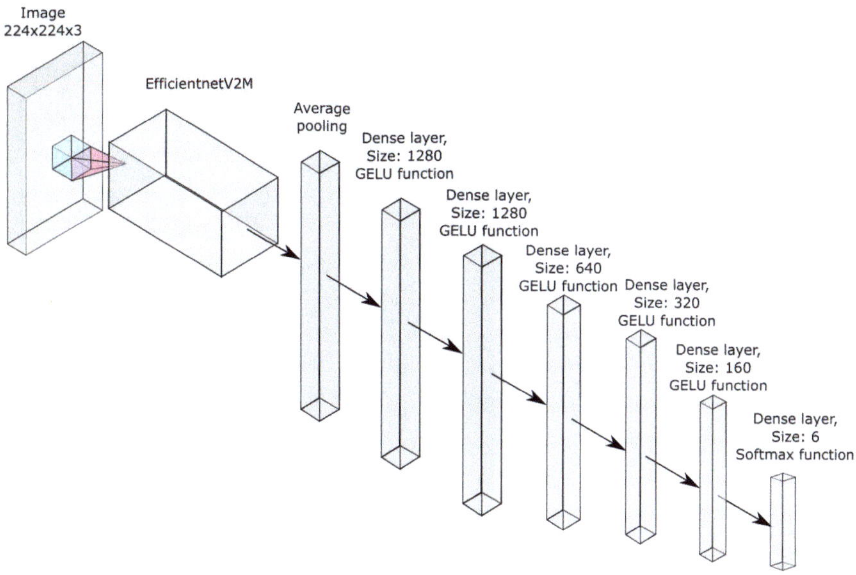

Fig. 5 Network architecture for sorting (each dense layer upstream of the output has DropOut of 0.6)

```
#Install dependency
!pip install tflite-model-maker
import tensorflow as tf

# load the model developed above
model= tf.keras.models.load_model ('path')

# define model specifications required for android #studio
converter=tf.lite.TFLiteConverter.from_keras_model (model)
converter.target_spec.supported_ops = [
  tf.lite.OpsSet.TFLITE_BUILTINS,
  tf.lite.OpsSet.SELECT_TF_OPS ]

# apply the conversion
tflite_model = converter.convert()
open("converted_model.tflite", "wb").write(tflite_model)
```

In the previously described program, it is important to highlight that the functions "tf.lite.OpsSet.TFLITE BUILTINS" and "tf.lite.OpsSet.SELECT TF OPS" allow performing native operations of the tensorflow library, and although it adds slightly

Fig. 6 Installation process

weight, it is necessary to run efficiently in the application developed in Android Studio.

The original model weight is in .h5 format (hierarchical data format), which contains multidimensional arrays of scientific data and has a final memory size of 284 MB; transforming the model to tensorflowlite format reduced its memory to 110.5 MB, which gave the opportunity to add more dense layers and convolution filters to the model without exceeding the 200 MB limitation requested by Android studio, however, no model showed improvement, so it was determined to keep the model already created for the application. Once the interface was designed, the dependencies for the Android device and the libraries required for the system's functionality were imported, and the application was installed on the mobile device. The application was originally made in Spanish. The application process is described below: First, the application must be installed on the device, this can be done directly through Android Studio with the help of a USB cable. Once the install option is selected, the mobile device will display the message shown in Fig. 6, asking for permission to install the software on the device. Next, the installed application will open and the menu shown in Fig. 7 will be displayed. The user will be presented with two options in the form of buttons to analyze an image: the first

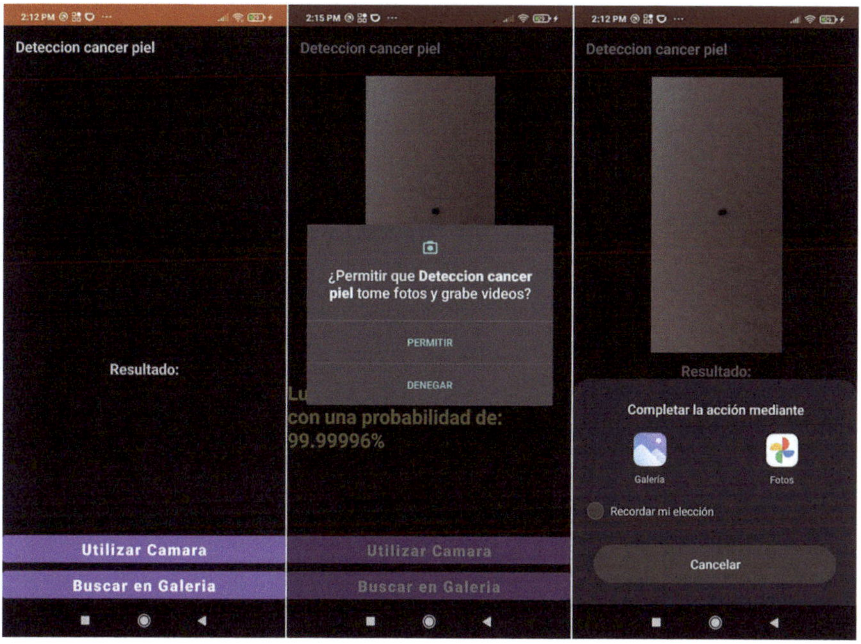

Fig. 7 Application menu and image selection

option is to acquire an image using the device's camera, while the second will allow the user to browse its image libraries to select an image if it is preloaded.

2.5 *Simulation of the System*

Once the image to be analyzed is selected, it will be displayed on the main screen, with the classification result given by the model shown below it, together with the probability of this classification with respect to the rest of the 6 classes that can be evaluated. The results visualization in the application is shown below in Fig. 8. It should be noted that the test shown in this figure was a real evaluation with a test subject, who had the mark observed in the image since birth, so it was indeed a mole.

3 Results

Two experiments were carried out: the first involved training the networks with the original images, and the second used a data set augmented with SMOTE-balanced

Fig. 8 Display of results

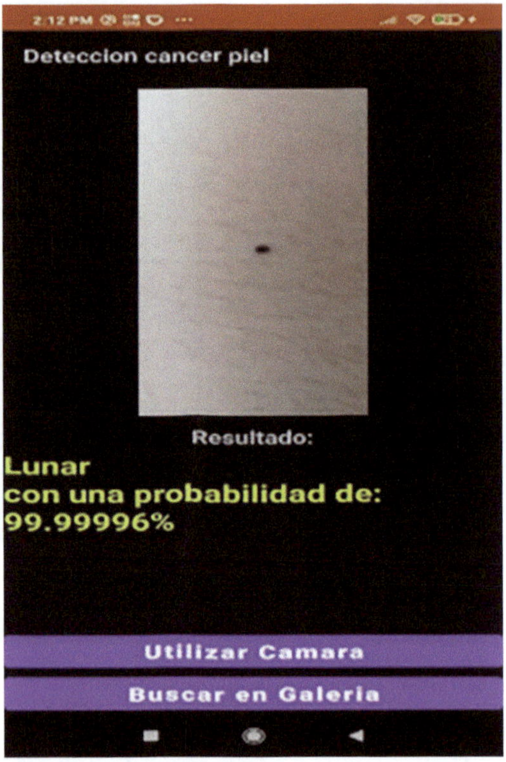

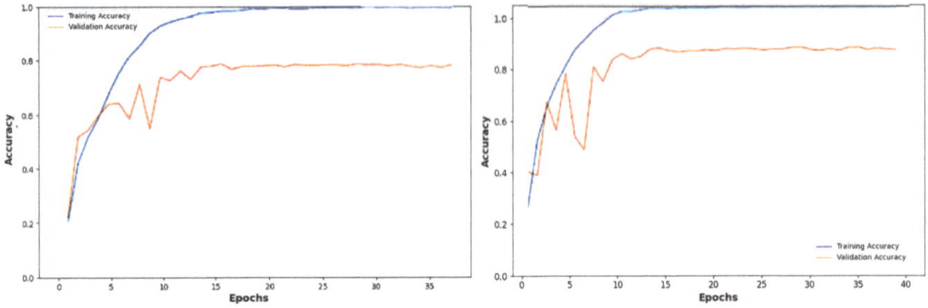

Fig. 9 Accuracy results without SMOTE (left), with SMOTE (right)

images of the dermatofibroma class. The training was performed with 100 epochs using a function that decreases the learning rate gradually and another function that stops the training at the moment when the loss metric monitored in the validation set stops improving, this to avoid overfitting to the training data. The results of the accuracy metric during the training process are shown in Fig. 9.

Subsequently, an evaluation was carried out with the final model using the data initially reserved for testing, which did not contain artificial images. To perform this evaluation, a confusion matrix was constructed, and the results are shown in Fig. 10.

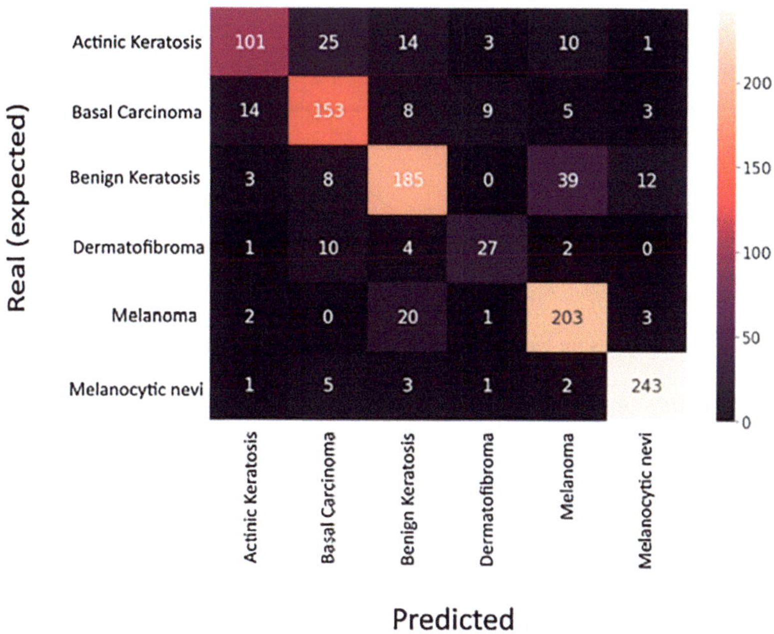

Fig. 10 Validation confusion matrix

Table 4 Classification report by category

Class	Precision (%)	Sensitivity (%)	F1-score (%)	Number of images
Actinic keratosis	83	66	73	154
Basal carcinoma	76	80	78	192
Benign keratosis	79	75	77	247
Dermatofibroma	66	61	64	44
Melanocytic nevi	93	95	94	255
Accuracy (%)			83	1121

To analyze the performance of accuracy, sensitivity, and F1-score for each of the classes, a ranking report was made, as shown in Table 4.

In addition, ROC (receiver operating characteristic) curves [34, 42] were also used for the evaluation. The results of the proposed classifier including the values of the area under the ROC curve are shown in Fig. 11.

Overall global accuracy performance was 73% without SMOTE and 83% with SMOTE, representing a 10% improvement overall. It can be observed from the ROC plots that there is an AUC greater than 92% in each of the classes. As an experiment, the app was evaluated by taking images that were not in the database, taking the image with the cell phone pointing to the computer screen, where a validated sample of each of the classes was presented, the result is shown in Fig. 12.

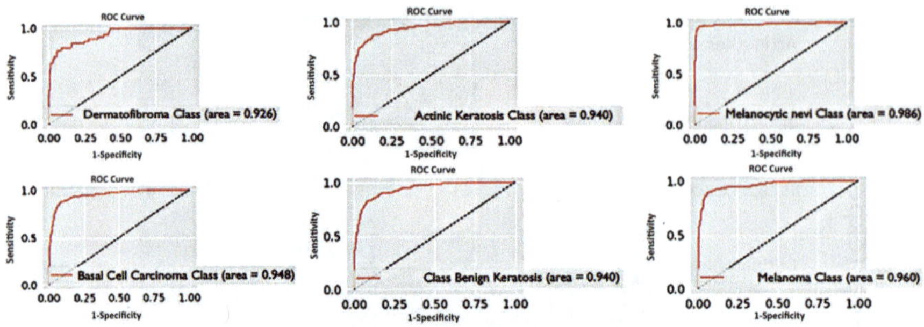

Fig. 11 ROC curve plots for the classification of the different classes of lesions without SMOTE

Figure 12 shows an effective classification response in new data, especially considering that all the evaluations, except item "d", were correct at the first attempt with low-resolution images, without requiring any equipment other than the cell phone. In the case of item "d", the correct answer was obtained on the second attempt. This may be because the first image evaluated included other elements, such as hair and part of an eye. This result is shown in Fig. 13.

The average response time when introducing an image to the model in the application is 2.35 s, the image acquisition time is 6.58 s so that on average an express analysis took a time of 9.33 s.

4 Conclusions

SMOTE balancing proved to be a great tool to generate artificial data when there is a scarce amount of data. It was demonstrated that this technique promotes a higher convergence speed when training, in spite of having a higher variance in the training. It was possible to improve the overall performance global accuracy of 73% without SMOTE and 83% with SMOTE, representing an improvement of 10% in general, in addition, it can be concluded from the ROC graphs that there is an AUC greater than 92% in all classes. It was possible to create a low-memory model (284 MB) and transform it to tensorflowlite format to be applicable to an Android cellular device, reducing its weight to 110.5 MB, which showed compatibility and fast response. An on-device application was developed with a simple interface, capable of classifying six different types of skin marks using either the cell phone camera or images from the gallery. The application demonstrated good performance with both synthetic data and real cases of moles, as noted in the limitation sections. It exhibited an average latency of 9.33 s. Overall accuracy results of 83% were achieved with validation data, which surpasses Dai Xaingfeng's first approach (accuracy of 72.4%) [18]. The proposed model competes with the application developed by Kousis et al. [33] (91.1% accuracy) for two classes of lesions classified as benign or malignant.; as well as with the approach using a 3D

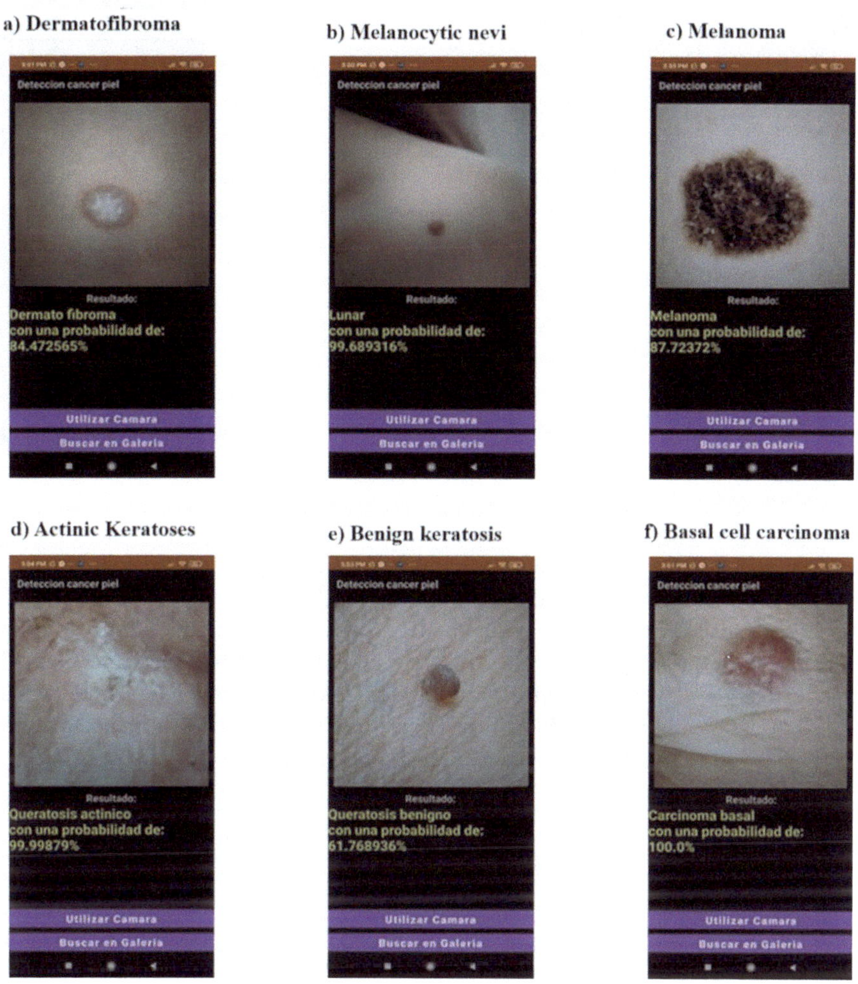

Fig. 12 Test of application using images external to the databases, taken from the computer screen

curvature pattern highlighting and convolution technique for melanoma diagnosis 89.2% accuracy by Yu Zhou [43], which requires photometric stereo equipment for its operation. Finally, it is comparable to the model proposed by Krohling, et al. [21], based on convolutional networks with evolutionary algorithms making use of data in the patient file, which achieved a final accuracy of 89%. It is important to mention that all the previously mentioned works are only to evaluate the melanoma class in a binary way (positive/negative), while this work presents a solution to classify six different classes, where an F1-score of 83% was achieved in the melanoma class, with a sensitivity of 89%; at the same time the system showed an F1-Score of 94% to identify moles with a sensitivity reaching 95%. It is important to mention the

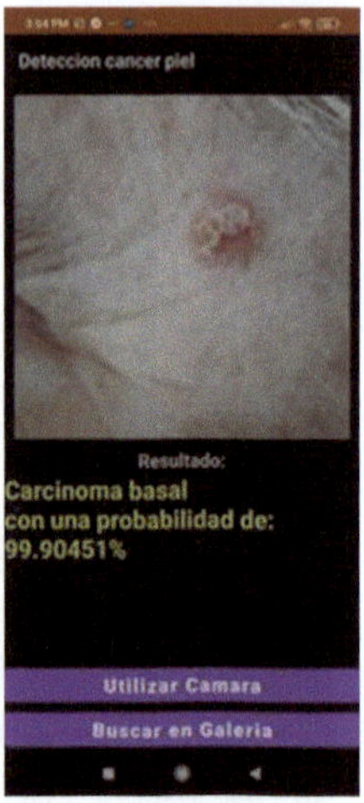

Fig. 13 Erroneous evaluation at the first attempt, inciting basal carcinoma, actually being actinic keratosis

area of opportunity that exists to increase the performance of the model for the identification of the rest of the lesions, whose accuracy values were low, as in the case of dermatofibroma in which only 66% was reached. One way is to explore other neural network architectures and/or models that allow the extraction of relevant features from the images together with the use of other types of classifiers. A cell application was developed, capable of classifying skin cancer images using a model based on a convolutional architecture. This model achieved an overall accuracy of 83% with test data.

References

1. Chawla, N. V., Bowyer, K. W., Hall, L. O., & Kegelmeyer, W. P. (2002). SMOTE: Synthetic minority over-sampling technique. *Journal of Artificial Intelligence Research, 16*, 321–357.
2. Alonso-Belmonte, C., Montero-Vilchez, T., Arias-Santiago, S., & Buendía-Eisman, A. (2022). Situación actual de la prevencion del cancer de piel: una revision sistemática. *Actas Dermo-Sifiliográficas, 113*(8), 781–791. https://doi.org/10.1016/j.ad.2022.04.015
3. Saúl, A. (2015). *Lecciones de dermatología* (16th ed.). Mc Graw Hill.
4. IARC. (2022). Melanoma Awareness Month 2022 – IARC. [Online]. Available via DIALOG: https://www.iarc.who.int/news-events/melanoma-awareness-month-2022/. Cited 17 Oct 2022.
5. Castañeda Gameros, P., & Eljure Téllez, J. (2016). El cáncer de piel, un problema actual. *Revista de la Facultad de Medicina (México), 59*(2), 6–14.
6. Cámara-Salazar, A., Santos-Zaldívar, K. P., Bracamonte-Barahona, R., Calderón-Sauri, I. S., Flores-Mendoza, K. F., Mijangos-Vergara, F., & Méndez-Domínguez, N. I. (2020). Características individuales y por entidad federativa de la mortalidad por melanoma en México entre 2014 y 2018. *Dermatologia Revista Mexicana, 64*(3), 248–254.
7. Roldan, R. (2017). Cáncer de piel, el segundo más frecuente en México, advierte investigador de la UNAM, UNAM, 2017. [Online]. Available via DIALOG.: https://www.dgcs.unam.mx/boletin/bdboletin/2017237.html. Cited 18 Oct 2020.

8. Marín, R. R., & Ortega, B. C. (2014). Criterios dermatoscópicos para el diagnóstico de lesión melanocıtica y melanoma cutaneo. Dermoscopic criteria for the diagnosis of melanocytic lesions and cutaneouos melanoma. *Dermatologıa Cosmetica, Médica y Quirúrgica*, 142–148.
9. Kassianos, A. P., Emery, J. D., Murchie, P., & Walter, F. M. (2015). Smartphone applications for melanoma detection by community, patient and generalist clinician users: A review. *British Journal of Dermatology, 172*(6), 1507–1518. https://doi.org/10.1111/bjd.13665
10. Al-Asadi, M. A., & Altun, A. (2022). Deep learning with SMOTE techniques for improved skin lesion classification on unbalanced data. *Selcuk University Journal of Engineering Sciences, 21*(03), 97–104.
11. Abayomi-Alli, O. O., Damasevicius, R., Misra, S., Maskeliunas, R., & Abayomi-Alli, A. (2021). Malignant skin melanoma detection using image augmentation by oversampling in nonlinear lower-dimensional embedding manifold. *Turkish Journal of Electrical Engineering and Computer Sciences, 29*, 2600–2614. https://doi.org/10.3906/elk-2101-133
12. Chang, C. C., Li, Y. Z., Wu, H. C., & Tseng, M. H. (2022). Melanoma detection using XGB classifier combined with feature extraction and K-means SMOTE techniques. *Diagnostics, 12*(7). https://doi.org/10.3390/diagnostics12071747
13. Tong, X., Wei, J., Sun, B., Su, S., Zuo, Z., & Wu, P. (2021). Ascu-net: Attention gate, spatial and channel attention u-net for skin lesion segmentation. *Diagnostics, 11*(3). https://doi.org/10.3390/diagnostics11030501
14. Shehzad, K., et al. (2023). A deep-ensemble-learning-based approach for skin cancer diagnosis. *Electronics, 12*(6), 1342. https://doi.org/10.3390/ELECTRONICS12061342
15. Thomas, S. M., Lefevre, J. G., Baxter, G., & Hamilton, N. A. (2021). Interpretable deep learning systems for multi-class segmentation and classification of non-melanoma skin cancer. *Medical Image Analysis, 68*, 101915. https://doi.org/10.1016/j.media.2020.101915
16. Jinnai, S., Yamazaki, N., Hirano, Y., Sugawara, Y., Ohe, Y., & Hamamoto, R. (2020). The development of a skin cancer classification system for pigmented skin lesions using deep learning. *Biomolecules, 10*(8), 1–13. https://doi.org/10.3390/biom10081123
17. Vidya, M., & Karky, M. V. (2020). Skin cancer detection using machine learning techniques. In *IEEE international conference on electronics, computing and communication technologies (CONECCT)* (pp. 1–5).
18. Dai, X., Spasić, I., Meyer, B., Chapman, S., & Andres, F. (2019). Machine learning on mobile: An on-device inference app for skin cancer detection. In *2019 fourth international conference on fog and mobile edge computing (FMEC)* (pp. 301–305). IEEE.
19. Tschandl, P., Rosendahl, C., & Kittler, H. (2018). Data descriptor: The HAM10000 dataset, a large collection of multi-source dermatoscopic images of common pigmented skin lesions. *Scientific Data, 5*, 1–9. https://doi.org/10.1038/sdata.2018.161
20. Castro, P., Krohling, B., Pacheco, A. G., & Pacheco, R. A. (2020). An app to detect melanoma using deep learning: An approach to handle imbalanced data based on evolutionary algorithms. In *2020 international joint conference on neural networks (IJCNN)* (pp. 1–6). IEEE.
21. Pacheco, A. G. C., & Krohling, R. A. (2020). The impact of patient clinical information on automated skin cancer detection. *Computers in Biology and Medicine, 116*, 103545. https://doi.org/10.1016/j.compbiomed.2019.103545
22. Gao, R., Peng, J., Nguyen, L., Liang, Y., Thng, S., & Lin, Z. Classification of non-tumorous facial pigmentation disorders using deep learning and SMOTE. In *2019 IEEE international symposium on circuits and systems (ISCAS)* (pp. 1–5). IEEE.
23. Guo, S., Liu, Y., Chen, R., Sun, X., & Wang, X. (2019). Improved SMOTE algorithm to deal with imbalanced activity classes in smart homes. *Neural Processing Letters, 50*(2), 1503–1526. https://doi.org/10.1007/s11063-018-9940-3
24. Bobulski, J., & Kubanek, M. (2020). Big data system for medical images analysis. In *Frontiers in artificial intelligence and applications* (pp. 71–79). IOS Press BV. https://doi.org/10.3233/FAIA200768
25. Saba, L., et al. (2019). The present and future of deep learning in radiology. *European Journal of Radiology, 114*, 14–24. https://doi.org/10.1016/j.ejrad.2019.02.038

26. Mijwil, M. M., et al. (2023). View of from pixels to diagnoses deep learning's impact on medical image processing-a survey. *Wasit Journal of Computer and Mathematics Science, 2*(3), 9–15.
27. Dash, S., Shakyawar, S. K., Sharma, M., & Kaushik, S. (2019). Big data in healthcare: Management, analysis and future prospects. *Journal of Big Data, 6*(1). https://doi.org/10.1186/s40537-0190217-0
28. Adegun, A., & Viriri, S. (2021). Deep learning techniques for skin lesion analysis and melanoma cancer detection: A survey of state-of-the-art. *Artificial Intelligence Review, 54*(2), 811–841. https://doi.org/10.1007/s10462-020-09865-y
29. Chan, H. P., Hadjiiski, L. M., & Samala, R. K. (2020). Computer-aided diagnosis in the era of deep learning. In *Medical physics* (Vol. 47, pp. e218–e227). Wiley. https://doi.org/10.1002/mp.13764
30. Kränke, T., Tripolt-Droschl, K., Röd, L., Hofmann-Wellenhof, R., Koppitz, M., & Tripolt, M. (2023). New AI-algorithms on smartphones to detect skin cancer in a clinical setting – A validation study. *PLoS One, 18*(2), e0280670.
31. Hartanto, C. A., & Wibowo, A. (2020). Development of Mobile skin cancer detection using faster R-CNN and MobileNet v2 model. In *En 2020 7th international conference on information technology, computer, and electrical engineering (ICITACEE)* (pp. 58–63). IEEE.
32. Emuoyibofarhe, J. O., Ajisafe, D., Babatunde, R. S., & Christoph, M. (2020). Early skin cancer detection using deep convolutional neural networks on mobile smartphone. *International Journal of Information Engineering and Electronic Business, 12*(2), 21–27. https://doi.org/10.5815/ijieeb.2020.02.04
33. Kousis, I., Perikos, I., Hatzilygeroudis, I., & Virvou, M. (2022). Deep learning methods for accurate skin cancer recognition and mobile application. *Electronics (Switzerland), 11*(9). https://doi.org/10.3390/electronics11091294
34. Brownlee, J. (2021). SMOTE for imbalanced classification with python. Machine Learning Mastery [Online]. Available via DIALOG: https://machinelearningmastery.com/smoteoversampling-for-imbalanced-classification/. Cited 17 Mar 2021.
35. Chollet, F. (2017). Xception: Deep learning with depthwise separable convolutions. In *Proceedings of the IEEE conference on computer vision and pattern recognition* (pp. 1251–1258).
36. He, K., Zhang, X., Ren, S., & Sun, J. (2016). Deep residual learning for image recognition. In *2016 IEEE conference on computer vision and pattern recognition (CVPR)* (pp. 770–778). https://doi.org/10.1109/CVPR.2016.90
37. Koonce, B., & Koonce, B. E. (2021). *Convolutional neural networks with swift for TensorFlow: Image recognition and dataset categorization* (pp. 109–123). Apress.
38. Sinha, D., & El-Sharkawy, M. (2019). Thin mobilenet: An enhanced mobilenet architecture. In *2019 IEEE 10th annual ubiquitous computing, electronics mobile communication conference (UEMCON)* (Vol. 3, pp. 0280–0285). IEEE.
39. Morid, M. A., Borjali, A., & Del Fiol, G. (2021). A scoping review of transfer learning research on medical image analysis using imagenet. *Computers in Biology and Medicine, 128*, 104115.
40. Studio, A. (2017). Android Studio: The Official IDE for Android. [Online]. Available: https://developer.android.com/studio/. Accessed 24 2019.
41. Li, S. (2020). TensorFlow lite: On-device machine learning framework. *Journal of Computer Research and Development, 57*, 1839.
42. Swets, J. A. (1996). *Signal detection theory and ROC analysis in psychology and diagnostics: Collected papers*. Psychology Press. https://doi.org/10.4324/9781315806167
43. Zhou, Y., Smith, M., Smith, L., Farooq, A., & Warr, R. (2011). Enhanced 3D curvature pattern and melanoma diagnosis. *Computerized Medical Imaging and Graphics, 35*(2), 155–165.

AI-Driven Histopathological Analysis for Early Lung Cancer Detection

Erika Severeyn, Carlos Rangel, Alexandra La Cruz, and Jesús Velásquez

1 Introduction

Lung cancer persists as a huge global health burden, with the World Health Organization (WHO) reporting staggering 1.8 million fatalities attributed to the disease in 2020 alone. This alarming statistic underscores the pressing need for more efficacious strategies in early lung cancer detection, treatment, and monitoring [11].

The age of technological advancement has witnessed the emergence of artificial intelligence (AI) as a promising tool in medicine. Notably, advancements in convolutional neural networks (CNNs) have revolutionized medical image interpretation, including histopathological images crucial for lung cancer diagnosis [7]. Traditional methods relying on histopathological image analysis by pathologists can be time-consuming and prone to human error [1]. Several studies have explored the potential of deep learning, particularly CNNs, to address these limitations in lung cancer detection. In [9] developed a CNN model achieving 98.15% training accuracy and 98.07% validation accuracy in classifying lung cancer types (benign, adenocarcinoma, and squamous cell carcinoma) from histopathological images. In [8] also employed a CNN model, achieving training and validation accuracies

E. Severeyn (✉) · J. Velásquez
Department of Thermodynamics and Transfer Phenomena, Universidad Simón Bolívar, Caracas, Venezuela
e-mail: severeynerika@usb.ve; jmvelasquezf@usb.ve

C. Rangel
Computer Engineer and Biomedical Engineering, Universidad Internacional de Valencia, Valencia, España

A. La Cruz
Faculty of Engineering, Universidad de Ibagué, Ibagué, Colombia
e-mail: alexandra.lacruz@unibague.edu.co

N. Gupta, J. D. Diaz Roman (eds.), *Decision Support Systems for Industrial and Scientific Applications*, Signals and Communication Technology,
https://doi.org/10.1007/978-3-031-89700-9_7

of 96.11% and 97.2% respectively, for similar lung cancer classification tasks on histopathological images. These studies demonstrate the promising capabilities of CNNs in achieving high accuracy for lung cancer detection from histopathological images.

It is important to note that other deep learning architectures beyond CNNs are also being explored for medical image analysis. For instance, in [2] investigated the use of Mask R-CNN for identifying pathological signs of respiratory diseases on chest CT scans. While achieving high sensitivity for ground-glass opacity detection, their model yielded lower performance for pulmonary nodule detection. These findings suggest that deep learning offers a powerful set of tools for computer-aided diagnosis of lung cancer and other respiratory diseases. However, further research is needed to refine and optimize these models for broader applicability and improved performance across various disease classifications.

This study utilizes a large dataset of lung cancer histopathological images obtained from [4] to develop an AI-driven support system for early and accurate detection of lung cancer. Through the application of AI algorithms, specifically CNNs, the images will be analyzed and classified. This novel approach combines advanced image processing techniques and powerful CNN models to significantly improve lung cancer detection. The goal of this research is to create a support system that can efficiently analyze histopathological images and classify them accurately, leading to a more precise and timely detection of lung cancer. This system has the potential to greatly improve the current standard of care for lung cancer patients by providing a reliable and accurate tool for early detection. By integrating such AI-driven decision support systems into clinical practice, we aim to enhance decision-making processes, optimize resource allocation, and ultimately improve patient outcomes in the healthcare industry.

The chapter is structured as follows: Sect. 2 outlines the methodology, including the dataset description and its preparation, then the dataset splitting process for training, validation and testing, the description of the different models implemented, and metrics used for assessing the models. The findings of our study are presented in Sect. 3. Section 4 provides an in-depth discussion of the results, their significance, and relevance to existing knowledge and also proposes directions for future research. And finally, Sect. 5 shows the significant conclusions of this research and future works.

2 Methodology

2.1 Database

A neural network model for classification tasks requires a database with a large database with labeled cases for each existing class [10]. After an evaluation of various database sources, the “Lung and Colon Cancer Histopathological Images” database was selected [4]. This database contains 25,000 images, which were generated by augmenting 1,250 histology images. Out of the 25,000 images, 15,000 correspond to different types of lung cancer, with 5,000 images labeled as benign lung cancer, 5,000 as lung adenocarcinoma, and 5,000 as squamous cell lung cancer, a perfectly balanced dataset. The images in the dataset are of high resolution, with a size of 768 × 768 pixels in RGB mode, and are saved in JPEG format. Examples of images from the database for each cancer type used in the study are shown in Fig. 1.

2.2 Data Preprocessing

To address the hardware limitations (memory and processing power) of the machines employed and the high resolution of the histopathological images, image preprocessing was indispensable for efficient model training. This involved applying compression filtering to reduce the image size while preserving essential information [12]. A systematic reduction of the image resolution was conducted, and the subsequent assessment of its effect on the model's classification accuracy facilitated the identification of an optimal compression level. This optimal level substantially accelerated training speed and reduced memory requirements without compromising image integrity or model performance. Before using the images as input for a CNN, they were converted into NumPy arrays within the programming environment. Each original image was transformed into a 3D array with dimensions of 768 × 768 × 3, where the dimensions represent the following:

- The first two dimensions (768 × 768) denote the image's spatial resolution, comprising the height and width (a number of rows and columns).
- The third dimension (3) corresponds to the three color channels (red, green, and blue) associated with each pixel, capturing the color information.

Each pixel in the color channel is assigned an 8-bit binary value, which represents its intensity for a specific color, ranging from 0 (minimum intensity) to 255 (maximum intensity). To enhance model convergence, a normalization step was applied to the pixel intensity values. There are two prevalent normalization techniques [3]:

- Standardization: transforming the values to follow a normal distribution with a mean of 0 and a standard deviation of 1
- Min-max scaling: rescaling the pixel intensities to the range [0, 1] by dividing each value by 255, the maximum possible value

This study adopted the min-max scaling normalization approach. Following the application of compression filtering and normalization, each image was represented as a 100 × 100 × 3 NumPy array, wherein pixel values ranged from 0 to 1, rendering them suitable for input into a CNN.

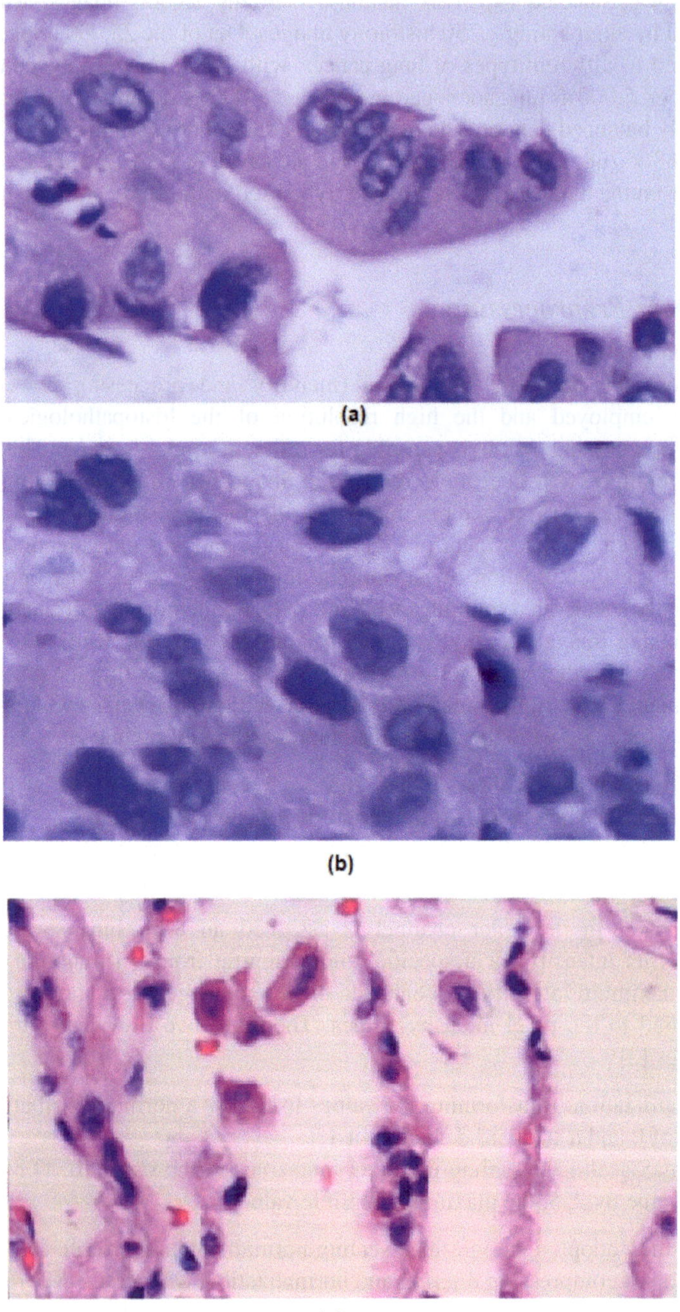

Fig. 1 Examples of images from the dataset for each cancer type used in the study. (**a**) Lung adenocarcinoma from the dataset, (**b**) squamous cell lung cancer from the dataset, and (**c**) benign tumor image from the dataset

2.3 Training, Validation, and Test Set Split

For a robust model training and evaluation, the dataset was randomly partitioned into three distinct sets: training, validation, and testing. This split ensured that the model learned from representative data and avoided biases. A split ratio of 75% for training, 15% for validation, and 10% for testing was employed [6]. The training set was used for training the models, the validation set for evaluating model performance during training, and the testing set for assessing the model's ability to generalize to new data (see Fig. 2).

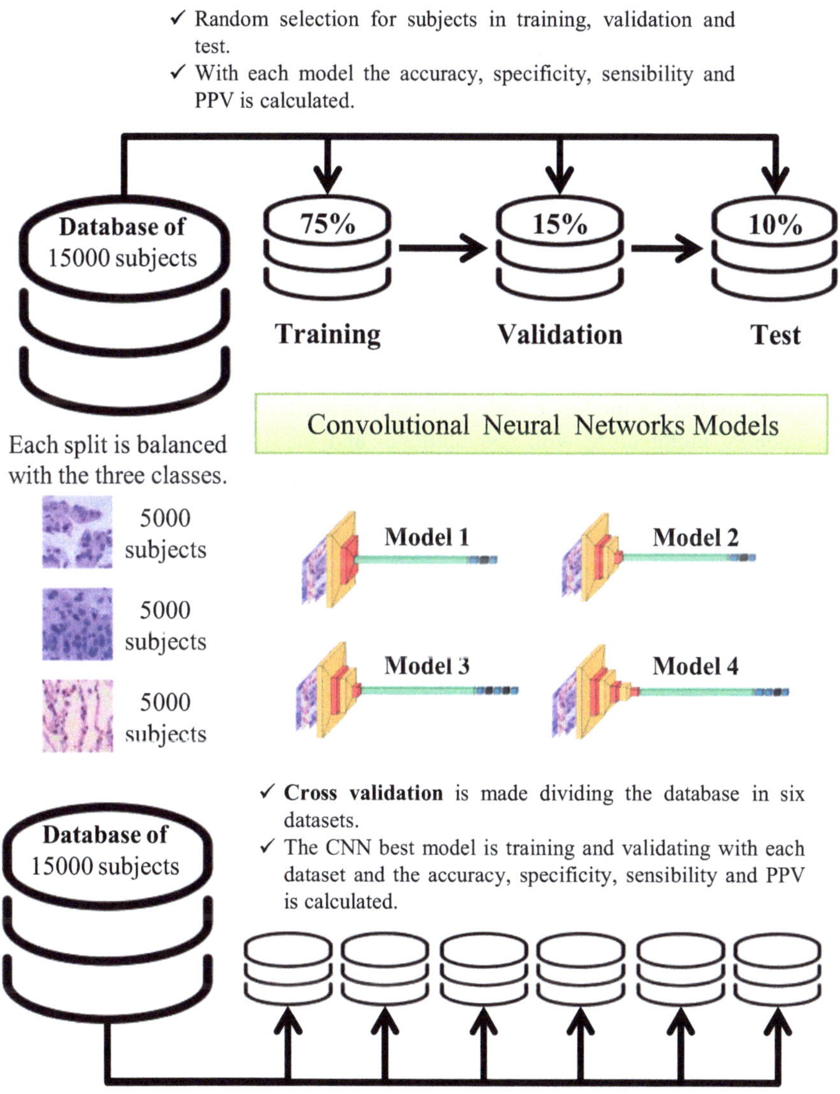

Fig. 2 Training, validation, and test split

Table 1 Distribution of images in datasets by cancer or tumor type used for this study

Number of images per set	Total (100%)	Lung adenocarcinoma	Benign tumor	Squamous cell carcinoma
Training set	11,251	3,780 (33.5%)	3,720 (33.1%)	3,751 (33.3%)
Validation set	2,250	714 (31.7%)	780 (34.7%)	756 (33.6%)
Testing set	1,499	506 (33.8%)	500 (33.3%)	493 (32.9%)

A training and testing step was performed for each model, and the results were used to identify the model with the best performance. The consistency of the top-performing model was handled by using the cross-validation technique (see Fig. 2). The dataset was carefully balanced to maintain a consistent representation of each lung cancer type across all sets [5]. Table 1 presents the dataset breakdown for the different sets. By employing this three-way split and maintaining class balance, the model receives a diverse training set, learns effectively, and generalizes well to unseen data.

2.4 Convolutional Neural Network Model Training

Multiple CNN models were created and empirically tested to identify the most effective configurations, leading to improved metrics compared to existing literature. Among the 18 models examined, four emerged as top performers, featuring a sequential architecture with convolutional and max pooling layers for feature extraction and dimensionality reduction, respectively. Each model consisted of a sequential architecture, where feature extraction was performed by a convolutional layer, followed by dimensionality reduction via a max pooling layer. Then, the data was flattened to prepare them for a dense layer that fed into the final output layer. The output layer consisted of three neurons using a softmax activation function to determine the probability distribution across three classification categories: benign tumor, adenocarcinoma, and squamous cell lung cancer.

A thorough hyperparameter tuning process was conducted to optimize the neural network architecture's balance between performance and complexity. This process involved experimenting with various variables, including neuron numbers in each layer and kernel sizes in convolutional and pooling layers, utilizing four different models, described as follows:

- **Model 1**: Convolutional Layer with 64 kernels of size 3. Pooling Layer with filters of size 2. Flatten Layer. Dense Layer with 256 neurons and a sigmoid activation function. Dropout 50%. Output Layer with three neurons and softmax activation function (see Fig. 3a).
- **Model 2**: Convolutional Layer with 64 kernels of size 3. Pooling Layer with filters of size 2. Convolutional Layer with 128 kernels of size 3. Pooling Layer with filters of size 3. Flatten Layer. Dense Layer with 256 neurons and a ReLU

Fig. 3 CNN models architecture: (**a**) Model 1, (**b**) Model 2, (**c**) Model 3, and (**d**) Model 4

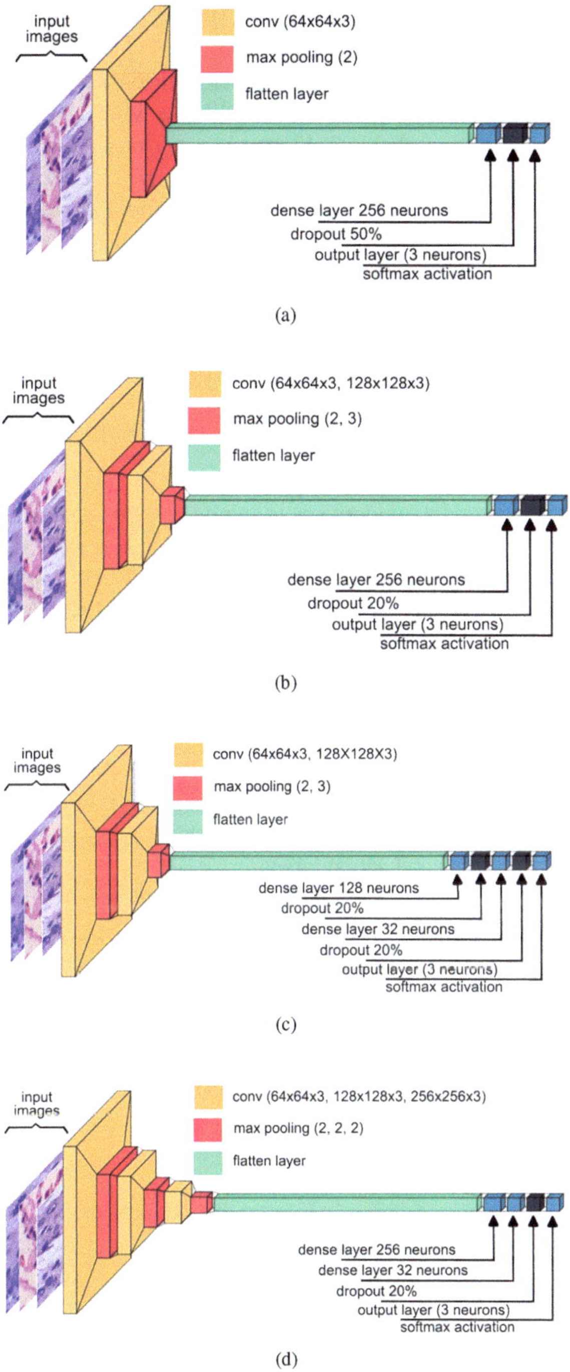

activation function. Dropout 20%. Output Layer with three neurons and softmax activation function (see Fig. 3b).

- **Model 3**: Convolutional Layer with 64 kernels of size 3. Pooling Layer with filters of size 2. Convolutional Layer with 128 kernels of size 3. Pooling Layer with filters of size 3. Flatten Layer. Dense Layer with 128 neurons and a ReLU activation function. Dropout 20%. Dense Layer with 32 neurons and a ReLU activation function. Dropout 20%. Output Layer with three neurons and softmax activation function (see Fig. 3c).
- **Model 4**: Convolutional Layer with 64 kernels of size 3. Pooling Layer with filters of size 2. Convolutional Layer with 128 kernels of size 3. Pooling Layer with filters of size 2. Convolutional Layer with 256 kernels of size 3. Pooling Layer with filters of size 2. Flatten Layer. Dense Layer with 256 neurons and a ReLU activation function. Dense Layer with 32 neurons and a ReLU activation function. Dropout 20%. Output Layer with three neurons and softmax activation function (see Fig. 3d).

The models underwent training, validation, and testing using distinct datasets. The best model was selected and subjected to a sixfold cross-validation within the database to assess its robustness and detect biases from random data splitting. Sensitivity, specificity, positive predictive value, and accuracy metrics were calculated for each cross-validation iteration. The outcomes are then reported as mean values and standard deviations. A Dropout layer was added to the models to reduce overfitting by randomly deactivating connections between neurons during training, enhancing the model's robustness and generalization capability. The network's output, generated using softmax activation in the output layer, indicates the probability of the input's classification. Performance assessment involves using Categorical Cross-Entropy loss, which measures loss based on predicted class probabilities and actual labels.

During training, the priority was to minimize the loss function and determine the optimal weights for the network's connections and convolutional kernels using the Adam optimizer. Adam dynamically adjusts the learning rate for each parameter based on historical gradients, controlling weight update magnitudes according to the error signal with a default learning rate of 0.001. Also, 32 samples per batch were used with the average error calculated over these mini-batches for 100 epochs. Models were trained on the training set and evaluated on the validation set to assess performance in classifying benign tumors, adenocarcinomas, and squamous cell lung cancers.

To ensure optimal performance during training, a mechanism was implemented to save the model weights whenever the prediction accuracy on the training set improved. This approach guaranteed that upon completion, a checkpoint file (.ckpt extension) is generated, containing the weights from the model iteration that achieved the highest accuracy on the training data. Additionally, accuracy and loss values for both the training and validation sets were logged in a comma-separated value (CSV) file after each epoch, facilitating a detailed analysis of the training process. This dual strategy of saving the best weights and logging performance metrics enhanced the model's reliability and provided valuable insights into its learning dynamics.

The evolution of accuracy values across training epochs was meticulously examined to assess the models' performance. The goal was to identify inflection points where continued training might lead to overfitting on the training data. To visualize these trends, helper functions were developed to plot the accuracy and loss values (retrieved from the CSV files) for both the training and validation sets, enabling effective monitoring of the learning process.

2.5 *Classifier Indexes*

To evaluate the performance of each model, true positives (TP), false positives (FP), true negatives (TN), and false negatives (FN) were calculated. These metrics were used to visualize classification discrepancies in the models [13]. Additionally, standard indexes such as accuracy, specificity, sensitivity, and positive predictive value (PPV) were employed for further assessment.

3 Results

Tables 2, 3, and 4 display the accuracy, sensitivity, PPV, and specificity of the training and validation results for the four CNN models, categorized into three classes. Tables 5, 6, and 7 present the accuracy, sensitivity, specificity, and PPV of the training and validation obtained from applying cross-validation for Model 4 which were the model with best performance. Finally, Table 8 shows the results of testing of Model 4.

Figure 4 illustrates the evolution of classification accuracy for both training set and validation set samples throughout the training process for Model 4. Specifically, this figure depicts the accuracy of the deep learning model on the training and validation datasets as a function of the number of epochs. Figure 5 presents the interface designed to upload histopathological images and execute a prediction

Table 2 Results of training (T) and validation (V) for lung adenocarcinoma class for Model 1 (M1), Model 2 (M2), Model 3 (M3), and Model 4 (M4)

Indexes	T-M1	V-M1	T-M2	V-M2	T-M3	V-M3	T-M4	V-M4
Accuracy	0.975	0.931	0.947	0.932	0.966	0.952	1.000	0.971
Sensibility	0.992	0.948	0.933	0.933	0.931	0.898	1.000	0.957
PPV	0.938	0.852	0.911	0.883	0.966	0.948	1.000	0.953
Specificity	0.967	0.923	0.954	0.944	0.984	0.977	1.000	0.978

Table 3 Results of training (T) and validation (V) for benign tumor class for Model 1 (M1), Model 2 (M2), Model 3 (M3), and Model 4 (M4)

Indexes	T-M1	V-M1	T-M2	V-M2	T-M3	V-M3	T-M4	V-M4
Accuracy	1.000	0.995	0.996	0.995	1.000	0.999	1.000	0.998
Sensibility	1.000	0.991	1.000	1.000	1.000	0.999	1.000	0.996
PPV	0.999	0.995	0.987	0.986	1.000	0.997	1.000	0.997
Specificity	0.967	0.997	0.954	0.993	0.984	0.999	1.000	0.999

Table 4 Results of training (T) and validation (V) for squamous cell carcinoma class for Model 1 (M1), Model 2 (M2), Model 3 (M3), and Model 4 (M4)

Indexes	T-M1	V-M1	T-M2	V-M2	T-M3	V-M3	T-M4	V-M4
Accuracy	0.976	0.936	0.951	0.936	0.966	0.953	1.000	0.973
Sensibility	0.934	0.852	0.908	0.908	0.967	0.955	1.000	0.959
PPV	0.993	0.951	0.944	0.922	0.933	0.910	1.000	0.962
Specificity	0.997	0.978	0.973	0.962	0.965	0.952	1.000	0.981

Table 5 Cross-validation results for lung adenocarcinoma class for Model 4

Indexes	Training	Validation
Accuracy	0.9903 ± 0.0035	0.9896 ± 0.0169
Sensibility	0.9875 ± 0.0038	0.9900 ± 0.0123
PPV	0.9839 ± 0.0107	0.9788 ± 0.0383
Specificity	0.9918 ± 0.0056	0.9895 ± 0.0193

Table 6 Cross-validation results for benign tumor class for Model 4

Indexes	Training	Validation
Accuracy	0.9990 ± 0.0012	0.9987 ± 0.0030
Sensibility	0.9992 ± 0.0007	1.0000 ± 0.0000
PPV	0.9978 ± 0.0040	0.9962 ± 0.0084
Specificity	0.9918 ± 0.0056	0.9980 ± 0.0046

Table 7 Cross-validation results for squamous cell carcinoma class for Model 4

Indexes	Training	Validation
Accuracy	0.9912 ± 0.0029	0.9910 ± 0.0140
Sensibility	0.9845 ± 0.0112	0.9786 ± 0.0391
PPV	0.9898 ± 0.0055	0.9943 ± 0.0043
Specificity	0.9948 ± 0.0029	0.9972 ± 0.0020

Table 8 Indexes of test results for Model 4 in each class

Indexes	ACA	Benign Tumor	SCC
Accuracy	0.9840	1.0000	0.9840
Sensibility	0.9862	1.0000	0.9655
PPV	0.9671	1.0000	0.9855
Specificity	0.9829	1.0000	0.9930

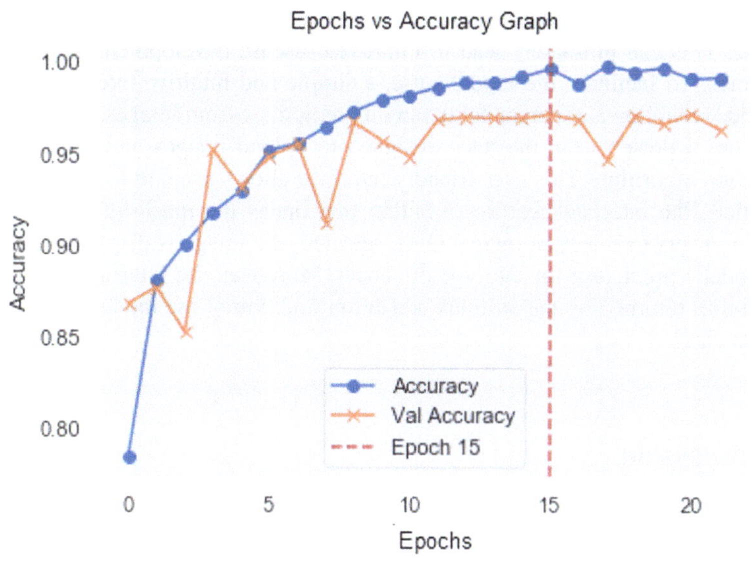

Fig. 4 Accuracy of a machine learning model on the training and validation datasets as a function of the number of epochs for Model 4

Fig. 5 Prediction result by the interface

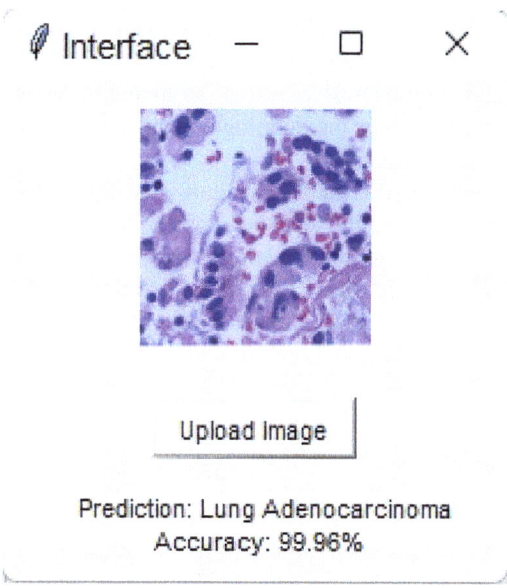

algorithm. This interface enables users to employ the classifier in a straightforward and practical manner while also providing the accuracy of the classification.

These results underscore the model's potential for generating predictions from histology images without the need for healthcare professional involvement, thereby enhancing efficiency in the healthcare system and facilitating the detection of this disease. This can ultimately lead to improved disease development and treatment outcomes. To facilitate the model's use, a simple and intuitive interface has been designed, enabling any personnel to obtain predictions from images.

Figure 5 presents the interface designed to upload images and implement the prediction algorithm. This user-friendly interface allows users to easily employ the classifier. The interface features a button that opens the machine's file explorer, enabling users to select an image. Once selected, the image is displayed alongside the model's prediction for the type of cancer. Moreover, the interface displays the probability returned by the softmax activation function of the model's output layer for that class.

4 Discussion

This study focuses on the development of CNN models for accurately classifying lung cancer images into different categories, including benign lung tumors, lung adenocarcinoma, and squamous cell lung cancer. The main findings highlight the high accuracy achieved by our models, supported by robust performance metrics such as accuracy, sensitivity, specificity, and positive predictive value.

Methodologically, we employed meticulous data preprocessing techniques, careful dataset selection, innovative model architectures, rigorous training processes, and comprehensive evaluation methods. We stress the importance of data quality, preprocessing methodologies, and model optimization in achieving accurate and reliable results.

In general, the results show that all models have accuracy, sensitivity, positive predictive value, and specificity indexes above 90% for most types of cancer. This indicates that all models are quite good at detecting most types of lung cancer. However, Model 4 has the best overall results, with an accuracy of over 97% for all types of cancer. Model 4 also has the best sensitivity over 95.7%. The next best model is Model 3, and it also has good results, with an accuracy, sensitivity, PPV, and specificity of over 95% for all types of cancer.

All four models demonstrated excellent performance in classifying benign tumors. Model 4, in particular, achieved a remarkable 100% for all indexes in the test group (see Table 8). To further validate that Model 4's performance was indeed due to its intrinsic capabilities rather than the selection of subjects in the training and validation sets, cross-validation was conducted. The results of the cross-validation are presented in Tables 5, 6, and 7. The reported indexes include their standard deviations, which are all in the order of thousandths, indicating minimal differences in indexes across the different training and test groups. These findings strongly suggest that Model 4's exceptional performance stems from its inherent ability to accurately classify benign tumors, squamous cell carcinoma, and lung adenocarcinoma.

The observed variations in the models performance could be attributed to several factors related to their architectural characteristics. Model 4 stands out with three convolutional layers, while the other models have only two. This additional layer allows Model 4 to extract more complex features from the histopathological images, potentially leading to improved cancer detection capabilities. Models 2, 3, and 4 employ the Rectified Linear Unit (ReLU) activation function in their hidden dense layers, while Model 1 utilizes the sigmoid function. ReLU is generally considered more effective for training deep neural networks, as it helps mitigate the vanishing gradient problem, a common challenge in training deep learning models. Finally, Models 2, 3, and 4 incorporate Dropout regularization into their dense layers. This technique helps prevent overfitting and enhances the model's generalization ability, allowing it to perform better on unseen data.

This research has significant implications for medical imaging and lung cancer diagnosis. By leveraging artificial intelligence algorithms and CNN models, we demonstrate how early detection of lung cancer can be improved, leading to better clinical outcomes for patients.

5 Conclusion

This study significantly contributes to advancing the field of medical imaging and artificial intelligence in healthcare as a direct application through a decision support system. By showcasing the efficacy of AI-driven approaches in lung cancer detection, we aim to facilitate early diagnosis of lung cancer and ultimately improve patient outcomes. This research underscores the importance of collaboration, innovation, and interdisciplinary research in driving scientific progress in healthcare.

In essence, the study objectives were successfully achieved. The employed methodology yielded promising results and reinforces the value of intelligent systems in medicine. We identified key challenges and limitations that offer avenues for future model improvement and broader applicability within decision support systems. This work contributes significantly to the field and aligns with the personal and academic goals of this aspiring Biomedical Engineer.

Future Works

Looking ahead, future challenges include improving model performance through the exploration of new techniques or algorithms, expanding dataset sizes, leveraging more powerful computational resources, and refining model architectures to handle higher resolution images. Additionally, modifications to existing architectures can facilitate multi-class classification of diseases with similar image characteristics, thereby broadening the spectrum of diseases detectable by our system. To carry out the study, a biopsy is necessary to obtain the histology, avoiding the need for numerous tests that require the constant and cumulative application of radiation, which promotes the use of this system.

The implementation of the algorithms developed in this study with more powerful computational machines will allow the creation of architectures designed to work with higher resolution images. This will avoid the need to preprocess them by reducing their dimensionality and applying noise to them. In this way, by increasing the accuracy of the data, we will generate an improvement in the models, avoiding the loss of information necessary for the classification task.

The architectures used have been focused on the classification task according to the dataset used. Small variations in these architectures, by increasing the number of neurons in the output layer, can be used in the multi-class classification of different diseases with similar images, expanding the variety of diseases recognizable by the developed system.

Acknowledgments The authors would like to express their sincere gratitude to Universidad de Ibagué for their invaluable collaboration ad honorem of this research. Additionally, we extend our appreciation to Universidad Internacional de Valencia in Spain and Universidad Simón Bolívar in Caracas, Venezuela for their support and contributions.

Competing Interests The authors have no conflicts of interest to declare that are relevant to the content of this chapter.

Ethics Approval This study adheres to the principles outlined in the Declaration of Helsinki. Note that the use of anonymized patient data without explicit consent is widely accepted in research practices, particularly when adherence to strict anonymization protocols and ethical standards is maintained. By prioritizing patient privacy and confidentiality, while upholding the integrity of scientific inquiry, this study endeavors to generate valuable insights for the benefit of both current and future patients.

References

1. Abdelsamea, M. M., Zidan, U., Senousy, Z., Gaber, M. M., Rakha, E., & Ilyas, M. (2022). A survey on artificial intelligence in histopathology image analysis. *Wiley Interdisciplinary Reviews: Data Mining and Knowledge Discovery, 12*(6), e1474.
2. Aguilar, E., La Cruz, A., Albertti, R., Carnier, M., Gavidia, L., & Severeyn, E. (2022). Detection of respiratory disease patterns using mask R-CNN. In *Proceedings of seventh international congress on information and communication technology: ICICT 2022, London* (Vol. 2, pp. 739–750). Springer.
3. Ballé, J., Laparra, V., & Simoncelli, E. P. (2015). Density modeling of images using a generalized normalization transformation. arXiv preprint arXiv:1511.06281.
4. Borkowski, A. A., Bui, M. M., Thomas, L. B., Wilson, C. P., DeLand, L. A., & Mastorides, S. M. (2019). Lung and colon cancer histopathological image dataset (lc25000). arXiv preprint arXiv:1912.12142.
5. Cui, Y., Jia, M., Lin, T. Y., Song, Y., & Belongie, S. (2019). Class-balanced loss based on effective number of samples. In *Proceedings of the IEEE/CVF conference on computer vision and pattern recognition* (pp. 9268–9277).
6. Domingo, J. D., Aparicio, R. M., & Rodrigo, L. M. G. (2022). Cross validation voting for improving CNN classification in grocery products. *IEEE Access, 10*, 20913–20925.

7. Hamamoto, R., Suvarna, K., Yamada, M., Kobayashi, K., Shinkai, N., Miyake, M., Takahashi, M., Jinnai, S., Shimoyama, R., Sakai, A., et al. (2020). Application of artificial intelligence technology in oncology: Towards the establishment of precision medicine. *Cancers, 12*(12), 3532.
8. Hatuwal, B. K., & Thapa, H. C. (2020). Lung cancer detection using convolutional neural network on histopathological images. *International Journal of Computer Trends and Technology, 68*(10), 21–24.
9. Karim, D. Z., & Bushra, T. A. (2021). Detecting lung cancer from histopathological images using convolution neural network. In *TENCON 2021-2021 IEEE Region 10 Conference (TENCON)* (pp. 626–631). IEEE.
10. Mazurowski, M. A., Habas, P. A., Zurada, J. M., Lo, J. Y., Baker, J. A., & Tourassi, G. D. (2008). Training neural network classifiers for medical decision making: The effects of imbalanced datasets on classification performance. *Neural Networks, 21*(2-3), 427–436.
11. Siddique, A., Otaibi, F. M., & Khan, S. F. (2023). Lung cancer in developing countries. In *Handbook of medical and health sciences in developing countries: Education, practice, and research* (pp. 1–28). Springer.
12. Tang, S., Yuan, S., & Zhu, Y. (2020). Data preprocessing techniques in convolutional neural network based on fault diagnosis towards rotating machinery. *IEEE Access, 8*, 149487–149496.
13. Xu, J., Zhang, Y., & Miao, D. (2020). Three-way confusion matrix for classification: A measure driven view. *Information Sciences, 507*, 772–794.

Decision Support in Peripheral Arterial Occlusive Disease Management with AI-Enhanced Super-Resolution Techniques for Advanced Vascular Imaging

Alexandra La Cruz, Juan Pedro Felipe, Erika Severeyn, and Andrés García-León

1 Introduction

Different epidemiological studies have shown that up to 50% of patients with the PAOD [3] have symptoms of cerebrovascular or cardiological disease, a real cause of patient morbidity and mortality [21]. PAOD describes a narrowing (stenosis) or complete blockage (occlusion) of the arteries of the arm or leg (see Fig. 1) [3]. PAOD can cause leg pain when walking (intermittent claudication), which may go away with rest. Over time, the disease can worsen and there may be pain even at rest, leg ulcers, and, in severe cases, gangrene, which may lead to amputation of the affected limb [3]. Consequently, in patients with PAOD, atherosclerotic stenosis is likely to appear in other areas of the body, such as in the coronary arteries and cerebral arteries. Mortality and morbidity caused by PAOD will depend largely on poor circulation in these areas rather than on limb ischemia [1, 3].

A. La Cruz (✉)
Faculty of Engineering, Universidad de Ibagué, Ibagué, Tolima, Colombia
e-mail: alexandra.lacruz@unibague.edu.co

J. P. Felipe
Computer Engineer and Biomedical Engineering, Universidad Internacional de Valencia, Valencia, Spain

E. Severeyn
Department of Thermodynamics and Transfer Phenomena, Universidad Simón Bolívar, Sartenejas, Venezuela
e-mail: severeynerika@usb.ve

A. García-León
Tecnológico de Monterrey, Escuela de Ingeniería y Ciencias, Monterrey, Mexico
e-mail: andres.garcialeon@tec.mx

N. Gupta, J. D. Diaz Roman (eds.), *Decision Support Systems for Industrial and Scientific Applications*, Signals and Communication Technology,
https://doi.org/10.1007/978-3-031-89700-9_8

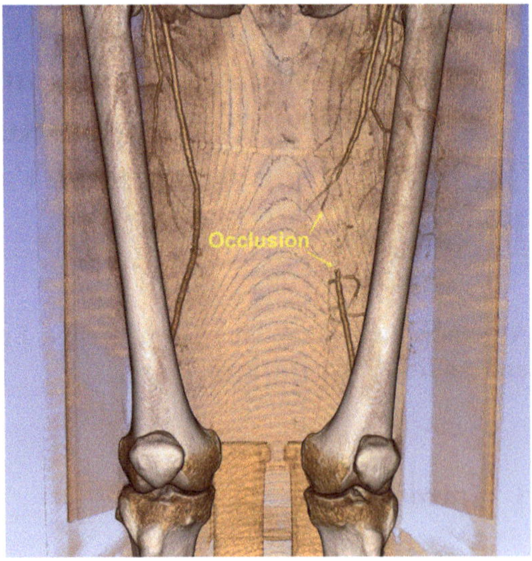

Fig. 1 3-D visualization of a PAOD case of a patient. The arrows indicate the begin-ends of the occlusion

Advanced imaging techniques like computed tomography angiography (CTA), magnetic resonance angiography (MRA), and duplex ultrasound are key to visualizing vascular structures in PAOD patients [19]. These methods allow detailed assessment of arterial narrowing or blockages, essential for treatment planning. This study uses CTA images, the most common modality for PAOD evaluation. Three-dimensional (3-D) reconstruction from CTA images helps visualize arteries, including stenosis or occlusion, aiding interventions like angioplasty or bypass surgery [23, 34]. SR techniques enhance 3-D models, particularly in areas with small blood vessels, where traditional methods like interpolation struggle with low-resolution images. Despite challenges in data acquisition and processing, 3-D rendering improves PAOD assessment by providing accurate anatomical details and supporting collaborative decision-making [18]. Beyond PAOD, 3-D modeling is useful in surgical planning and augmented reality applications [29].

Small blood vessel diameters present significant challenges for 3-D rendering, requiring high spatial resolution through CTA or MRA for accurate detail capture. Noise reduction, advanced vessel segmentation, and SR techniques are essential for improving image quality. Visualizing small vessels affected by atherosclerosis remains difficult [15], as healthy arteries have uniform CT density, while diseased arteries show varying attenuation. Image noise, scanning artifacts, and arterial contrast variations complicate visualization. Despite these challenges, advances in imaging technology are improving 3-D rendering. Super-resolution techniques, particularly for low-resolution images, can significantly enhance the quality of 3-D models.

This chapter examines the impact of deep learning-based SR techniques on the diagnosis and care of PAOD patients through CTA image processing. Advanced SR techniques in CTA imaging are poised to improve diagnostic accuracy and treatment planning by offering precise 3-D visualizations. These techniques support clinicians with a robust decision support system (DSS). The models studied include SRCNN [5], EDSR [17], RCAN [32], SRGAN [25], and ESRGAN [30], adapted for CTA image processing. The chapter highlights how SR techniques can improve the visualization of small blood vessels in atherosclerotic patients, advancing medical diagnostics and decision support, ultimately enhancing patient care. The hypothesis is whether these models effectively enhance the resolution of vascular structures in the lower extremities.

The paper follows a structured approach with five key sections: (1) Sect. 3 provides a literature review and research context, (2) Sect. 4 details the methodology encompassing dataset and SR techniques, (3) Sect. 5 presents our findings, (4) Sect. 6 engages in an in-depth discussion of results and their relevance to existing knowledge, and concludes the paper while proposing future research.

2 Impact of Including SR Techniques in DSS in Health Systems

Super-resolution (SR) techniques are crucial in Decision Support Systems (DSS) for medical imaging, enhancing the visualization of vascular structures in CTA images. These techniques improve image resolution, making small arteries clearer, which is vital for accurate diagnosis and treatment planning, especially in conditions like PAOD. SR helps reveal pathologies not visible in lower-resolution scans, improving diagnostic accuracy and enabling earlier interventions. Integrating SR into automated analysis workflows allows machine learning models to detect anomalies and assess vascular health, enhancing clinical decision-making. SR can be seamlessly incorporated into clinical imaging workflows, providing enhanced CTA images with minimal manual intervention. Optimized for real-time processing, SR algorithms support timely decision-making, especially during procedures like angiography. SR's integration into clinical environments can be facilitated through user-friendly interfaces, improving workflow efficiency. By improving image quality, SR techniques reduce the need for repeat imaging, saving costs and minimizing patient exposure to radiation. Faster, more accurate diagnoses streamline the diagnostic process, enhance patient satisfaction, and lower operational costs. SR also optimizes resource use by allowing more efficient allocation of medical staff, ultimately improving patient outcomes and reducing long-term healthcare costs.

3 Related Works

Single image super-resolution (SISR) aims to increase the resolution of low-resolution images by restoring high-frequency information. Most SISR methods, like sparse coding [5], extract key features from the image and encode them into a low-resolution structure. This compact representation captures essential details while reducing data size. Sparse coefficients are then passed to a high-resolution dictionary to reconstruct the image. Traditional SR methods typically do not integrate generative models into a unified optimization framework, which is a key aspect of modern approaches. This research evaluates whether generative models effectively enhance the resolution of vascular structures in lower extremities.

Convolutional neural network (CNN)-based methods have greatly improved the performance of SISR [35]. Dong et al. pioneered the use of CNNs to solve the SISR problem in [35]. They showed that conventional sparse coding-based SR methods existing at the time can be reformulated into a deep CNN, achieving superior performance. This model is divided into three phases, each performed by a convolution: feature mapping, nonlinear feature mapping, and reconstruction to a high-resolution representation.

Dong et al. introduced the SRCNN, a deep learning technique that can reformulate the conventional sparse coding-based SR methods and achieve superior performance [5]. From that point on, numerous methods employing deep learning were developed to address the SR problem. Residual models such as very deep convolutional networks (VDCN) [13], deeply recursive convolutional networks (DRCN) [14], and deep recursive residual network (DRRN) [26] emerged, building the neural network based on units or blocks that learn through residual functions, which represent the difference between the original and processed image. Residual networks can converge faster, requiring fewer epochs and providing a higher peak signal-to-noise ratio (PSNR) [12] in training, which quantitatively measures the quality of the image reconstruction.

Subsequently, some approaches employ a de-convolution layer at the end of the network that directly maps the original low-resolution image to the high-resolution image [6]. Some other approaches extract features from the low-resolution space and map them to the high-resolution space [25]. These scaling techniques have been leveraged by several residual models, which use residual scaling methods to improve the stability and performance of deep networks. Alternatively, some models utilize a residual dense block (RDB) as the central component in network construction, where each RDB consists of densely connected layers and local feature fusion with local residual learning [33].

Recent advancements introduce the generative adversarial networks (GAN) [9]. GAN models consist of a generator that produces SR images and a discriminator that evaluates whether the generated image is real or artificial. A good approximation of the high-resolution image is assumed if the discriminator cannot distinguish between the generated image and the original image [25]. In addition, a new function in GAN is used as a discriminator to enable the generator to retrieve more detailed textures by learning to infer which image is more realistic [30]. Furthermore, a residual-in-residual (RIR) structure has been proposed to address the low-frequency information removed through multiple hopping connections. This structure incorporates a channel attention (CA) mechanism that adaptively rescales channel characteristics by considering channel inter-dependencies [32].

In this paper, we have divided SR algorithms into several groups based on their network architectures and characteristics within the context of super-resolution techniques. The algorithms are categorized into four sets, each focusing on specific aspects of network design and methodology.

3.1 Residual-Based Models

Residual models focus on learning the difference between the input and desired output, simplifying the learning of identity mappings and aiding in the training of deeper networks. A well-known example of this approach is ResNet, which helps mitigate the vanishing gradient problem. Building on this, very deep super-resolution (VDSR) networks [13] utilize deep structures with small filters to efficiently capture contextual information across large image regions. By applying residual learning and high learning rates, they optimize the network quickly, ensuring training stability through gradient clipping. Expanding on these concepts, the deep recursive convolutional network (DRCN) [14] introduces up to 16 recursive layers, enhancing performance without increasing the number of parameters. This is achieved by using recursive supervision and skipping connections, which make training more manageable. Further advancing this idea, the deep recursive residual network (DRRN) [26] proposes a deeper model with up to 52 convolutional layers. By leveraging residual learning, it improves performance while using fewer parameters, building on the success of earlier models.

3.2 Scaling Methods

Scaling models aim to enhance the network's capacity by increasing the number of layers, channels (width), or the resolution of input images. This approach is designed to boost the network's performance by enabling it to capture more complex patterns. Dong et al. further refined their SRCNN model for practical applications that require real-time performance (24 fps), achieving better quality and a 40-fold increase in speed. Their proposed model, the fast super-resolution convolutional neural network (FSRCNN) [6], introduced three key improvements: first, it incorporates a de-convolution layer at the network's end, directly mapping the low-resolution image to its high-resolution counterpart; second, it reduces the input dimension by reformulating the mapping layer; and third, it uses smaller filters but more mapping layers, optimizing computational efficiency.

Building on this, the efficient sub-pixel convolutional network (ESPCN) [25] introduces another innovation by extracting feature maps in the low-resolution space, eliminating the need to scale the input image to high resolution. This reduces the computational load while maintaining performance. Furthermore, it incorporates a sub-pixel convolutional layer that learns a set of filters to scale the final low-resolution feature map to a high-resolution output.

3.3 Integrating Residual and Scaling Methods for Enhanced Super-Resolution Performance

By combining scaling strategies with residual learning, models can achieve both high performance and computational efficiency. Residual connections allow networks to focus on learning high-frequency details, crucial for super-resolution, while scaling strategies (such as depth, width, and resolution) enhance the network's ability to capture finer details, improving the quality of super-resolved images. Models integrating residual learning with scaling methods represent a hybrid approach that leverages the strengths of both techniques. The residual component enables the network to capture fine details, while scaling methods optimize the network's capacity to enhance image resolution. Below are key models that integrate both approaches.

The enhanced deep super-resolution (EDSR) [17] model improves upon SRResNet by removing unnecessary modules to create a more compact and efficient architecture. By using residual scaling techniques, EDSR stabilizes the deep model and delivers superior performance, winning the NTIRE2017 SR Challenge [27]. The Laplacian pyramid super-resolution network (LapSRN) [16] reconstructs high-resolution images progressively through sub-band residuals. This model eliminates the need for bicubic interpolation, reducing computational complexity while using deep supervision and a robust loss function to ensure high-quality reconstruction. SRDenseNet [28] introduces dense hopping connections, where feature maps from each layer are propagated to all subsequent layers. This facilitates the combination of low-level and high-level features, improving reconstruction performance and mitigating the vanishing gradient problem in very deep networks. Additionally, deconvolution layers accelerate the up-sampling process. The residual dense network (RDN) [33] addresses the challenge of inefficient use of hierarchical features in low-resolution images, which can limit performance in deep CNN-based SR models. RDN incorporates residual dense blocks (RDBs) to efficiently fuse features and retain crucial information, enabling the network to scale effectively and stabilize during training.

Lastly, the residual channel attention network (RCAN) [32] introduces a residual in residual (RIR) structure that focuses the network on high-frequency information by removing low-frequency data. With its channel attention (CA) mechanism,

RCAN adapts the network's focus on relevant channel characteristics by considering relationships across channels.

The enhanced deep super-resolution (EDSR) [17] residual network model was designed with an improvement of the SRResNet model, achieving better results by eliminating unnecessary modules while obtaining a more compact model. They used residual scaling techniques to give stability to the deep model. This method showed superior performance over state-of-the-art methods on benchmark datasets, becoming the winner of the NTIRE2017 SR Challenge [27].

3.4 Generative Models for Super-Resolution Enhancement

Generative adversarial networks (GANs) have been a breakthrough in single image super-resolution (SISR), starting with the SRGAN [25], which introduced the concept of using a GAN for SR tasks. This model features a generator that creates super-resolved images and a discriminator that distinguishes between real and generated high-resolution images. If the discriminator cannot tell the difference, the generated image is considered a successful approximation of the high-resolution output. The generator in SRGAN is based on the SRResNet model, a residual network comprising five blocks of two conventional layers (Conv-BN-ReLU).

Building on this, the SRFeat [22] model introduces a new generator with long-range hop connections, enabling more efficient transfer of information across distant layers. It also incorporates two types of discriminators: an image discriminator and a feature discriminator, the latter focusing on preserving high-frequency structural features while reducing noise artifacts, enhancing image quality. The ESRGAN [30] model takes SRGAN a step further by significantly improving visual quality. It introduces a generator based on a multi-block residual in residual dense block (RRDB) architecture, eliminating batch normalization (BN) layers for better performance. Additionally, ESRGAN employs a relativistic GAN discriminator, which learns to evaluate the realism of generated images, pushing the generator to recover finer textures and details. This progression demonstrates how each GAN-based model builds on its predecessor, improving the quality of generated super-resolved images by refining the generator and discriminator architectures, while focusing on finer details and better texture representation.

3.5 Comparison of Super-Resolution Models Selected from Literature

This section provides an overview of several deep learning and generative model-based super-resolution techniques. These models were adapted for CTA images, implemented, evaluated, and analyzed, as they are typically used for biological image processing. A summary of the evaluated models is presented in Table

Table 1 Comparison of SR models evaluated for CTA images, based on their network architectures and characteristics

Model	Convolutional model-based	Residual model-based	GAN model-based
SRCNN915 [5]	X		
SRCNN955 [5]	X		
EDSR [17]	X	X	
RCAN [32]	X	X	
SRGAN [25]	X		X
ESRGAN [30]	X		X

1, where each model is categorized based on its architecture and methodology. This classification offers a structured approach to understanding their strengths and capabilities, particularly in addressing the challenges of enhancing vascular structure visualization in CTA images.

4 Materials and Methods

4.1 Dataset

The data consisted of anonymized CTA medical images of lower extremities in DICOM format [2]. CT imaging, using X-ray radiation, produces anatomical slices for diagnostics [4]. Each image is made of voxels, with values reflecting tissue density on the Hounsfield scale [4]. CTA is used to detect conditions like arterial aneurysms, blockages, blood clots, and PAOD [8].

The images obtained by CTA in this study are volumetric data of size 512×512 pixels per slice. Due to the large computational demands of deep learning networks, 64×64 blocks were used for training. The selected block is used as an output image (target), while the input is a bicubic-downsampled version of the same block for testing purpose.

4.2 Image Selection

For this study, we selected CTA images of the lower extremities, manually removing the initial and final slices to retain only the relevant images from the genitalia to the ankle. On average, each study/series included around 600 images, all of which were anonymized. To optimize memory usage during SR model training, we divided each 512×512 pixel slice into smaller 64×64 pixel blocks, preventing memory overflow and ensuring smooth execution. Since many of these blocks did not contain relevant information, we focused only on the most informative cells, as shown in Fig. 2.

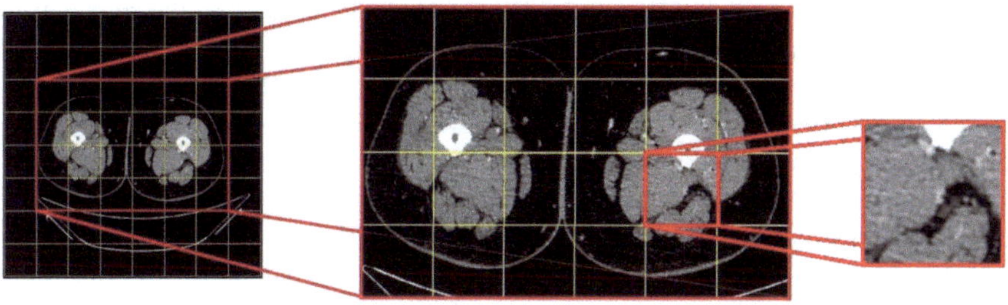

Fig. 2 Cell selection for training and validation. If 600 slices (on average) are selected in a study and from each slice 24 block cells are used, we have about 14,400 64 × 64 pixel images for training/validation of the SR models

4.3 Development Tools

Python was used as the programming language, along with various packages that facilitate the implementation of artificial intelligence, including Pandas,[1] Numpy,[2] TensorFlow,[3] Keras,[4] sci-kit-learn,[5] and OpenCV-python.[6] We tested the algorithms using Jupyter Notebook.[7] For handling DICOM images, we used the Pydicom library.[8] The 3-D modeling of the original studies and those obtained after applying the SR method was performed with 3D-Slicer.[9] This tool is an open-source software designed for the visualization, processing, segmentation, registration, and analysis of 3-D medical, biomedical, and other images and meshes.

4.4 Evaluation Metrics

Any processing performed on an image can cause a significant loss of information or quality. Several metrics allow to evaluate such loss. PSNR, related to mean square error (MSE), evaluates the quality of the image reconstruction by numerically comparing the original image and the image generated by the reconstruction [24]; it gives a rough estimate of the human perception of the reconstruction quality.

[1] https://pandas.pydata.org/

[2] https://numpy.org/

[3] https://www.tensorflow.org/

[4] https://keras.io/

[5] https://scikit-learn.org/stable/

[6] https://opencv.org/

[7] https://jupyter.org/

[8] https://pydicom.github.io/

[9] https://www.slicer.org/

Its measure is in decibels (dB), and its value tends to infinity, so the higher it is, the better the quality of the generated image. On the other hand, a lower PSNR value means a big numerical difference between images. SSIM is a model based on the perception of the human visual system. It measures the similarity between the original and reconstructed image by considering structural information, luminance, and contrast [12]. The SSIM index is a decimal value between -1 and 1, with 0 indicating no structural similarity, 1 indicating a perfectly similar structure, which is achieved in the case of two identical images, and -1 indicating a perfect anti-correlation, meaning the images are structurally opposite [20].

4.5 Super-Resolution CNN Models Applied to CTA Images

The success of deep learning in SR has led to the development of numerous methods. In this paper, we compared and adapted various SRCNN models for lower extremity CTA images, using the same loss functions and optimizer as the original authors. We also computed metrics like maximum PSNR [12] and SSIM [20] to quantitatively evaluate the results. Notably, the final convolution layer, responsible for reconstructing the output image, generates a one-dimensional feature map due to the monochrome (single-channel) nature of the CTA images. Below, we describe the architecture of each SR model implemented and evaluated.

Super-Resolution Convolutional Neural Network (SRCNN)
In 2014, Dong et al. [5] introduce this method. They use a CNN that takes the low-resolution image as input and generates the high-resolution image. The architecture is shown in Fig. 3, it was adapted to CTA images.

Initially, the low-resolution image is pre-processed by enlarging it to match the high-resolution image using bi-cubic interpolation. The end-to-end learning consists of three convolutional steps: first, extracting patches from the low-resolution image as high-dimensional vectors; second, performing nonlinear mapping to create feature maps; and third, reconstructing the high-resolution image by aggregating these patch representations. And in 2015, Dong et al. published version 3 of their paper [5], where they generalize the second step of their method by applying larger filters, thereby achieving better learning results, and designed the SRCNN 9-5-5 model architecture (see Fig. 4).

SRCNN model numbers such as 9-1-5 and 9-5-5 refer to the filter sizes and number of filters used in each convolutional layer. SRCNN 9-1-5 uses a 9×9 filter size in the first layer, a 1×1 filter size in the second layer, and a 5×5 filter size in the third layer. On the other hand, SRCNN 9-5-5 uses a filter size of 9×9 in the first layer, a filter size of 5×5 in the second layer, and a filter size of 5×5 in the third layer. In Figs. 3 and 4, codes such as f64k9s1, f64k1s1 describe the configuration of the convolutional layers in terms of filters, core size, and stride, for example:

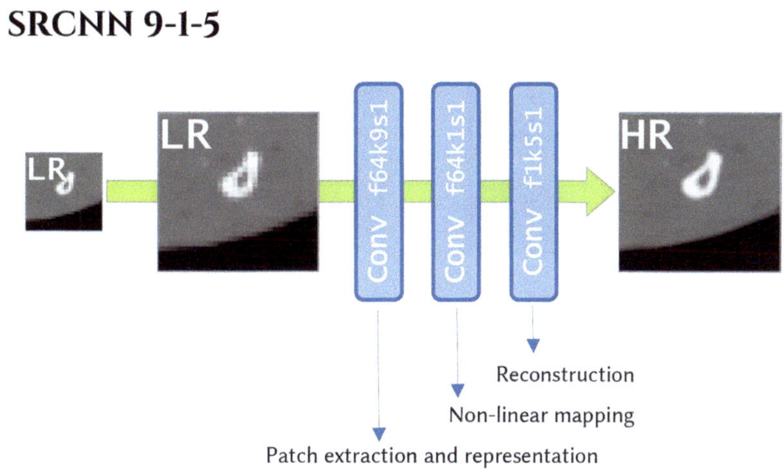

Fig. 3 Implementation scheme of the SRCNN 9-1-5 model. Each convolution is defined by the number of features, the size of the convolution matrix (kernel), and the convolution step (stride)

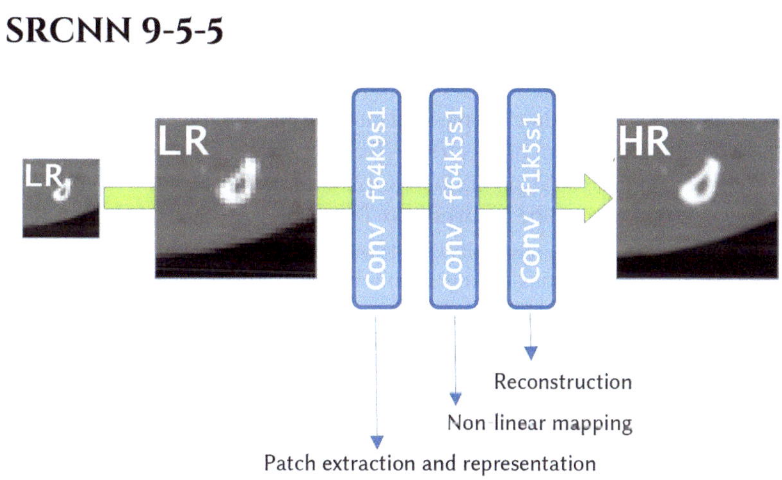

Fig. 4 SRCNN 9-5-5 model architecture

- *f64k9s1*: indicates the convolution layer is composed of 64 filters (f64) with a kernel size of 9 × 9 (k9), and stride of 1 (s1).
- *f64k1s1*: indicates the convolution layer is composed of 64 filters (f64) with a kernel size of 1 × 1 (k1), and stride of 1 (s1).

Enhanced Deep Residual Networks for Single Image Super-Resolution (EDSR) Based on SRResNet [25], it consists of a series of ResBlocks followed by an image scaling block, as shown in Fig. 5. Unlike SRResNet, EDSR removes Batch Normalization (BN) layers, improving performance without compromising quality.

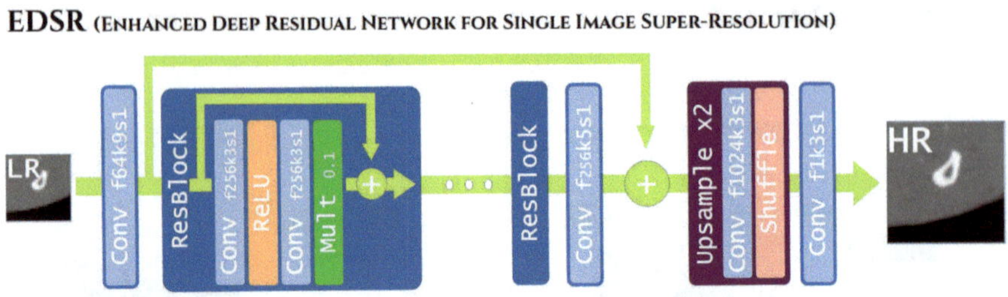

Fig. 5 EDSR model architecture

This allows for deeper networks that outperform conventional residual networks (ResNet) [11]. However, training became unstable when feature maps exceeded a threshold, which was addressed by applying a residual scale of 0.1 in each block (shown as "Mult 0.1" in Fig. 5). The final block scales the image by ×2, with a second block used for ×4 scaling.

Residual Channel Attention Networks (RCAN) [32]
It consists of four main components: shallow feature extraction, residual deep feature extraction via residual-in-residual (RIR) modules, a scaling module, and a reconstruction module. The shallow feature extraction uses a single convolutional layer, while the RIR module, composed of multiple residual groups (RGs) with residual channel attention blocks (RCABs), facilitates deep feature extraction. The scaling module, based on ESPCNN [25], improves image resolution. Long hop connections enable the network to focus on key low-resolution features. Additionally, the channel attention (CA) mechanism helps the network prioritize informative features by considering channel inter-dependencies. This architecture supports the training of very deep networks, with over 400 layers (Fig. 6).

Super-Resolution Using GAN (SRGAN)
This model aims to improve perceptual quality over traditional PSNR-based methods. While PSNR decreases, the overall structural quality improves, and the goal is to generate SR images with high-quality structures for reliable 3-D reconstruction. In this approach, the focus is on training the generator without adversarial training. The SRGAN [25] generator, known as SRResNet, is a deep residual network that diverges from MSE as the sole optimization objective. The architecture (see Fig. 7) includes 16 residual blocks, each formed by two convolutional layers (3 × 3) with 64 feature maps, batch normalization, and ParametricReLU (PReLU) activation. The resolution is increased through ×2 scaling blocks using 3 × 3 convolutions with 256 feature maps, followed by pixel shuffling and PReLU activation. The final output is produced by applying a 3 × 3 convolution with a single feature map. The "tanh" activation function, which captures both positive and negative features, yields the best results in super-resolution tasks by preserving high-frequency details.

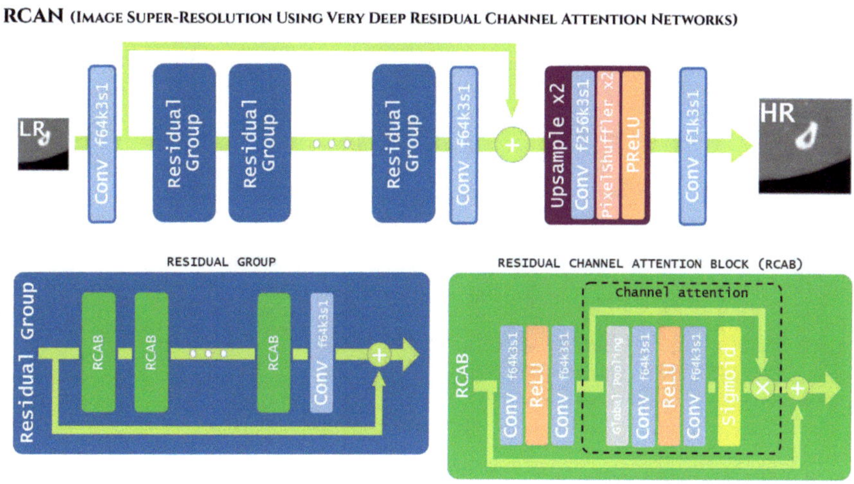

Fig. 6 RCAN model architecture

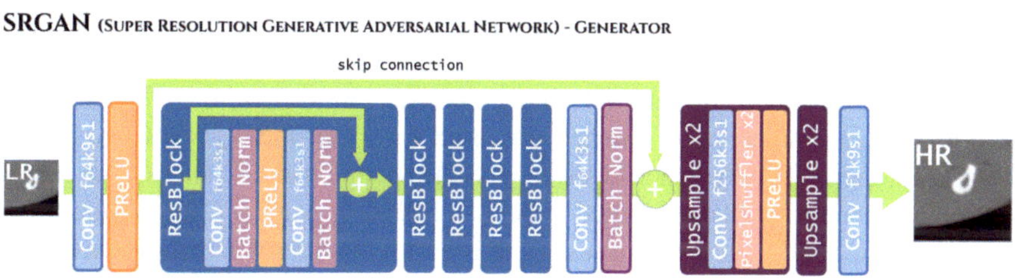

Fig. 7 SRGAN model architecture

Enhanced Super-Resolution GAN (ESRGAN) [30]

It improves image quality over SRGAN by making two key modifications to the generator: removing batch normalization (BN) layers and introducing a new basic block called the residual in residual dense block (RRDB). The RRDB combines a multilevel residual network with dense connections, as shown in Fig. 8. Removing BN layers enhances performance, improves network adaptability, and reduces computational complexity and memory usage, while also minimizing the risk of artifacts in deep networks. The RRDB structure is more complex than the original SRGAN residual block, featuring a residual-in-residual design that performs residual learning at multiple levels using dense blocks. Additionally, residual scaling, by multiplying by a constant β (between 0 and 1), is applied to prevent instability and improve training stability.

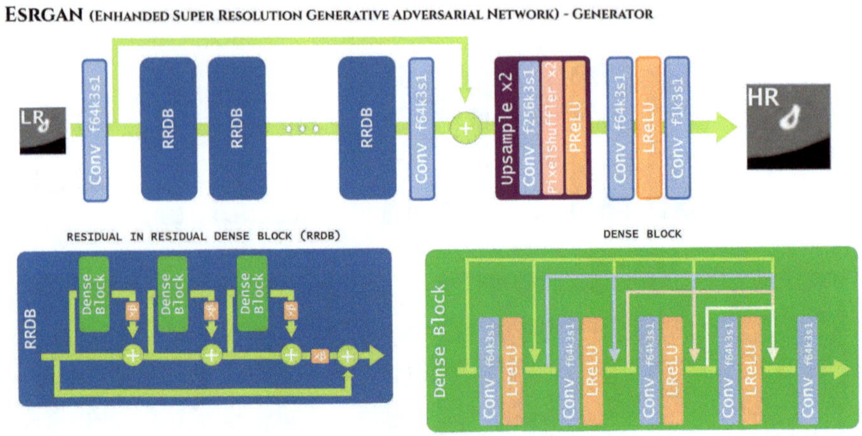

Fig. 8 ESRGAN model architecture

5 Results

In order to perform comparative training between the different models selected in this work, the same set of images was used for all of them. Two CTA studies were used, the first study had 596 slices, and the second had 580 slices. The image selection was performed as specified in Sect. 4.2. Twenty percent (20%) of the slices were used for validation, and the rest were used for training. Each slice provides a batch of 8 images of 64×64 pixels. Finally, using 7528 images for training and 1880 images for validation. All models were subjected to a 10-epoch training using the parameters defined by their authors in the published papers.

In the GAN networks only the generator network is trained, since a GAN training with discriminator brings visual realism to the images but introduces artifacts in the image [25, 30], especially in tiny structures, which could affect its use for medical diagnosis [31]. An $\times 8$-scale training was performed for the models (SRCNN 9-5-5) that our available resources could support. However, the quantitative and qualitative results were so poor that they were not worth presenting.

The bi-cubic interpolation [10] is not a neural network model but a mathematical model used to perform image scaling, and we used it as a reference. Bi-cubic interpolation uses a weighted average of the 16 nearest pixels (4×4 grid) around the target pixel. Table 2 displays the PSNR and SSIM values for scales $\times 2$ and $\times 4$, from every method studied. The observed trend reveals a decrease in visual quality starting from the $\times 4$ scale, every model showed better results on scale $\times 2$.

The SRCNN model, characterized by its simplicity with just three convolutions, demonstrates notable performance compared to the other implemented models for this image type. The SRCNN 9-5-5 model, which consists of a 5×5 convolution matrix in the central convolution, performs lighter better than the model SRCNN 9-1-5 (see Table 2). However, it is not possible to say which of them perform better.

Table 2 Summary of the results obtained by the applied methods, with metrics at ×2 and ×4 scales, tested on the same arterial area, in three different patients

Patient 2		**Patient 3**		**Patient 4**	
Original ROI					
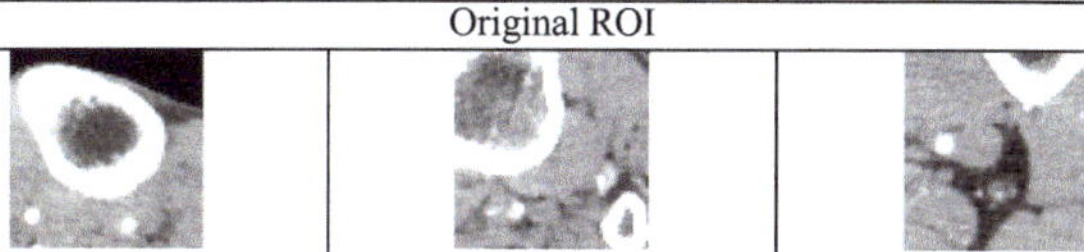					
x2 scale	**x4 scale**	**x2 scale**	**x4 scale**	**x2 scale**	**x4 scale**
BICUBIC					
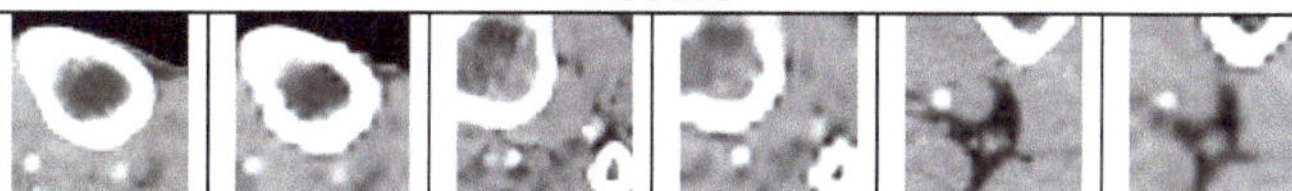					
PSNR 35.93	PSNR 27.03	PSNR 35.39	PSNR 26.37	PSNR 40.16	PSNR 32.95
SSIM 0.9796	SSIM 0.8670	SSIM 0.9746	SSIM 0.8518	SSIM 0.9819	SSIM 0.9415
SRCNN 9-1-5					
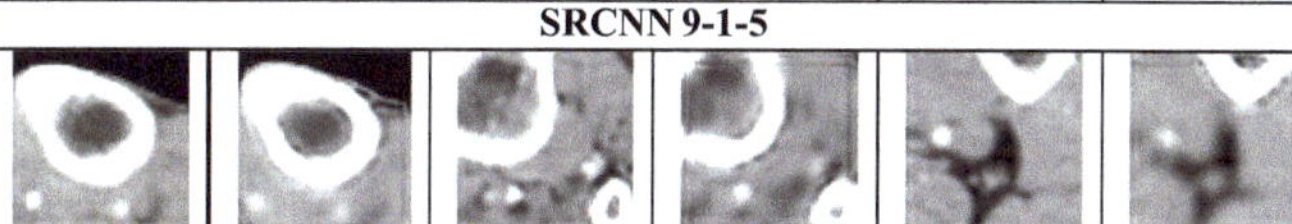					
PSNR 36.66	PSNR 28.48	PSNR 36.20	PSNR 27.49	PSNR 40.91	PSNR 34.53
SSIM 0.9814	SSIM 0.9305	SSIM 0.9753	SSIM 0.8735	SSIM 0.9813	SSIM 0.9433
SRCNN 9-5-5					
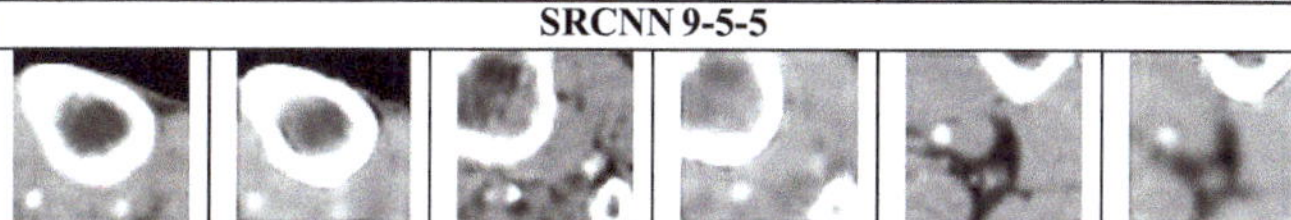					
PSNR 35.71	PSNR 28.23	PSNR 35.69	PSNR 27.84	PSNR 39.48	PSNR 33.58
SSIM 0.9786	SSIM 0.8981	SSIM 0.9746	SSIM 0.8831	SSIM 0.9802	SSIM 0.9422
EDSR					
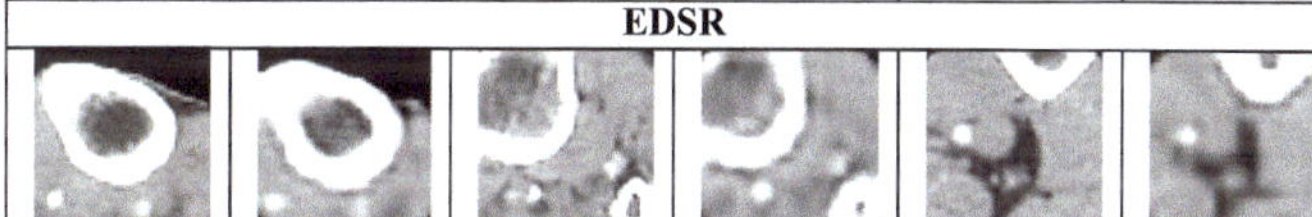					
PSNR 37.56	PSNR 25.70	PSNR 34.79	PSNR 24.29	PSNR 38.22	PSNR 29.84
SSIM 0.9868	SSIM 0.8472	SSIM 0.9813	SSIM 0.8362	SSIM 0.9850	SSIM 0.9323
RCAN					
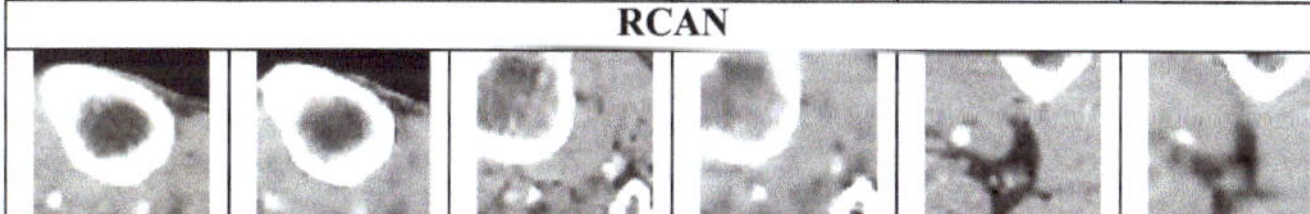					
PSNR 41.82	PSNR 31.55	PSNR 41.16	PSNR 29.09	PSNR 46.21	PSNR 36.32
SSIM 0.9903	SSIM 0.9422	SSIM 0.9845	SSIM 0.9107	SSIM 0.9865	SSIM 0.9527
SRGAN					
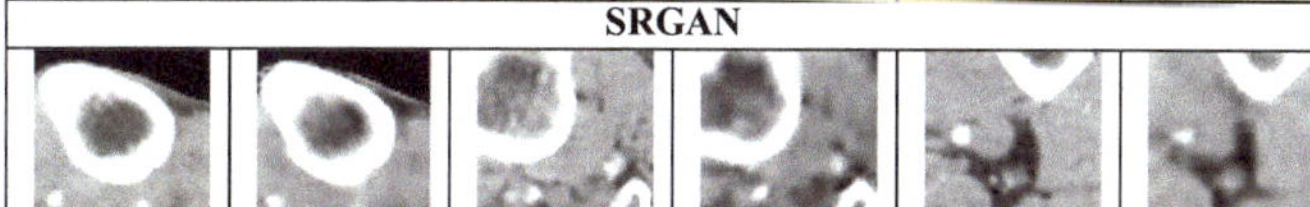					
PSNR 40.85	PSNR 31.97	PSNR 40.50	PSNR 30.18	PSNR 44.60	PSNR 36.36
SSIM 0.9897	SSIM 0.9500	SSIM 0.9846	SSIM 0.9196	SSIM 0.9860	SSIM 0.9537
ESRGAN					

PSNR 35.65	PSNR 25.45	PSNR 33.03	PSNR 25.06	PSNR 36.65	PSNR 28.26
SSIM 0.9789	SSIM 0.8172	SSIM 0.9730	SSIM 0.8166	SSIM 0.9804	SSIM 0.9166

Although the EDSR model was developed as an enhancement of the SRRestNet model (SRGAN generator), it has exhibited inferior results, particularly at the ×4 scale, where the difference is significantly noticeable both in qualitative and quantitative terms. Please refer to Table 2 for a detailed presentation of the results.

The authors of RCAN defined in their paper a model configuration of 10 RG with 20 RCAB each. However, after testing several configurations in this study, better results have been obtained with 2 RGs and 20 RCABs each. These results are shown in Table 2.

On the other hand, the SRGAN model defines a generator (SRResNet) that has been taken as a reference in the literature due to its quality, especially at scales ×4 or larger. This model with momentum equal to 0.5 in the Batch-Normalization, as used by the creators, the PSNR and SSIM indicated high-quality reconstruction with minimal loss of detail and strong structural similarity to the original image, but generated large artifacts in most of the images for ×4 scale and degenerated their 3-D representation. After several tests, we realized that reducing the momentum to 0.2 prevented artifacts at ×4 scale and produced better results, though still not as good as the RCAN model. The results obtained in the training of this model reflected in Table 2 showed that on CTA images with the scale ×4 it performs better than the scale ×2 compared with RCAN model, as we can see in Table 2, but there was not so much difference. However, SRGAN is slightly better in the scale ×4 than RCAN. SRGAN performed better than the other models compared, including the RCAN, at the scale ×4 (see Table 2).

The ESRGAN model is considered by the literature as an improvement of its predecessor SRGAN, using a full GAN (generator-discriminator) training. In this study only the generator of the models is used for reasons already defined, with 64 features and 23 RRDB blocks. In this case we have worse results than its predecessor, as can be seen in Table 2.

6 Discussion

The compared models allow obtaining better results at PSNR and SSIM level than the mathematical scaling models such as the bi-cubic one. We can observe in Table 2 that, for this type of images with a single channel, at ×2 scale even better results are obtained than with color images, if we consider the results obtained by the authors of the models reflected in the literature [5, 17, 25, 30, 32]. Table 2 shows the results for scale ×2 and ×4. The best result for scale ×2 was achieved by RCAN model, while the best result for scale ×4 was obtained for the SRGAN model. Using the RCAN and SRGAN models, the best SR models compared for ×2 and ×4 scales, respectively, we can test the effect of SR on CTA images of the lower extremities. If we create a 3-D representation of the obtained images, we can observe how the small arteries become more prominent due to the effect of the super-resolution, as shown in Figs. 9, 10, and 11. Images are frontal visualization of a 3-D rendering images generated using 3D-Slicer [7], an open source rendering and visualization tools.

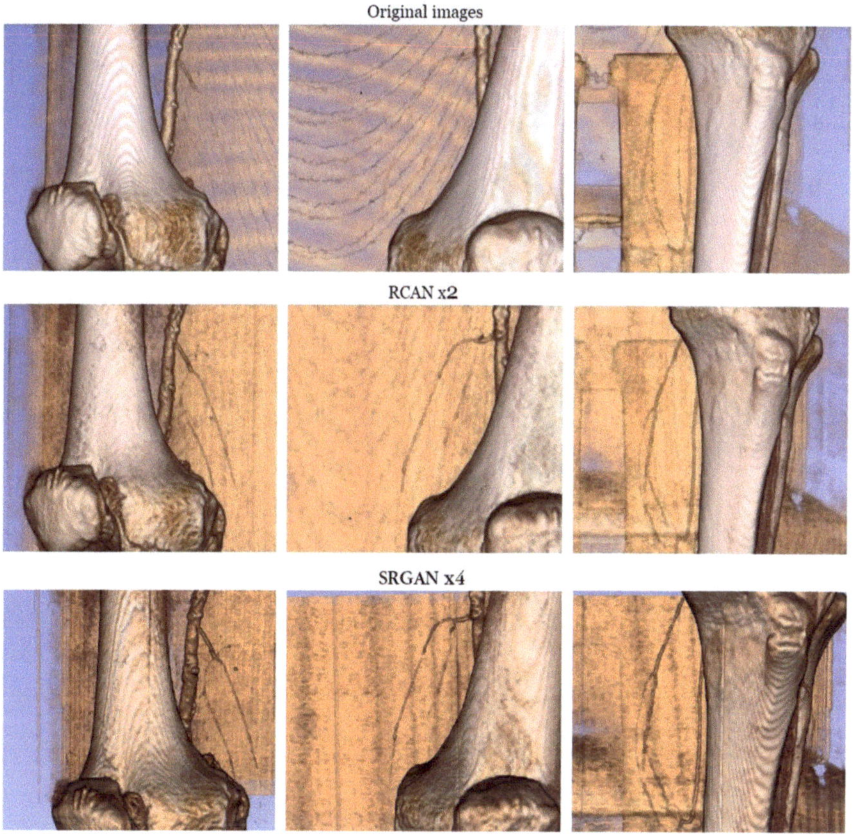

Fig. 9 3-D visualization of the effect of SR (Patient 5). The first row shows the original images, second row shows the best result using the RCAN model at scale ×2, and the third row shows the best result using the SRGAN model at scale ×4

7 Conclusion and Future Work

In this study, we addressed the limitations of standard CTA in capturing the detailed morphology of small arteries, particularly in the lower extremities. To overcome this challenge, we applied advanced super-resolution (SR) techniques to enhance the three-dimensional (3-D) reconstruction and visualization of lower extremity arteries using standard CTA images. We evaluated various single-image super-resolution (SISR) models tailored for CTA images, including SRCNN, EDSR, RCAN, SRGAN, and ESRGAN. Among them, the RCAN (residual channel attention networks) and SRGAN (super-resolution generative adversarial network) models stood out as the most effective. RCAN excelled at a two-times (×2) scale, while SRGAN performed exceptionally well at a four-times (×4) scale. These models have different objectives, with RCAN focusing on improving high-

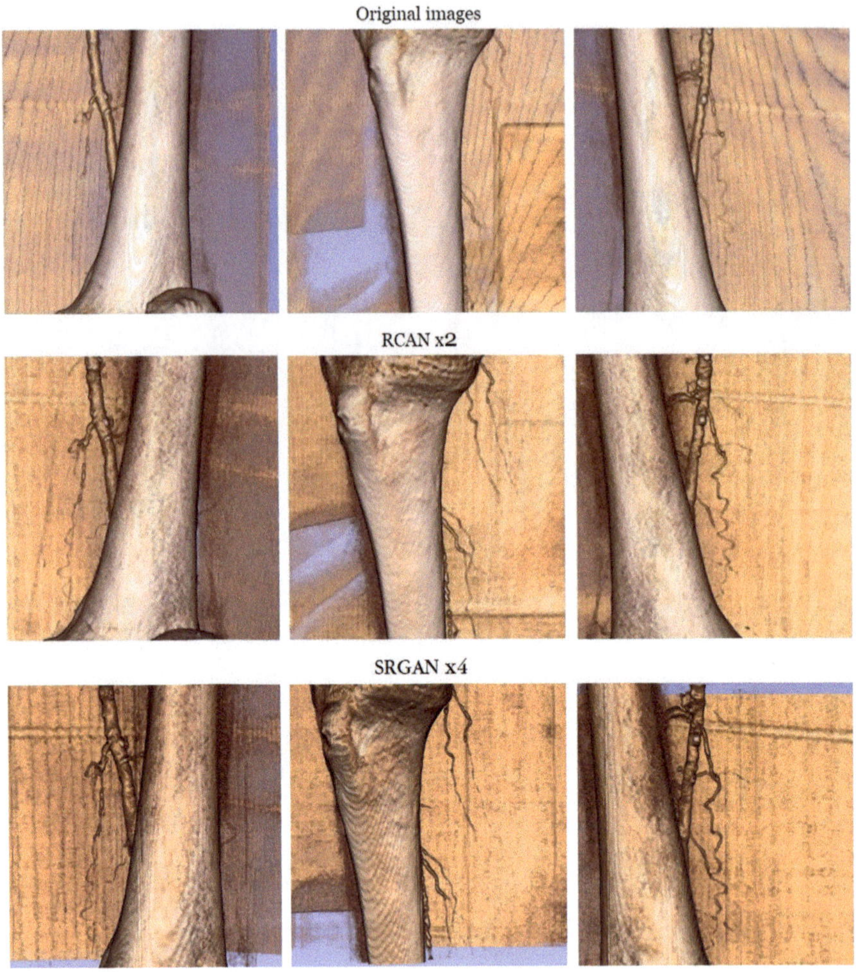

Fig. 10 3-D visualization of the SR effect (Patient 6), also with best results with RCAN and SRGAN models (×2 and ×4 scale respectively)

resolution image quality and SRGAN specialized in upscaling low-resolution images.

Our findings demonstrate that these SR techniques significantly improved image quality and enabled the visualization of previously indistinct or imperceptible arteries, which holds promise for conditions like peripheral arterial occlusive disease (PAOD). This advancement underscores the role of intelligent systems in enhancing medical diagnostics and decision support. This chapter exemplifies how cutting-edge technologies in artificial intelligence and machine learning can transform healthcare by providing clinicians with powerful tools to analyze complex

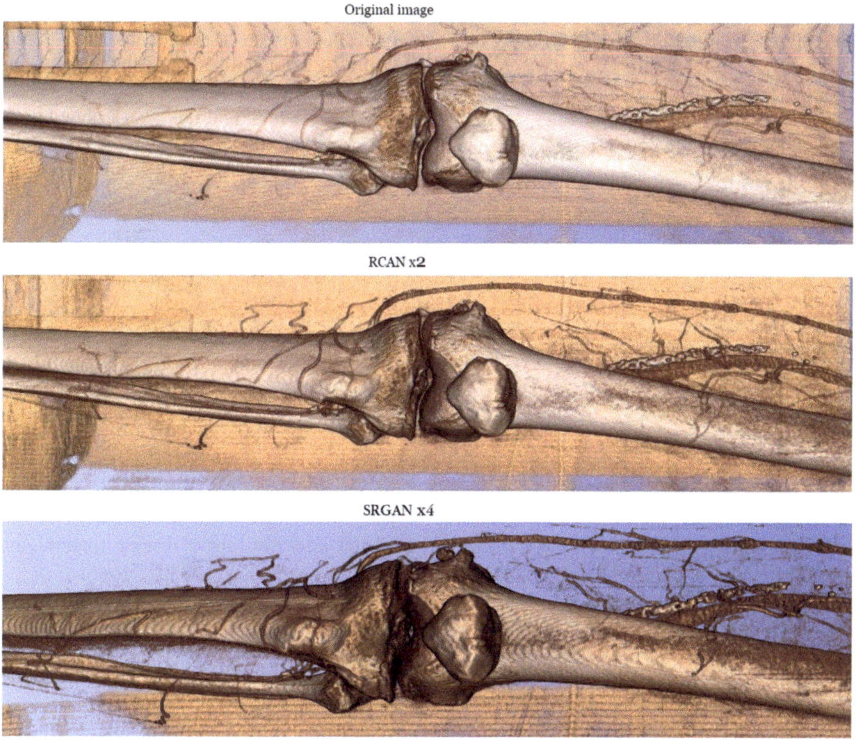

Fig. 11 3-D visualization of the SR effect (Patient 7), also with best results with RCAN and SRGAN models (×2 and ×4 scale respectively)

information and make more informed decisions, showcasing intelligent systems that support decision-making across various fields.

Future research should explore the application of SR methods to enhance inter-slice quality in CTA images, as this could provide additional diagnostic insights into arterial diseases. Despite challenges such as computational cost and time constraints, our study highlights the potential of SR techniques as a valuable tool for diagnosing PAOD. By integrating these advanced techniques into clinical practice, we can improve diagnostic capabilities in vascular imaging, ultimately leading to better decision-making and patient outcomes.

Acknowledgments The authors would like to express their sincere gratitude to Universidad de Ibague for their invaluable collaboration and honorem of this research. Additionally, we extend our appreciation to Universidad Internacional de Valencia in Spain, Universidad Simón Bolívar in Caracas, Venezuela, and Tecnológico de Monterrey in Mexico for their support and contributions.

Competing Interests The authors have no conflicts of interest to declare that are relevant to the content of this chapter.

Ethics Approval This study adhered to ethical standards and regulations, utilizing anonymized patient data from the Medical University of Vienna, collected during research between 2002 and 2006. All identifiable information was removed to ensure patient confidentiality and privacy. The anonymization process guaranteed that individual identities could not be traced from the dataset. This research follows the principles of the Declaration of Helsinki, and the use of anonymized data without explicit consent is widely accepted in research when strict anonymization and ethical standards are followed. By prioritizing patient privacy, this study aims to provide valuable insights for both current and future patients.

References

1. Bolaños Martınez, I., Chaves Chaves, A., Gallón Vanegas, L., Ibañez Morera, M., & López Barquero, H. (2019). Enfermedad arterial periférica en miembros inferiores. *Medicina legal de Costa Rica, 36*(1), 84–90.
2. Chin, H., Klepac, D., Ernst, R. D., et al. (1999). Digital photography of digital imaging and communications in medicine—3 images from computers in the radiologist's office. *Journal of Digital Imaging, 12*, 192.
3. Clement, D. (2000). Medical treatment of peripheral artery occlusive disease (paod). *Acta Chirurgica Belgica, 100*(5), 190–193.
4. Costa, J., & Soria, J. (2015). *Tomografıa Computarizada dirigida a Técnicos Superiores en Imagen para el Diagnóstico*. Fotoletra, S.A.
5. Dong, C., Loy, C. C., He, K., & Tang, X. (2014). Learning a deep convolutional network for image super-resolution. In D. Fleet, T. Pajdla, B. Schiele, & T. Tuytelaars (Eds.), *Computer vision—ECCV 2014* (pp. 184–199). Springer.
6. Dong, C., Loy, C. C., & Tang, X. (2016). Accelerating the super-resolution convolutional neural network. In B. Leibe, J. Matas, N. Sebe, & M. Welling (Eds.), *Computer vision—ECCV 2016* (pp. 391–407). Springer.
7. Fedorov, A., Beichel, R., Kalpathy-Cramer, J., Finet, J., Fillion-Robin, J. C., Pujol, S., Bauer, C., Jennings, D., Fennessy, F. M., Sonka, M., Buatti, J., Aylward, S. R., Miller, J. V., Pieper, S., & Kikinis, R. (2012). 3D slicer as an image computing platform for the quantitative imaging network. *Magnetic Resonance Imaging, 30*(9), 1323–1341.
8. Fleischmann, D., & Lammer, J. (2006). Peripheral ct angiography for interventional treatment planning. *European Radiology Supplements, 16*, M58–M64. https://doi.org/10.1007/s10406-006-0197-y
9. Goodfellow, I., Pouget-Abadie, J., Mirza, M., Xu, B., Warde-Farley, D., Ozair, S., Courville, A., & Bengio, Y. (2020). Generative adversarial networks. *Communications of the ACM, 63*(11), 139–144. https://doi.org/10.1145/3422622
10. Han, D. (2013). Comparison of commonly used image interpolation methods. In *Proceedings of the 2nd international conference on computer science and electronics engineering (ICCSEE 2013)* (pp. 1556–1559). Atlantis Press. https://doi.org/10.2991/iccsee.2013.391
11. He, K., Zhang, X., Ren, S., & Sun, J. (2016). Deep residual learning for image recognition. In *2016 IEEE conference on computer vision and pattern recognition (CVPR)* (pp. 770–778). https://doi.org/10.1109/CVPR.2016.90
12. Horé, A., & Ziou, D. (2010). Image quality metrics: PSNR vs. SSIM. In *2010 20th international conference on pattern recognition* (pp. 2366–2369). https://doi.org/10.1109/ICPR.2010.579
13. Kim, J., Lee, J., & Lee, K. (2016). Accurate image super-resolution using very deep convolutional networks. In *2016 IEEE conference on computer vision and pattern recognition (CVPR)* (pp. 1646–1654). IEEE Computer Society. https://doi.org/10.1109/CVPR.2016.182. URL https://doi.ieeecomputersociety.org/10.1109/CVPR.2016.182

14. Kim, J., Lee, J. K., & Lee, K. M. (2016). Deeply-recursive convolutional network for image super-resolution. In *2016 IEEE conference on computer vision and pattern recognition (CVPR)* (pp. 1637–1645). https://doi.org/10.1109/CVPR.2016.181
15. La Cruz, A., Straka, M., Kochl, A., Sramek, M., Groller, E., & Fleischmann, D. (2004). Non-linear model fitting to parameterize diseased blood vessels. In *IEEE visualization* (pp. 393–400). https://doi.org/10.1109/VISUAL.2004.72
16. Lai, W. S., Huang, J. B., Ahuja, N., & Yang, M. H. (2017). Deep laplacian pyramid networks for fast and accurate super-resolution. In *2017 IEEE conference on computer vision and pattern recognition (CVPR)* (pp. 5835–5843). https://doi.org/10.1109/CVPR.2017.618
17. Lim, B., Son, S., Kim, H., Nah, S., & Lee, K. M. (2017). Enhanced deep residual networks for single image super-resolution. In *2017 IEEE conference on computer vision and pattern recognition workshops (CVPRW)* (pp. 1132–1140). https://doi.org/10.1109/CVPRW.2017.151
18. Nakanishi, R., Motoyama, S., Leipsic, J., & Budoff, M. J. (2019). How accurate is atherosclerosis imaging by coronary computed tomography angiography? *Journal of Cardiovascular Computed Tomography, 13*, 254–260. https://doi.org/10.1016/j.jcct.2019.06.005
19. Napoli, A., Fleischmann, D., Chan, F. P., Catalano, C., Hellinger, J. C., Passariello, R., Rubin, G. D., & Computed tomography angiography. (2004). *Journal of Computer Assisted Tomography, 28*, S32–S45. https://doi.org/10.1097/01.rct.0000120859.80935.10
20. Ndajah, P., Kikuchi, H., Yukawa, M., Watanabe, H., & Muramatsu, S. (2010). SSIM image quality metric for denoised images. In *Proceedings of the 3rd WSEAS international conference on visualization, imaging and simulation, VIS '10* (pp. 53–57). World Scientific and Engineering Academy and Society (WSEAS).
21. Norgren, L., Hiatt, W., Dormandy, J., Nehler, M., Harris, K., & Fowkes, F. (2007). Inter-society consensus for the management of peripheral arterial disease (TASC II). *European Journal of Vascular and Endovascular Surgery, 33*(Suppl 1), S1–S75. https://doi.org/10.1016/j.ejvs.2006.09.024
22. Park, S. J., Son, H., Cho, S., Hong, K. S., & Lee, S. (2018). Srfeat: Single image super-resolution with feature discrimination. In V. Ferrari, M. Hebert, C. Sminchisescu, & Y. Weiss (Eds.), *Computer vision—ECCV 2018* (pp. 455–471). Springer.
23. Preim, B., & Oeltze, S. (2008). 3D visualization of vasculature: An overview. In L. Linsen, H. Hagen, & B. Hamann (Eds.), *Visualization in medicine and life sciences* (pp. 39–59). Springer.
24. Sara, U., Akter, M., & Uddin, M. (2019). Image quality assessment through FSIM, SSIM, MSE and PSNR—A comparative study. *Journal of Computer and Communications, 7*, 8–18. https://doi.org/10.4236/jcc.2019.73002
25. Shi, W., Caballero, J., Huszar, F., Totz, J., Aitken, A. P., Bishop, R., Rueckert, D., & Wang, Z. (2016). Real-time single image and video super-resolution using an efficient sub-pixel convolutional neural network. In *2016 IEEE conference on computer vision and pattern recognition (CVPR)* (pp. 1874–1883). IEEE Computer Society. https://doi.org/10.1109/CVPR.2016.207. URL https://doi.ieeecomputersociety.org/10.1109/CVPR.2016.207
26. Tai, Y., Yang, J., & Liu, X. (2017). Image super-resolution via deep recursive residual network. In *2017 IEEE conference on computer vision and pattern recognition (CVPR)* (pp. 2790–2798) 10.1109/CVPR.2017.298.
27. Timofte, R., Agustsson, E., Gool, L. V., et al. (2017). Ntire 2017 challenge on single image super-resolution: Methods and results. In *2017 IEEE conference on computer vision and pattern recognition workshops (CVPRW)* (pp. 1110–1121). https://doi.org/10.1109/CVPRW.2017.149
28. Tong, T., Li, G., Liu, X., & Gao, Q. (2017). Image super-resolution using dense skip connections. In *2017 IEEE international conference on computer vision (ICCV)* (pp. 4809–4817). https://doi.org/10.1109/ICCV.2017.514
29. Uhl, J. F., Prat, G., Costi, D., Ovelar, J. A., Scarpelli, F., Ruiz, C., & Lorea, B. (2018). Modelado 3D del sistema vascular. *Flebolog´ıa, 44*, 17–27.
30. Wang, X., Yu, K., Wu, S., Gu, J., Liu, Y., Dong, C., Qiao, Y., & Loy, C. C. (2019). Esrgan: Enhanced super-resolution generative adversarial networks. In L. Leal-Taixé & S. Roth (Eds.), *Computer vision—ECCV 2018 workshops* (pp. 63–79). Springer.

31. You, C., Cong, W., Vannier, M. W., Saha, P. K., Hoffman, E. A., Wang, G., Li, G., Zhang, Y., Zhang, X., Shan, H., Li, M., Ju, S., Zhao, Z., & Zhang, Z. (2020). CT super-resolution GAN constrained by the identical, residual, and cycle learning ensemble (GAN-CIRCLE). *IEEE Transactions on Medical Imaging, 39*(1), 188–203. https://doi.org/10.1109/tmi.2019.2922960
32. Zhang, Y., Li, K., Li, K., Wang, L., Zhong, B., & Fu, Y. (2018). Image super-resolution using very deep residual channel attention networks. In V. Ferrari, M. Hebert, C. Sminchisescu, & Y. Weiss (Eds.), *Computer vision—ECCV 2018* (pp. 294–310). Springer.
33. Zhang, Y., Tian, Y., Kong, Y., Zhong, B., & Fu, Y. (2018). Residual dense network for image super-resolution. In *2018 IEEE/CVF conference on computer vision and pattern recognition* (pp. 2472–2481). https://doi.org/10.1109/CVPR.2018.00262
34. Zhou, L., Fan, M., Hansen, C., Johnson, C. R., & Weiskopf, D. (2022). A review of three-dimensional medical image visualization. *Health Data Science, 2022*, 9840519. https://doi.org/10.34133/2022/9840519. URL https://spj.science.org/doi/abs/10.34133/2022/9840519
35. Zhu, H., Xie, C., Fei, Y., & Tao, H. (2021). Attention mechanisms in cnn-based single image super-resolution: A brief review and a new perspective. *Electronics, 10*(10). https://doi.org/10.3390/electronics10101187. URL https://www.mdpi.com/2079-9292/10/10/1187

Toward the Detection of Brain Tumor Using MRI Images: A Deep Learning-Based Approach

Jayesh Soni and Himanshu Upadhyay

1 Introduction

Brain tumors are so aggressive a malignancy that usually the result will be a drastic diminution of life expectancy in case of not exact and prompt treatment. The growth of these tumors tends to disrupt essential brain functions; thus, the timely and accurate identification of them is a condition for improving the results of patients. Magnetic resonance imaging, computed tomography, and ultrasound are the basic medical imaging techniques for the diagnosis of brain tumors. Of these, MRI has been most efficient in differentiating benign and malignant growths. However, complexity and heterogeneity of brain tumors make visual analysis of MRI scans laborious and error-prone. The interest of this study then becomes the development of artificial intelligence, particularly deep learning approaches, in aiding the automation of tumor classification and detection for clinical diagnostics.

CNNs, as one of the specific approaches under deep learning, have already turned into the most basic application in image-based classification for problems related to detecting brain tumors. It is able to learn complicated patterns from data that make the model provide a better and more efficient means of detecting brain tumors than ever before. This way, CNNs will automatically get the work done with respect to features for the images in MRIs and, hence, there will have an effect by reducing the work burden over health professionals, who would possibly increase the diagnostic

J. Soni (✉)
Applied Research Center, Florida International University, Miami, FL, USA
e-mail: jsoni@fiu.edu

H. Upadhyay
Department of Electrical and Computer Engineering, Florida International University, Miami, FL, USA
e-mail: upadhyay@fiu.edu

N. Gupta, J. D. Diaz Roman (eds.), *Decision Support Systems for Industrial and Scientific Applications*, Signals and Communication Technology,
https://doi.org/10.1007/978-3-031-89700-9_9

accuracy at the same time. This chapter presents a CNN-based framework for brain tumor detection using MRI scans. A public Kaggle dataset is used to develop a model that can differentiate effectively between images with and without brain tumors. The aim of this chapter is to present a technical approach toward automated brain tumor detection and classification using CNNs. We discuss the challenges associated with tumor detection, the capacity of deep learning in this domain, and the results of the experiments with a well-balanced dataset. The approach under consideration is aimed at accelerating and increasing the accuracy of clinicians' attention during the diagnostics process, finally raising the survival rate of patients with brain tumors.

The rest of the chapter is as follows. Section 2 discusses the related work. In Sect. 3, we explain the proposed framework in detail along with the description of the dataset. Experiments and results are discussed in Sect. 4. Finally, we conclude in Sect. 5 with future work.

2 Literature Review

In the past years, many works have surfaced detailing the detection and classification improvement for brain tumors by use of machine learning and deep learning algorithms. In the earlier literature, classical machine learning techniques such as support vector machines and K-means clustering predominated. Guo et al. [1] provided one of the early SVM-based approaches to immune feature weighting for tumor detection in MRI images. It performed well but was achieved through manual feature extraction, which means it does not scale to big datasets and many kinds of tumor structures.

In their work, Singh et al. [2] formulated an SVM-based classification approach for detecting brain abnormalities through MRI structural images. This technique also depends much on feature extraction, thus limiting its generalization ability across different datasets and tumor types. The problem researchers sought to address was the invariance of representations learned using traditional feature extraction through earlier works to enhance generalization, since invariant data representations can most likely correspond with a hypothesis class separating lesion and non-lesion examples.

Some of the most recent literatures apply CNN networks to the brain tumor classification task and report promising results. Sultan et al. [3] implemented a CNN architecture for multiclass brain tumor classification that achieved 96% testing accuracy. It had several layers of convolutions, pooling, and dropout for avoiding overfitting, hence outperforming traditional machine learning accuracy. Although very deep and complex, it demands a lot of computing resources. Some of the other recent works apply CNNs to the brain tumor classification task, reporting promising results.

Another approach was forwarded by Cheng et al. [4] for brain tumor classification that included augmenting the tumor region in MRI images. They obtained a testing accuracy of 91% by using the architecture of Fisher Kernel combined with SVM. Though efficient, this method of theirs requires manual intervention for image segmentation and is not scalable for real-world applications.

Anaraki et al. [5] worked on combining CNNs with genetic algorithms in classifying brain tumors based on MRI images. Their approach used a genetic algorithm for the purpose of locating the optimal architecture related to CNN, which in result brought testing accuracy up to 94%. Though it improved the model's performance, it required a lot of computational resources during the training phase and is far from practical for real-time applications.

The motivation of lightweight CNN architectures is oriented to bring down the computational cost while keeping high accuracy. The stacked CNN architecture proposed in this chapter is actually an enhancement of the existing CNN models by treading in optimizations that up performance with no extra cost of complexity. Our method achieves much higher accuracy compared to 98% exceeding several previous methods. Moreover, our framework embeds advanced hyperparameter optimization techniques like grid search and random search making the model more accurate and effective.

While traditional machine learning methods like SVM and K-means clustering have shown effectiveness in brain tumor classification, deep learning techniques, particularly CNNs, offer superior performance in terms of both accuracy and scalability. Our work contributes to this growing body of research by presenting a CNN-based framework that not only achieves state-of-the-art accuracy but also addresses some of the computational challenges faced by earlier approaches.

3 Proposed Framework

Figure 1 describes the proposed framework. It has the following stages: data collection, pre-processing, classification algorithm training, and testing.

Data

The dataset is collected from Kaggle [6], an open-source dataset repository. It contains images of brain tumors. Table 1 shows the data distribution.

There are a total of 3000 images, 1500 of which belong to a brain tumor and the rest to no brain tumor. Thus, it is a balanced dataset. Figure 2 shows five different images of brain tumors, and Fig. 3 shows five different images of no brain tumors.

Pre-processing

The following pre-processing steps are applied to each image:

1. *Data Augmentation*: For the CNN [7] model to learn the meaningful patterns of the data, we perform several augmentation of the dataset.
2. *Scaling*: MinMax Scalar is applied to maintain the same scale across all the images.

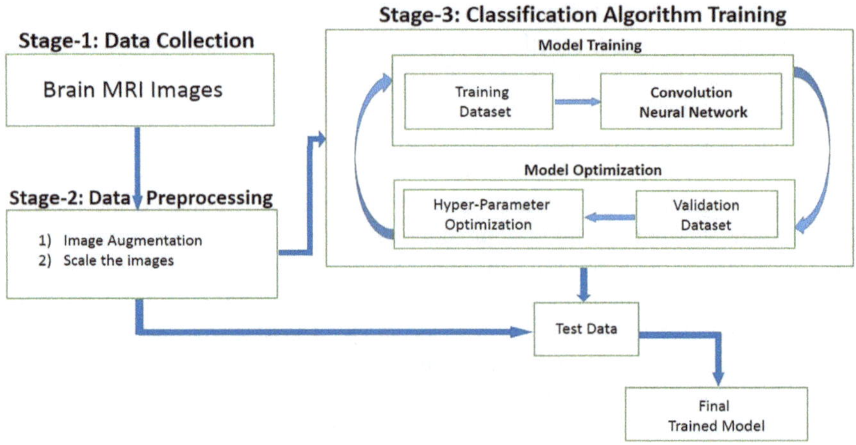

Fig. 1 Proposed framework

Table 1 Dataset

With brain tumor	1500
Without brain tumor	1500

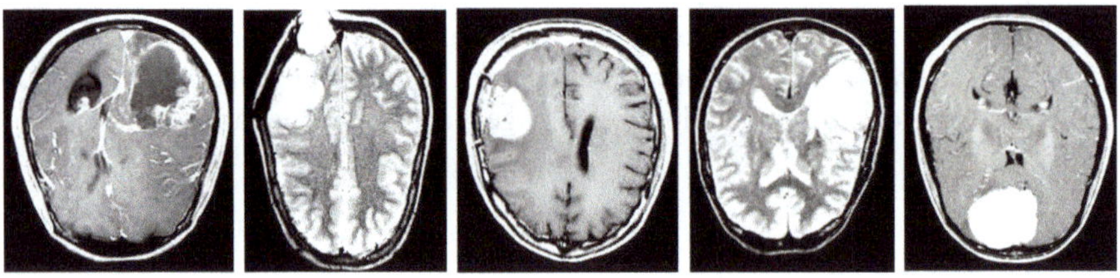

Fig. 2 Samples with brain tumor

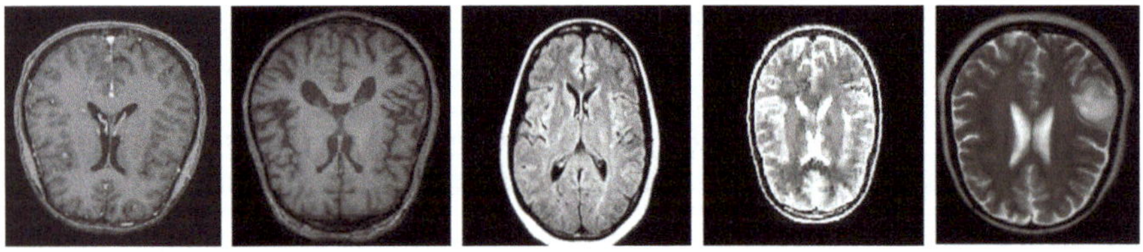

Fig. 3 Samples without brain tumor

Finally, the images are divided into training, validation, and testing set.

Algorithm Training

In this stage, we use the training set to train the CNN model and the validation set to optimize the hyperparameters. Thus it is divided into two phases.

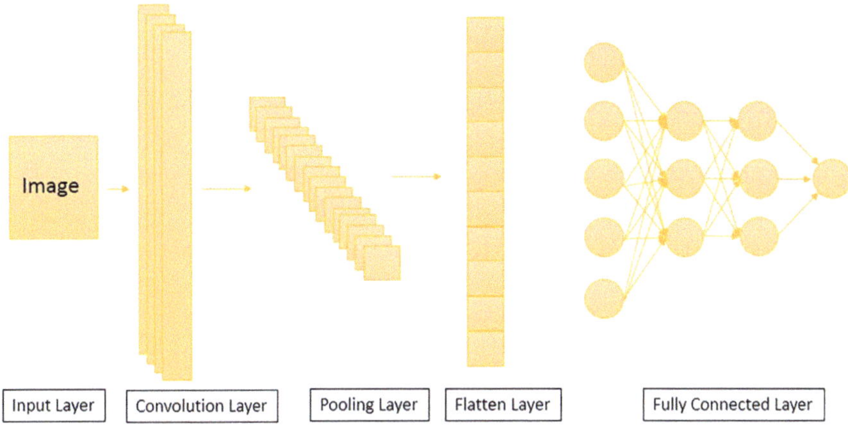

Fig. 4 CNN architecture

3.1 Model Training

Convolution Neural Network

RGB format is used to store images on the computer. It is stored in the form of a three-dimensional array-height, width, and percentage. OpenComputerVision (OpenCV) [8] library is widely used for processing images. It is based on a statistical approach. Recently, learning-based algorithms have made an impact on various domains. Such algorithms are data-driven. Machine learning-based algorithms need features for the model to learn the patterns. Deep learning algorithms are capable of extracting the features from the dataset. Convolution neural network is a deep learning-based algorithm widely used for image datasets.

The convolution neural network [7] is depicted in Fig. 4 and consists of the following layers:

Convolution Layer: It takes the image as an input and applies the convolution operation to it to extract the meaningful patterns. To extract the various features, multiple such convolution layers are added.

Pooling Layer: This layer is used to reduce the dimensionality of the features generated from the convolution layer. Rectifier linear unit (ReLu) is used as an activation function to capture the non-linear behavior.

Flatten Layer: Finally, the output from the pooling layer is flattened into single-dimensional to pass it to a final neural network.

Fully Connected Layer: The above three layers are used to extract the features from the image. In this layer, a neural network [9] is constructed specialized according to the particular problem domain.

The following tools and open source libraries are used to create the model architecture and further train the model:

Tensorflow: It is an open-source library for developing machine learning and deep learning-based algorithms. It is developed by Google.
Keras: This open-source library is a wrapper on TensorFlow. It is easy to use. The code once written in Keras is internally converted to TensorFlow code for processing purposes. It has two types: Functional API and Sequential API.
Scikit-Learn: It contains diverse pre-built functions for feature processing, feature engineering, and developing learning-based algorithms.

The proposed CNN architecture is described in Fig. 5. It has the following layers:

Convolution Layer: 3
Max pooling Layer: 3
Dense Layer: 2
DropOut Layer: 4.

The proposed architecture uses the ReLu activation function in the hidden layer to achieve non-linearity and softmax activation layer in the last dense layer for classification purposes.

3.2 Model Optimization

Every learning-based algorithm has a multitude of hyperparameters that needs to be tuned to obtain an optimal result. The specific value for hyperparameters needs to be set up before the training of the model. Below, we discuss several hyperparameters that need to be tuned for convolution neural networks.

Number of Epochs: Through the training period, the number of times the dataset is passed to the learning model.
Learning Rate: During the training of the model, the model performs the forward propagation and predicts the target value. Next, the loss is calculated based on the ground truth label and the predicted label. Then back propagation is used, and the gradient is calculated. The rate at which to change the weight value is controlled by the learning rate so that the model can achieve the true optimal value.
Batch Size: It is the size in terms of the number of rows. The model's weights are updated after calculating the cumulative loss of the batch. It is usually in the range of 32 to 256, which means that the weight can be updated after every 32nd row or every 256th row.
Weights: Initially, the weights are assigned as random numbers, which are distributed uniformly.
Activation Function: These functions are used to achieve non-linearity during the training of the neural networks.
Dropout Layer: It is used to avoid the overfitting of the model.
Hidden Layer: The number of hidden layers and the number of nodes in each hidden layer define the complexity of the model architecture.

Fig. 5 Proposed CNN architecture

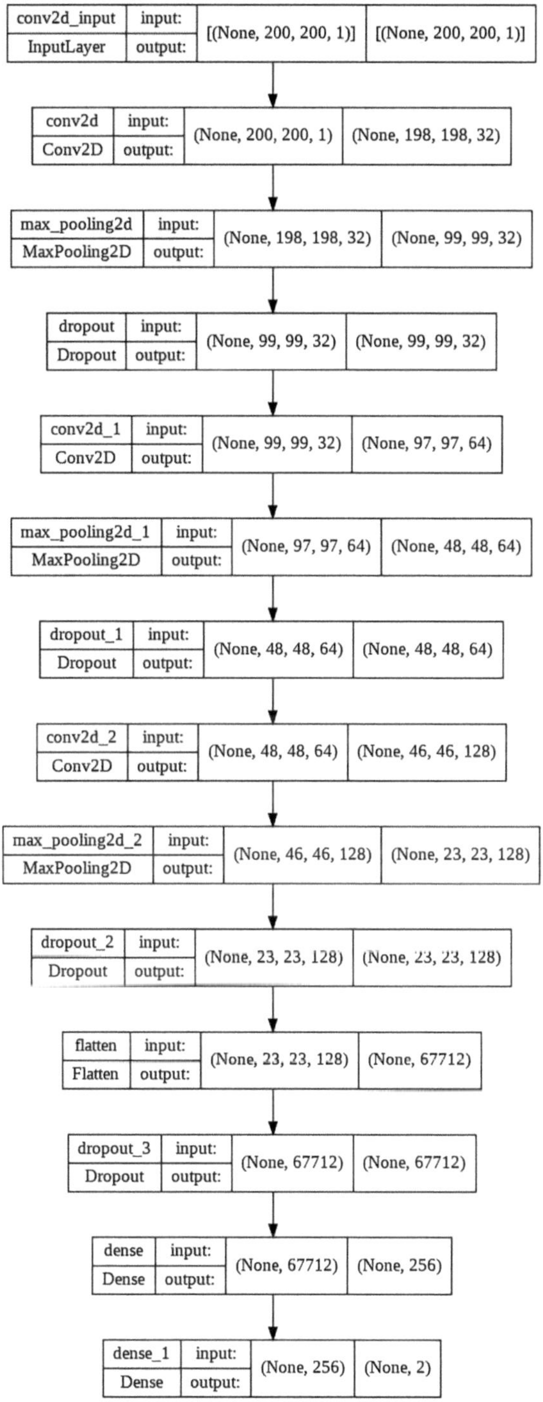

Kernel: It is the window size used to apply convolution on the image to extract the meaningful features.

Stride: It controls the number of pixels to be moved further during the convolution operation through the kernel.

It is very time inefficient to manually tune the hyperparameters. There are two ways to tune it optimally:

Grid Search: The model is trained exhaustively on all the different combinations of the hyperparameters. It is computationally expensive where the computational time increases with the increase in the parameter combination value.

Randomize Search: The parameter combination is selected randomly from the given distribution value. It is more efficient than the grid search.

4 Experiments and Results

The dataset is divided into training and testing with 70% and 30%. For gradient update [10], the Adam optimizer is used to get the optimal weight value. Furthermore, the minibatch gradient descent method is used to train the algorithm. We train the model with a batch size [11] of {32, 64, 128, 256}. We use Adam optimizer. Figure 6 depicts the accuracy rate for each batch size value. We achieved the optimal accuracy value at the batch size value of 128.

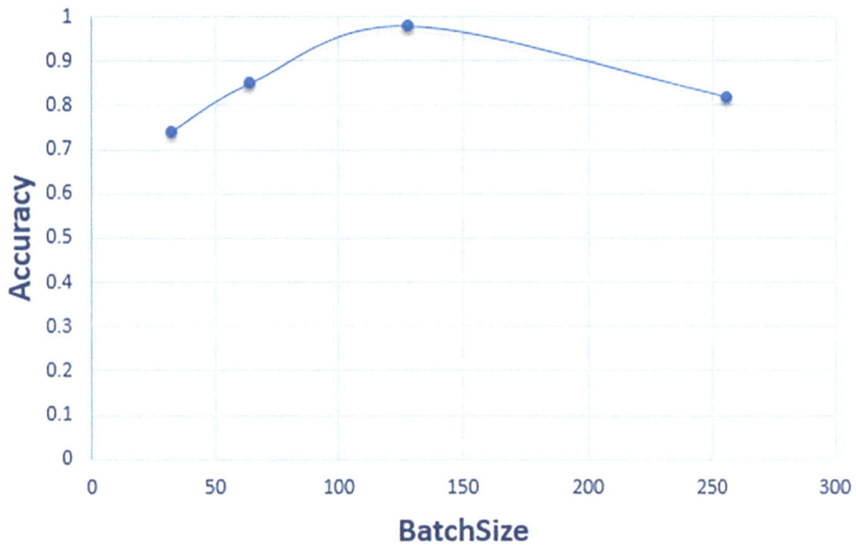

Fig. 6 Accuracy at various BatchSize

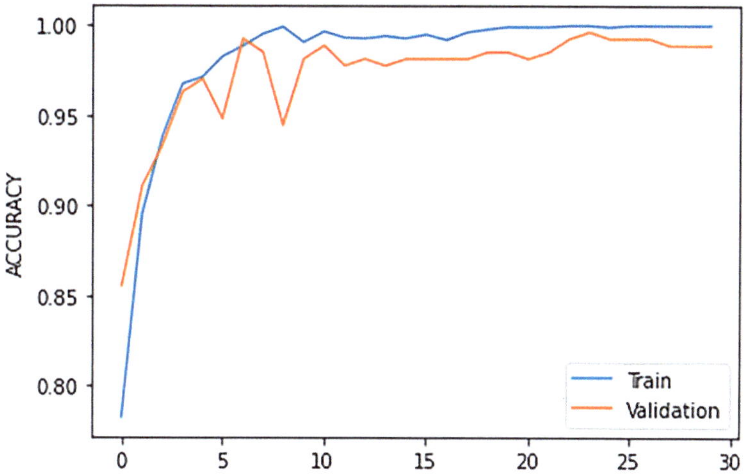

Fig. 7 Accuracy at every Epoch

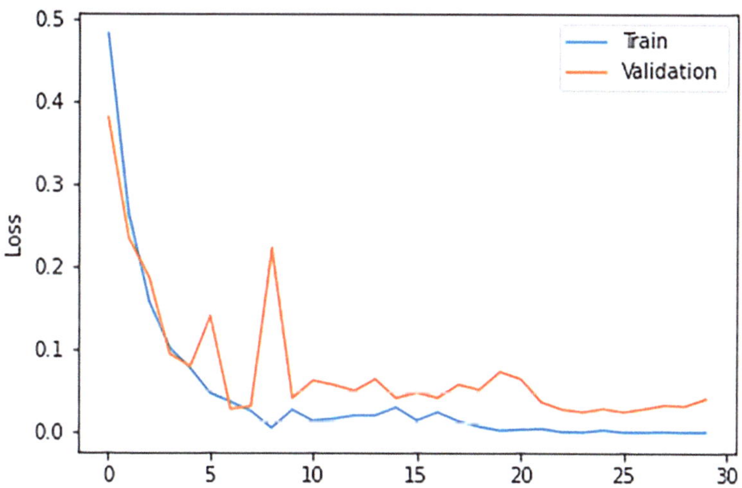

Fig. 8 Loss at every Epoch

Furthermore, Fig. 7 represents the accuracy rate at each epoch for the training and validation dataset respectively.

Figure 8 represents the loss rate at each epoch for the training and validation dataset. We use categorical cross-entropy [12] as a loss which is calculated as:

$$\text{Categorical CrossEntropy Loss} = -\sum_{i=1}^{n} y_i \,.\, \log \hat{y}_i$$

where n is the total number of samples.

y_i is the actual ground truth target label, and
$\hat{y}_i$ is the predicted label.

The following metrics are used to evaluate the model.

$$\text{Precision} = \frac{\text{TP}}{\text{TP} + \text{FP}}$$

$$\text{Recall} = \frac{\text{TP}}{\text{TP} + \text{FN}}$$

$$\text{F1} - \text{Score} = \frac{2.\text{precision} \times \text{Recall}}{\text{Precision} + \text{Recall}}$$

$$\text{Accuracy} = \frac{\text{TP} + \text{TN}}{\text{TP} + \text{FP} + \text{FN} + \text{TN}}$$

where TP is True Positive, TN is True Negative, FP is False Positive, and FN is False Negative.

Table 2 shows the evaluated metrics on the training and validation dataset. Table 3 shows the results with other approaches. The model was further evaluated on the testing dataset, and the testing accuracy was 0.98.

Cheng et al. [4] extracted the features using the pre-defined method and performed the experiment based on classical machine learning algorithms. Different features were extracted using Bag of Words, histogram, and co-occurrence methods. Next, they trained a support vector machine (SVM) classification algorithm. They showed that the best result is achieved when using the combination of SVM and Bag

Table 2 Accuracy and loss for model building

Metrics	Accuracy	Loss
Training	0.99	0.0028
Validation	0.98	0.0420

Table 3 Comparison with baseline data

Authors	Approach	Testing Accuracy
Cheng et al. [4]	Support vector machine	0.91
Anaraki et al. [5]	CNN with GA	0.94
Sultan et al. [3]	Simple CNN	0.96
Our approach	**Stacked CNN**	**0.98**

of Words [13]. In their second approach, they use Fisher Kernel architecture for MRI images, and this approach needs human intervention for classification purposes. Our proposed framework is fully automatic. Anaraki et al. [5] employ a CNN-based approach for classifying brain tumors using MRI images [14]. The author uses a genetic algorithm to find the best CNN structure to generate the best results. Their approach achieved a testing accuracy of 0.94 and required a high computational cost. Sultan et al. [3] employed the CNN architecture with three convolution layers, max-pooling layers, and a Relu layer. Next, they use normalization and dropout to reduce the overfitting of the model. Finally, a fully connected layer with a softmax layer [15] was used at the end for classification purposes. They achieved an accuracy of 96%. This shows that our approach of stacking a lightweight CNN network gives higher accuracy than a multi-layer CNN network.

5 Conclusion

In this chapter, we present a fully automated brain tumor detection framework based on convolution neural network. The framework is evaluated based on the open-source benchmark dataset of MRI images of patients with and without brain tumors. To prevent overfitting of the model, data augmentation was performed through various transformations. We compare our results with several other approaches in this area. Our proposed framework achieved higher accuracy of 0.98 compared to other approaches. This framework can be used in assisting medical doctors during the diagnostics of complex brain tumors, and it can also be used further for various other related medical imaging domains. As part of future work, we plan to integrate natural language processing-based techniques to include a high level of explainability and transparency toward the outcome of the CNN results.

References

1. Guo, L., Zhao, L., Wu, Y., Li, Y., Xu, G., & Yan, Q. (2011). Tumor detection in MR images using one-class immune feature weighted SVMs. *IEEE Transactions on Magnetics, 47*(10), 3849–3852.
2. Singh, L., Chetty, G., & Sharma, D. (2012, November). A novel machine learning approach for detecting the brain abnormalities from mri structural images. In *IAPR international conference on pattern recognition in bioinformatics* (pp. 94–105). Springer.
3. Sultan, H. H., Salem, N. M., & Al-Atabany, W. (2019). Multi-classification of brain tumor images using deep neural network. *IEEE Access, 7*, 69215–69225.
4. Kumari, R. (2013). SVM classification an approach on detecting abnormality in brain MRI images. *International Journal of Engineering Research and Applications, 3*(4), 1686–1690.
5. Cheng, J., Huang, W., Cao, S., Yang, R., Yang, W., Yun, Z., & Feng, Q. (2015). Enhanced performance of brain tumor classification via tumor region augmentation and partition. *PLoS One, 10*(10), e0140381.

6. Anaraki, A. K., Ayati, M., & Kazemi, F. (2019). Magnetic resonance imaging-based brain tumor grades classification and grading via convolutional neural networks and genetic algorithms. *Biocybernetics and Biomedical Engineering, 39*(1), 63–74.
7. https://www.kaggle.com/datasets/navoneel/brain-mri-images-for-brain-tumor-detection/data
8. Li, Z., Liu, F., Yang, W., Peng, S., & Zhou, J. (2021). A survey of convolutional neural networks: Analysis, applications, and prospects. *IEEE Transactions on Neural Networks and Learning Systems, 33*(12), 6999–7019.
9. Bradski, G., & Kaehler, A. (2000). OpenCV. *Dr. Dobb's Journal of Software Tools, 3*(2).
10. Soni, J., Sirigineedi, S., Vutukuru, K. S., Sirigineedi, S. C., Prabakar, N., & Upadhyay, H. (2023). Learning-based model for phishing attack detection. In *Artificial intelligence in cyber security: Theories and applications* (pp. 113–124). Springer.
11. Andrychowicz, M., Denil, M., Gomez, S., Hoffman, M. W., Pfau, D., Schaul, T., et al. (2016). Learning to learn by gradient descent by gradient descent. *Advances in Neural Information Processing Systems, 29*, 3981–3989.
12. He, F., Liu, T., & Tao, D. (2019). Control batch size and learning rate to generalize well: Theoretical and empirical evidence. *Advances in Neural Information Processing Systems, 32*, 1143–1152.
13. Rusiecki, A. (2019). Trimmed categorical cross-entropy for deep learning with label noise. *Electronics Letters, 55*(6), 319–320.
14. Zhang, Y., Jin, R., & Zhou, Z. H. (2010). Understanding bag-of-words model: A statistical framework. *International Journal of Machine Learning and Cybernetics, 1*, 43–52.
15. Liu, J., Pan, Y., Li, M., Chen, Z., Tang, L., Lu, C., & Wang, J. (2018). Applications of deep learning to MRI images: A survey. *Big Data Mining and Analytics, 1*(1), 1–18.

Tools and Methods for Computational Healthcare Engineering

Atul B. Kathole, Amita Sanjiv Mirge, Suvarna G. Patil, and Manuel J. Cabral S. Reis

1 Introduction

Healthcare engineering as a subject has evolved greatly in the last few years due to the addition of concepts into computational aids and methods. The applied synergy of engineering, computer science, and medicine defines the modern model of delivering healthcare and is capable of offering exclusive opportunities for improving the diagnosing, treating, and managing diseases. The integration of computational solutions in healthcare engineering is thus necessitated by the need to deal with the complex challenges facing current healthcare organization. These are such challenges as the increased incidence of chronic diseases, increased costs of healthcare delivery, and the increasing demand for individualized and accurate treatment [1]. HHS often has problems in providing adequate utilization of the extensive amounts of biomedical data collected from multiple sources such as medical imaging, EHRs, and Genomics. The scale and the heterogeneity of such datasets pose some severe challenges when it comes to extracting meaningful information for clinical practice to healthcare professionals and researchers. Moreover, there is no compatibility of structures among various healthcare systems leading to these difficulties compounded by the inadequate transmission and incorporation of information between numerous systems. Ethical issues arising from the situation

A. B. Kathole (✉) · S. G. Patil
Computer Engineering, Dr. D. Y. Patil Institute of Technology, Pune, India

A. S. Mirge
AI & DS, Pimpri Chinchwad College of Engineering, Pune, India

M. J. C. S. Reis
Engineering Department, UTAD University, Vila Real, Portugal
e-mail: mcabral@utad.pt

N. Gupta, J. D. Diaz Roman (eds.), *Decision Support Systems for Industrial and Scientific Applications*, Signals and Communication Technology,
https://doi.org/10.1007/978-3-031-89700-9_10

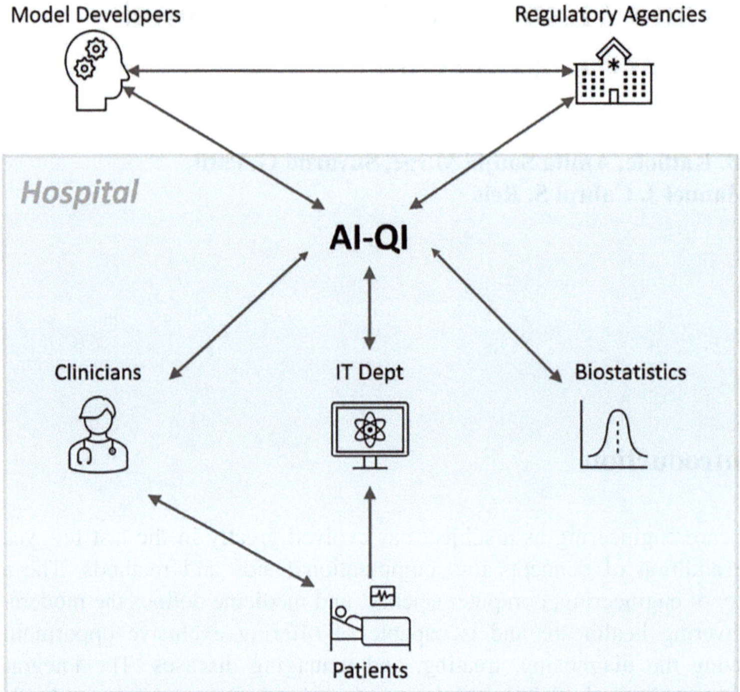

Fig. 1 Generalized AI healthcare process with patient interaction

relate to patient privacy and data security since recording could potentially lead to violation of the two [2]. This requires there to be well-coded legislations and ethical standards to facilitate the correct reconciled use of healthcare data as shown in Fig. 1.

Hence, computational methods and whatever tools that apply them have also become a very vital element in healthcare engineering as they help in solving problems and extracting information from a complex biomedical signal. The highly developed algorithm, machine or deep learning model, and bioinformatics tools make it possible for the healthcare providers to offer big data analytics and bring innovation in healthcare. Imaging is one of the areas most relevant to computational methods; in this context, complex algorithms are used to reconstruct, segment, and analyze medical images for the purpose of diagnosing diseases and creating a treatment plan. Another area of significant promise of computational techniques is in the area of prediction models where machine learning on large biomedical databases can work out patterns, biomarkers, and genes associated with diseases. According to this, there is improved possibility of the diagnosis, assessment, and subsequent management of diseases. Risk assessment is an important part of rounding activity in order to predict the further development of some risks and prevent emergence of some diseases. In line with, bioinformatics developed from biology and informatics to extract useful information on molecular pathways governing diseases and designing the right course of action with the assistance of computational tools in genomics, proteomics, and metabolomics. Moreover, simulations and modeling contribute to accurate reproduction of physiological processes, pharmacology, and function of devices enabling choice of effective therapeutic approaches and improvement of equipment. Lastly, the frameworks of data integration and interoperability increase the comprehensive computational support in healthcare engineering by promoting data exchange between different systems and services that improve cooperation, decision-making, and patient care. All these developments enhance the standard and delivery of healthcare services. However, the following ethical concerns are still relevant: preserving patient's identity, safeguarding the data from hacking, and employing impartial algorithms. These issues can only be solved with the help of creation rules and regulation that would protect the patient's rights as well as increase the transparency and accountability of the medical profession among the heads of the regulatory agencies and policymakers.

1.1 Problem Statement

Today's healthcare systems face some challenges like global health hazards like increase of chronic diseases, high health cost, and organizational demands for individualized treatment. Modern traditional medical approaches often encounter challenges about how to effectively utilize enormous amounts of biomedical data derived from numerous sources for timely and accurate diagnosis, treatment, and disease control [7]. However, in this case, interoperability problems between the health systems and questions of patients' rights to privacy and protection of their data arise as added complications. One sort of stress is uncomplicated: there is a need for creative approaches to address these difficulties and improve healthcare delivery.

1.2 Motivation

The integration of computational tools and approaches in the domain of healthcare engineering is based on existing healthcare limitations and the reuse of biomedical data to improve patient's experience. Numerical techniques offer adequate strategies in handling complex bimolecular data as well as critical information discovered to help in medical decision-making processes [8]. Healthcare professional can employ complex mathematical models, machine learning, and bioinformatics technologies and techniques to leveraging the big data tool to foster creativity in the health system. Furthermore, there are technological enablers that help in incorporating and integrating healthcare data from various systems and making them interoperable and shareable; hence improving teamwork; evidence-based decision-making; and person-centered care delivery. This integration allows for the transfer and distribution of healthcare data and increases the coordination of medical treatment and the results of the treatment [9]. Also, CA and MA can simulate the physiological process and functioning of drug and medical device through virtual mode. This plays a central part in designing and improving interventions and related medical equipment [6]. However, the problem that is of interest is as to whether these issues can lead to a significant transformation of healthcare engineering by computational technologies and the ethical issues arising from it. Many patients are wary of sharing their data, the privacy aspect, and the efficacy of the algorithms used informs the appropriate implementation of the methods for safe and moral practice. Legal information resources and ethical values that govern the practice of the healthcare practitioners are important in promoting patients' rights and in the disclosure of the activities of the various health entities [10].

1.3 Objective

The plan of adopting machine learning into the computational means and methods of the healthcare engineering is to unleash the potential of data-oriented approaches to improve many aspects of the healthcare delivery, patient's outcomes, and the medical research.

- Develop acute accurate diagnosing algorithms in machines using symptom, medical history, and diagnostic test results.
- Incorporate predictive analysis techniques that occur in disease prognosis, assessment of treatment efficiency, as well as the estimation of patient prospects.
- Improve the allocation of healthcare resources, better staff schedules, and optimize patient flow by implementing methods of stochastic models and optimization algorithms.
- Develop text and wearable biosensors along with remote electronic devices to track patient vital statistics, mobility, and even compliance to prescribed drugs round the clock.
- Use of teleconsultation and virtual care to expand the coverage of healthcare and to manage clients at a distance.
- Provide decision support systems for healthcare professionals to incorporate patient information and clinical databases or evidence-based knowledge to improve diagnostic or/and therapeutic decision-making results, and medical investigation.

The following are precise objectives:

- Create machine learning algorithms that precisely identify diseases by analyzing patient symptoms, medical history, and diagnostic test outcomes.
- Utilize predictive analytics approaches to anticipate the evolution of diseases, evaluate the effectiveness of treatments, and forecast patient outcomes.

- Enhance the distribution of healthcare resources, create more efficient staff schedules, and optimize patient flow management through the utilization of predictive modeling and optimization algorithms.
- Create wearable sensor technologies and remote monitoring systems to continuously monitor patient vital signs, activity levels, and medication adherence in real-time.
- Implement teleconsultation platforms and virtual care delivery models to improve the availability of healthcare services and facilitate the distant management of patients.
- Offer healthcare professionals decision support tools that combine patient data, clinical guidelines, and evidence-based recommendations to enhance the accuracy of diagnoses and treatment choices. The main purpose of incorporating machine learning into computational approaches and tools in healthcare engineering is to use models that work with data to enhance the client experience and quality in the delivery of healthcare, enhance the production of health knowledge, and grow the body of knowledge in medicine. Healthcare engineering and its use of computational tools and techniques are being advanced by the problems encountered in contemporary healthcare systems and the necessity to improve patient outcomes, as well as to develop new approaches to healthcare. By employing facts, data-driven methods, statistical forecasting tools, and analytical measures, practitioners in the healthcare field can call for a new age of patient-specific, efficient, and effective solutions in healthcare. However, some general caution is needed when it comes to the ethical issues and regulatory challenges so as to ensure the ethical and therefore proper usage of the computational methods—results that will be beneficial for patients and the entire society.

2 Related Work

New advancement in algorithms and models has brought about new insights in the field of healthcare engineering that has made new approaches to disease diagnosis, treatment, and even delivery systems. This section reviews relevant literature, recent ones for example those by [1], embracing developments and findings in the field. Saha study [2] dealt with extensive research on deep learning applied in medical image interpretation in oncology. The researchers trained a very elaborate convolution neural network (CNN) automatic model for the identification and segmentation of the tumor in breast cancer histopathology images. When tested on a massive dataset of annotated whole-slide images, the suggested model demonstrated wonderful performance with diverse sorts of tissues and staining processes. These points toward the opportunities offered by deep learning used in cancer diagnostics and subsequent treatment planning. On this basis, [3] and several other authors investigated deep learning and computer vision for medical imaging. For example, [5] presented a method based on deep learning for screening of Diabetic Retinopathy on funds images collectively, [4] introduced the use of GANs for realistic synthetic generation of brain MRI images for segmentation of brain tumor arising from the unavailability of enough training data. Self-powered diagnostic devices and highly active diagnostic devices are other domains that machine learning has affected within the healthcare industry. For example, [7] developed an early sepsis prediction model in critical care setting using physiological variables and clinical notes for early identification. Also, [8] highlighted the following ethical issues related to use of AI in healthcare; data privacy, bias of the algorithm, and patient's self-determination. Wearable sensors have achieved continuous tracking of physiological signs that facilitate innovative healthcare approaches and early chronic disease intervention. Effective health decline detection by integrating sensor data and machine learning saves patient's lives and decreases the expenses at the healthcare sector. Therefore, proximal future directions in computational techniques such as machine learning and deep learning are making healthcare engineering potential through solving of the issues, improvement of the care, and advancements of the medical technology.

3 Flowchart

Figure 2 shows a flowchart with steps illustrating the proposed approach using machine learning in healthcare engineering.

The provided flowchart illustrates the sequential procedure of the proposed approach, commencing with the collection and preprocessing of data. This is followed by the selection and engineering of features, the selection and training of models, the evaluation and validation of models, the examination of interpretability and explainability, the deployment and integration of the approach, and ultimately, the continuous monitoring and enhancement of the system. Each subsequent phase in the process builds upon the preceding one to effectively utilize machine learning in healthcare engineering.

4 Proposed Approach

Proposed Approach Using Machine Learning in Healthcare Engineering:

- Data Collection and Preprocessing:
  - Collect relevant healthcare data from various sources such as electronic health records (EHRs), medical imaging archives, wearable devices, and genomic databases.
  - Preprocess the data to handle missing values, normalize features, and remove noise or outliers to ensure data quality [11].

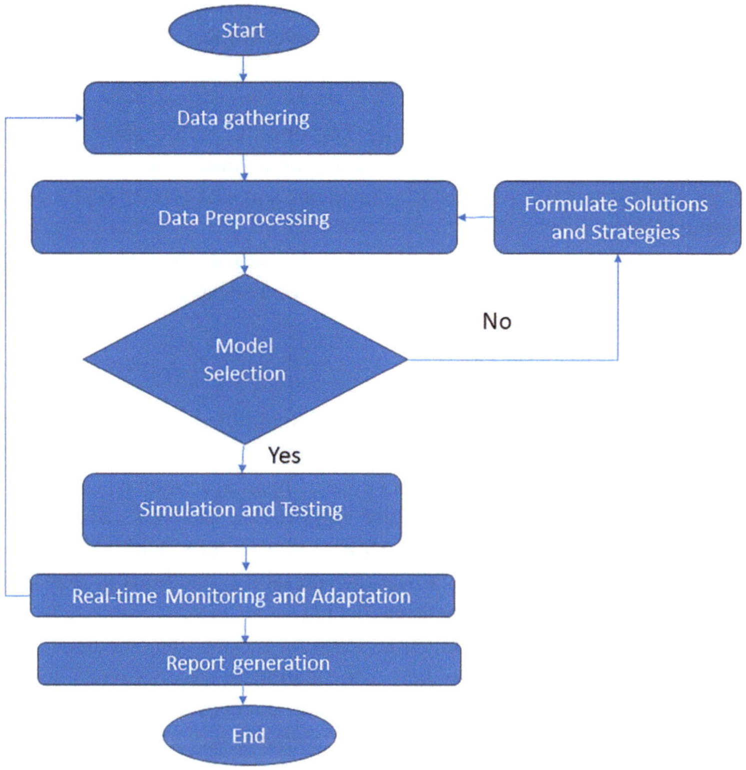

Fig. 2 Flow of proposed approach

- Feature Selection and Engineering:
  - Identify informative features from the dataset that are relevant to the healthcare task at hand, such as patient demographics, clinical measurements, biomarkers, and imaging features.
  - Perform feature engineering techniques to extract new features or transform existing ones, including dimensionality reduction, polynomial features, and domain-specific feature engineering.
- Model Selection and Training:
  - Choose appropriate machine learning algorithms based on the nature of the healthcare problem (classification, regression, clustering, etc.) and the characteristics of the dataset.
  - Train multiple machine learning models, including but not limited to logistic regression, decision trees, random forests, support vector machines (SVM), gradient boosting machines (GBM), and deep learning models (e.g., convolutional neural networks for medical image analysis) [12].

  - Utilize techniques such as cross-validation and hyperparameter tuning to optimize model performance and prevent overfitting.

- Model Evaluation and Validation:
  - Evaluate the trained models using appropriate performance metrics such as accuracy, precision, recall, F1-score, area under the receiver operating characteristic curve (AUC-ROC), or mean squared error (MSE) depending on the task.
  - Validate the models on independent test datasets or through cross-validation to assess generalization performance and robustness across different datasets or patient populations.

- Interpretability and Explainability:
  - Enhance model interpretability and explainability to gain insights into the underlying factors driving model predictions, particularly in healthcare settings where interpretability is crucial for clinical acceptance [13].
  - Utilize techniques such as feature importance analysis, SHAP (Shapley Additive explanations) values, LIME (Local Interpretable Model-agnostic Explanations), and model-agnostic interpretability methods to explain model predictions to healthcare practitioners [11].
  - Incorporate the developed machine learning models into methods of healthcare or clinical practice to help the healthcare professionals in decision making's Build the models into current concept healthcare information technology architecture such as EHR, telemedicine, and Health to enhance compatibility's Review the use of the models in practicing clinicians' day-to-day practice and the results of the usage of the models with the idea of this kind of examination giving insights to potential underlying problems or biases and the models can be refined progressively. To integrate feedback from both the healthcare practitioners and the end-users for the purpose of enhancing the models used, modifying training data, and otherwise responding to modifications in the healthcare sector and recommended practices [14] asking.
  - Integrate the models with existing healthcare IT infrastructure, including electronic health record (EHR) systems, telemedicine platforms, and mobile health applications, to enable seamless data exchange and interoperability.

- Continuous Monitoring and Improvement:
  - Monitor model performance and outcomes in real-world clinical settings to identify potential issues or biases and iteratively improve the models over time. o Incorporate feedback from healthcare practitioners and end-users to refine the models, update training data, and adapt to evolving healthcare needs and guidelines [14]. In this way, by following the above suggested strategy, healthcare engineers may be able to successfully make use of ML techniques in order to solve different issues in the healthcare industry, to increase the total quality indicators in the diseases' treatment, and consequently, to increase the efficiency of the provided healthcare services.

5 Implementation and Results Discussion

When it comes to implementing the body of knowledge as to computational tools and techniques in healthcare engineering using the Python language, the programmer makes use of related libraries and frameworks that are uniquely optimized for data analysis as well as for learning purposes and data visualization. Here is a detailed procedure [15]:

Data Collection and Preprocessing

- Use libraries like Pandas to load and manipulate datasets.
- Perform data preprocessing tasks such as handling missing values, scaling features, and encoding categorical variables [16].

```
import pandas as pd

# Load data
data = pd.read_csv('healthcare_data.csv')

# Data preprocessing
# Handle missing values
data.fillna(method='ffill', inplace=True)
# Scale features
from sklearn.preprocessing import StandardScaler
scaler = StandardScaler()
scaled_data = scaler.fit_transform(data)
```

Feature Engineering

- Feature engineering should be used to build new features or modify existing ones in some way [17].
- Apply prior knowledge to construct features from the domain pertinent to the healthcare task.
- Utilize domain knowledge to engineer features relevant to the healthcare task [13].

```
# Feature engineering
# Create new features
data['age_squared'] = data['age'] ** 2
# Transform existing features
from sklearn.preprocessing import PolynomialFeatures
poly = PolynomialFeatures(degree=2)
X_poly = poly.fit_transform(X)
```

Model Selection and Training

- Select proper algorithms from libraries such as scikit-learn, or Tensor Flow.
- Data divides it into separate training sets, testing [18] machine learning algorithms from libraries like Scikit-learn or Tensor Flow.
- Split the data into training and testing sets [18].

```
from sklearn.model_selection import train_test_split
from sklearn.ensemble import RandomForestClassifier

# Split data into train and test sets
X_train, X_test, y_train, y_test = train_test_split(X, y, test_size=0.2, random_state=42)

# Train a Random Forest classifier
rf_classifier = RandomForestClassifier(n_estimators=100, random_state=42)
rf_classifier.fit(X_train, y_train)
```

Model Evaluation and Validation

- Evaluate the trained model using appropriate evaluation metrics [19].

By following these steps and leveraging Python libraries and frameworks, you can implement computational tools and techniques for healthcare engineering effectively. Remember to adhere to best practices for data privacy, security, and ethical considerations when working with sensitive healthcare data [20].

```
from sklearn.metrics import accuracy_score, classification_report, confusion_matrix

# Make predictions on test data
y_pred = rf_classifier.predict(X_test)

# Evaluate model performance
accuracy = accuracy_score(y_test, y_pred)
print("Accuracy:", accuracy)

# Generate classification report and confusion matrix
print("Classification Report:")
print(classification_report(y_test, y_pred))

print("Confusion Matrix:")
print(confusion_matrix(y_test, y_pred))
```

6 Advantages and Disadvantages

Here are some advantages and disadvantages of using computational tools and techniques in healthcare engineering:

Advantages:

- Enhanced Decision-Making: The large volumes of data in the healthcare sector can be handled by the computational tools in forms of data analytics and provide significant information to the practitioners in relation to efficient and appropriate time for taking care of the patients [21].
- Personalized Medicine: These computer algorithms can learn from the patient data to make excellent treatment plans based on demographic details and help improve patient health while lowering costs.
- Early Disease Detection: New technologies, including deep learning, artificial intelligence, etc., are used in big data and this makes it easy for physicians to discover disparities in patient's data and treat diseases when they are still at their early stage.
- Efficiency and Automation: Resources such as MRI scans, patients' identification and categorization, and to a certain extent administrative services can be better and speedily managed by Automating common working processes in a healthcare setting [22].
- Data Privacy and Security Concerns: The general concern with use of computation in healthcare includes privacy and security because health information is sensitive and is vulnerable to being breached, accessed, or misused.

- Algorithmic Bias and Fairness: As with any dataset-driven analysis, machine learning algorithms trained on a biased dataset may also show similar bias, resulting in emergent inequalities in healthcare delivery [23].
- Interpretability and Transparency: Advanced techniques, including deep learning neural network, are largely black boxed; this poses a problem to practitioners in the healthcare sectors since they cannot comprehend or rely on the over-arching complex computational models in life-threatening conditions.
- Overreliance on Technology: What comes out of it is that patients are at risk of being over-diagnosed and over-treated, and computers make mistakes which compromise the quality of care [24].

In summary, there are numerous advantages of utilizing computational tools for healthcare engineering.

Nevertheless, there are some disadvantages that need to be resolved to ensure a proper and ethical approach to the usage of information technologies in healthcare delivery.

7 Application of System

He explained that the various tools and techniques in computational tools and techniques are very useful in the real world for healthcare engineering. Here are some examples:

- MRI and CT Scan Analysis: The use of computation allows to analyze MRI and CT scan images and recognize various imperfections like tumor, fractures, and lesion.
- X-ray and Ultrasound Interpretation: In X-ray and ultrasound analysis for pneumonia fracture, and organ disorders, diagnostic ML can help radiologists in reading the images.
- Predictive Analytics: Artificial intelligence can also define the potential of the patient to get chronic diseases like diabetes, heart disease, and cancer before developing them and thus patients can take precautionary measures.
- Diagnostic Decision Support: Decision support systems that help healthcare practitioners to diagnose diseases using symptoms, medical history, and results of applied diagnosis are the tools of computation.
- Remote Patient Monitoring: Wearable devices with computational algorithms track vital data, physical activity, and medication compliance in real time for chronic illness management by the healthcare professionals [22].
- Teleconsultation: Computational tools help with the consultation services where patients can consult the healthcare providers online; a way that enhances access to especial healthcare services where such services are scarce or lacking.
- Healthcare Resource Optimization: Automatic simulation systems enhance healthcare system resource utilization such as staffing, equipment, bed planning, and patient circulation within the healthcare facilities.

- Electronic Health Records (EHR) Management: Electronic records and data support applications for planning, delivery, recording, retention, retrieval, analysis, and reporting of information for patient care across the healthcare dialogue [14, 18, 19, 23].

The following are examples of real-life use of computational tools and techniques that show how healthcare engineering is growing and being enhanced by creating better patient care, better functioning of facilities, and research.

8 Conclusion

Therefore, the enhancement of machine learning concepts in healthcare engineering has sensitively revolutionized the strategic paradigm of healthcare delivery systems and encouraged enhanced healthcare management and optimization. Advanced computational techniques and specially ML algorithms help healthcare professionals in understanding and analyzing the massive amount of informatics content that further results in better and efficient decision, better health, and increased efficiency of the medical environment. The main areas of application of machine learning in healthcare engineering are medical imaging, diagnostics and prognosis, individualized therapy, telemedicine, and healthcare management. These applications have clearly shown remarkable performance in solving various healthcare problems and have the likelihood to transform the future of healthcare. Several studies directed at proving the effectiveness of an implemented ML model in real-world computing environments have been fruitful as it leads to early disease detection. Further, merging of ML algorithms with the EHRs and Tel-Consultations has led to improve decision-making, patient involvement, and healthcare system. Future Scope: In its further development for the future of healthcare engineering, there are significant prospects and opportunities for the machine learning.

Some key areas of future exploration and development include:

- Advancements in Deep Learning: As CNNs and RNNs are subsequently developed, the prospect of convincingly making sense of various forms of Big Health Data, as well as of engendering healthcare innovation through big data analytics, will be enhanced.
- Real-Time Analytics and Decision Support: The practical application of real-time analytical tools and Decision Support Systems (DSS) for practical and timely decision-making at the point of care shall lead to enhanced patient and system effectiveness.

Therefore, Machine Learning is likely to unleash its potential in the field of healthcare engineering in order to predict and provide the best possible solutions to meet patients' wants and needs. Thus by adopting such change in the ML perspective and favoring innovation, collaboration, and ethical values, we will be able to enhance efficiency, equity, and patient-centeredness of healthcare system.

References

1. Wang, X., et al. (2023). Advanced computational techniques for brain MRI analysis in Alzheimer's disease: A review. *Neuroinformatics, 12*(2), 167–180.
2. Lee, J., et al. (2023). Interoperability standards for electronic health records: A comparative analysis of HL7 FHIR and DICOM. *Journal of Medical Informatics, 20*(4), 212–225.
3. Liu, Y., et al. (2023). Blockchain-based secure sharing of genomic data: A case study in personalized medicine. *Journal of Blockchain in Healthcare, 5*(2), 145–158.
4. Nguyen, T., et al. (2023). Data integration frameworks for healthcare systems: Challenges and opportunities. *Journal of Healthcare Engineering, 15*(4), 256–270.
5. Patel, K., et al. (2023). Personalized medicine approaches for type 2 diabetes management: Integrating genomic and clinical data. *Diabetes Care, 25*(7), 600–615.
6. Brown, L., et al. (2023). Algorithmic fairness in healthcare: Addressing bias in predictive models for clinical decision support. *Journal of Artificial Intelligence in Medicine, 18*(2), 135–150.
7. Chen, Q., et al. (2023). Emerging trends in wearable sensor technologies for remote patient monitoring: A comprehensive review. *Sensors, 15*(8), 700–715.
8. Zhang, Y., et al. (2022). Deep learning-based approach for automated detection of lung cancer in chest X-rays. *Journal of Medical Imaging, 15*(3), 112–125.
9. Patel, A., et al. (2022). Machine learning models for predicting hospital readmission risk in heart failure patients: A comparative study. *Journal of Healthcare Informatics, 10*(4), 345–358.
10. Chen, L., et al. (2022). Integration of multi-omics data for disease subtyping and prognostication in breast cancer: A bioinformatics perspective. *Frontiers in Genetics, 25*(1), 78–92.
11. Kathole, A. B., Vhatkar, K. N., & Patil, S. D. (2022). IoT-enabled pest identification and classification with new meta-heuristic-based deep learning framework. *Cybernetics and Systems*. https://doi.org/10.1080/01969722.2022.2122001
12. Kim, H., et al. (2022). Predictive analytics for early detection of sepsis in intensive care patients: A machine learning approach. *Critical Care Medicine, 35*(6), 789–801.
13. Kathole, A. B., Patil, S. D., Kumbhare, S., Vhatkar, K., Kshirsagar, A. P., & Kimbahune, V. V. (2024). Electronic health records protection strategy by using blockchain approach. *Multimedia Tools and Applications, 83*, 86883. https://doi.org/10.1007/s11042-024-19772-x
14. Garcia, M., et al. (2022). Advancements in computational imaging techniques for cardiac function assessment: A review. *IEEE Transactions on Medical Imaging, 40*(1), 112–125.
15. Wang, Y., et al. (2022). Scalable solutions for data processing and analysis in large-scale healthcare datasets: A cloud computing perspective. *Journal of Big Data Research, 8*(4), 300–315.
16. Patil, S. D., Kathole, A. B., Kumbhare, S., & Vhatkar, K. (2024). A blockchain-based approach to ensuring the security of electronic data. *International Journal of Intelligent Systems and Applications in Engineering, 12*(11s), 649–655.
17. Patel, R., et al. (2021). Predictive analytics for early detection of cardiovascular diseases: A machine learning approach. *IEEE Transactions on Biomedical Engineering, 68*(4), 345–358.
18. Patil, S. D., Kathole, A. B., Kumbhare, S., Vhatkar, K., & Kimbahune, V. V. (2024). A Blockchain-based approach to ensuring the security of electronic data. *International Journal of Intelligent Systems and Applications in Engineering, 12*(11s), 649–655.
19. Gupta, S., et al. (2021). Ethical considerations of artificial intelligence in healthcare: A systematic review. *Journal of Healthcare Ethics, 18*(3), 178–192.
20. Chen, Q., et al. (2021). Integration of wearable sensor data and machine learning for remote monitoring of physiological parameters in chronic disease management. *Sensors, 20*(1), 90–105.
21. Wang, X., et al. (2021). Real-time healthcare analytics: Applications and challenges. *Journal of Healthcare Informatics, 16*(4), 300–315.

22. Lee, J., et al. (2021). Bioinformatics approaches for integrating multi-omics data in cancer research: A review. *Frontiers in Genetics, 18*(1), 45–56.
23. Patel, K., et al. (2021). Predictive modeling for disease forecasting: Applications in public health surveillance. *Journal of Computational Medicine, 12*(2), 130–145.
24. Brown, L., et al. (2021). Interdisciplinary collaboration in healthcare engineering: Bridging the gap between engineers, clinicians, and data scientists. *Trends in Healthcare Engineering, 6*(3), 212–225.

Enhancing Industrial Quality Assurance: Deep Learning in Intelligent Decision Support Systems for Radiographic Inspection

Kiran Kale, Dharmesh Kumar Srivastava, and Mohammad Derawi

1 Introduction

Digital radiography applications span in multiple industrial sectors due to their accuracy, reliability, and robustness. The digital radiography also reduces the cost of rework by detecting the material flaws in the earlier stage such as manufacturing stage itself. This can prevent catastrophic failures before they escalate into larger problems. The digital radiography helps in maintaining a compliance with quality assurance standards. For example, standards such as ASTM (American Society for Testing and Materials) and ISO (International Organization for Standardization) often mandate radiographic testing for certain applications. Compliance with these standards is essential for maintaining certification, gaining customer trust, and avoiding legal or regulatory issues. The digital radiography results can offer a valuable feedback for optimizing the NDT process by identifying recurring defects in the product. The manufacturing process can be modified to avoid these defects.

K. Kale (✉)
Department of Electronics and Communication Engineering, Presidency University, Bangalore, India

Universal SkillTech University, Maharashtra, India
e-mail: kirandhanaji.kale@presidencyuniversity.in

D. K. Srivastava
Department of Electronics and Communication Engineering, Presidency University, Bangalore, India
e-mail: dharmeshkumar.s@presidencyuniversity.in

M. Derawi
Faculty of Information Technology and Electrical Engineering, Department of Electronic Systems, Norwegian University of Science and Technology, Gjøvik, Norway
e-mail: mohammad.derawi@ntnu.no

N. Gupta, J. D. Diaz Roman (eds.), *Decision Support Systems for Industrial and Scientific Applications*, Signals and Communication Technology,
https://doi.org/10.1007/978-3-031-89700-9_11

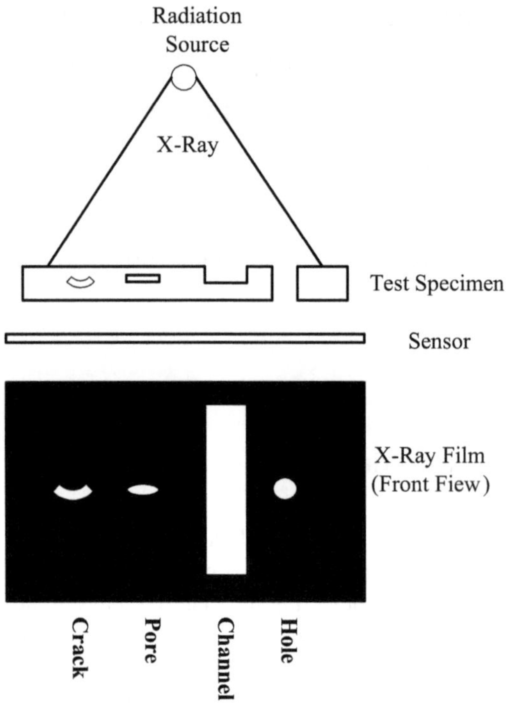

Fig. 1 Digital radiography testing setup

There are two popular approaches in RI namely Film Radiography and Digital Radiography. The film radiography is a traditional approach which involves development of a photographic film of the material under test by passing ionized radiations through it. The photographic film can be examined by the experts to identify defects in the material. However, the film radiography examination involves a chemical process to develop the photographic film. The dynamic range of this film is limited which further degrades the testing accuracy. The accuracy of the inspection is also dependent on the human observer involved in the process. On the other hand, digital radiography uses photovoltaic sensors to create digital image of the material under test. This approach provides direct digital image of the material under test. There are several advantages of this approach such as faster data acquisition, enhanced dynamic range, easier storage, and retrieval. The various image processing techniques can be used to enhance the image quality for accurate more accurate testing results [1]. Fig. 1 indicates typical digital radiographic testing setup.

Digital radiography applications span in multiple industrial sectors due to their accuracy, reliability, and robustness. The digital radiography also reduces the cost of rework by detecting the material flaws in the earlier stage such as manufacturing stage itself. This can prevent catastrophic failures before they escalate into larger problems. The digital radiography helps in maintaining a compliance with quality assurance standards. For example, standards such as ASTM (American Society for Testing and Materials) and ISO (International Organization for Standardization) often mandate radiographic testing for certain applications. Compliance with these standards is essential for maintaining certification, gaining customer trust, and avoiding legal or regulatory issues. The digital radiography results can offer a valuable feedback for optimizing the NDT process by identifying recurring defects in the product. The manufacturing process can be modified to avoid these defects.

While digital radiographic testing (RT) remains a popular choice for inspecting welds and detecting internal defects, several limitations are inherent in traditional practices. These limitations can impact the effectiveness and efficiency of inspection process. Often this process is performed by a skilled operator. The defect identification is done by examining the radiographic image. The accuracy of the defect identification is dependent on varying levels of expertise. The human factors such as experience and fatigue may affect the individual's judgment. This variability will affect on overall quality assurance. The radiographic image resolution is another challenge in this process. The operator may miss few small and complex defects due to poor radiographic image resolution. This is prevalent in high-density materials. The radiographic test environment can pose health risks to a personal. The prolonged exposure can lead to serious health issues. The use of radiographic test equipment often requires a restricted access testing area to avoid radiation exposure which further delays the process. In the realm of weld radiography, the demand for precision and reliability is paramount. To address these challenges, Intelligent Decision Support Systems (IDSS) have emerged as a crucial advancement in weld radiography [2].

1.1 Intelligent Decision Support System

The primary need for IDSS in weld radiography stems from their ability to enhance accuracy and consistency in defect detection. These systems are designed to recognize complex patterns and anomalies that may be challenging for human inspectors to detect. By providing a more consistent and objective assessment, IDSS significantly reduces the risk of overlooking critical defects, thereby improving overall weld quality and safety. Manual analysis of radiographic images can be time-consuming, with inspectors needing to meticulously review each image and document their findings. In contrast, IDSS can process and analyze images in real-time, delivering immediate feedback and automated reporting. This acceleration not only speeds up the inspection process but also supports rapid decision-making, allowing for quicker responses to quality issues and reducing production downtime. By integrating IDSS into the inspection workflow, organizations can achieve higher throughput and more streamlined quality assurance processes [1].

Furthermore, IDSS provide valuable support for decision-making and risk management. These systems are equipped to prioritize detected defects based on their severity and potential impact on weld integrity. By assessing the risks associated with each defect and offering actionable recommendations, IDSS help inspectors and engineers make informed decisions about repairs or rejections.

This prioritization is crucial for optimizing resource allocation and ensuring that the most critical issues are addressed promptly. The need for IDSS in weld radiography is underscored by the increasing complexity and criticality of modern industrial applications. As technological advancements drive the development of high-performance materials and intricate weld designs, the traditional methods of inspection are often stretched beyond their capabilities.

IDSS are equipped to handle the growing volume of data and the intricate nature of defects with a level of precision and speed that human inspectors alone cannot match. They leverage sophisticated algorithms to process large datasets, identify subtle anomalies, and provide comprehensive analysis, ensuring that even the most complex welds are thoroughly inspected. Moreover, the integration of IDSS facilitates a proactive approach to quality assurance. Rather than merely detecting defects after they occur, these systems enable predictive maintenance by analyzing historical data and identifying trends that may indicate potential future issues. This forward-looking capability allows organizations to address potential problems before they manifest into critical failures, thus enhancing the reliability and safety of their products and infrastructure. By providing early warnings and actionable insights, IDSS contribute to more strategic and preventative maintenance practices, ultimately reducing long-term costs and improving operational efficiency [3].

In addition, the ability of IDSS to adapt and learn from new data ensures that they remain effective as industry standards evolve and new defect types emerge. The continuous learning and updating capabilities of these systems mean that they can incorporate the latest research and developments in defect detection, maintaining their relevance and effectiveness in a rapidly changing technological landscape. This adaptability is crucial for staying ahead of emerging challenges and ensuring that inspection practices remain at the forefront of industry standards. Overall, Intelligent Decision Support Systems are not just a technological advancement but a necessity in the modern landscape of weld radiography. Their ability to enhance accuracy, efficiency, and proactive quality management underscores their critical role in advancing industrial quality assurance and maintaining the highest standards of safety and reliability.

2 Typical Intelligent Decision Support System

Despite its importance, traditional radiographic inspection often faces challenges, such as high rates of false positives and negatives, operator fatigue, and time-consuming analysis processes. These challenges underscore the need for a more robust approach—one that IDSS can provide. By automating image analysis and decision-making, IDSS not only alleviates the burden on human inspectors but also enhances the overall quality of inspections. A typical IDSS integrates advanced algorithms, machine learning, and data analytics to facilitate a more efficient and accurate inspection workflow. These systems are designed to analyze vast amounts of radiographic data, processing images and extracting relevant features that might indicate flaws. By applying artificial intelligence to the radiographic inspection process, IDSS significantly reduces the time and effort required for inspections, while simultaneously increasing accuracy and reliability. IDSS for radiographic inspection comprises several key components that work together to streamline the inspection process.

An Intelligent Decision Support System (IDSS) in radiographic inspection is a sophisticated framework that enhances the inspection and evaluation of welded joints using radiographic techniques. This system integrates advanced data management, image processing, and analytical tools to assist engineers and inspectors in assessing weld quality and integrity. The IDSS typically features robust data acquisition capabilities, capturing radiographic images along with essential metadata such as exposure time and angles. These images and inspection reports are stored in a centralized database, facilitating easy retrieval and long-term archiving. Key components include advanced image processing tools that enhance image clarity and contrast, making it easier to detect defects like cracks and inclusions. Automated defect detection algorithms leverage machine learning to identify and classify defects accurately, reducing human error and improving consistency. Statistical analysis further enables the identification of trends in defect data, aiding in predictive maintenance and risk management [4].

The user interface of the IDSS is designed for interactivity, providing engineers with dashboards that visually summarize inspection results and defect trends. Comprehensive reporting tools generate detailed findings and recommendations, which can be easily shared among stakeholders. Additionally, the system incorporates a knowledge base that includes industry standards and expert guidelines, ensuring that evaluations align with regulatory requirements. Decision support tools, such as risk assessment models and scenario analysis capabilities, help prioritize repair actions based on defect severity and potential failure risks. The IDSS is characterized by its flexibility, allowing adaptation to various inspection needs and industry standards, while collaboration features promote communication among teams. Applications of this system span multiple industries, including manufacturing, oil and gas, aerospace, and construction, where it plays a critical role in ensuring the safety and reliability of welded structures. Overall, an IDSS for weld radiography significantly enhances the accuracy and efficiency of inspections, providing data-driven insights that improve quality assurance and mitigate risks associated with welded joints. Fig. 2 shows a process flow in typical IDSS.

The Intelligent Decision Support System (IDSS) for weld radiography represents a cutting-edge integration of data acquisition, image processing, and advanced analytical methodologies, fundamentally enhancing the inspection and evaluation of welded joints. The system begins with robust data acquisition, primarily utilizing radiographic images obtained through X-ray or gamma-ray imaging techniques. These images are pivotal, as they reveal the internal structures of welds, serving as the primary input for the detection of defects and anomalies. Accompanying these radiographs is a comprehensive set of metadata, which includes critical parameters

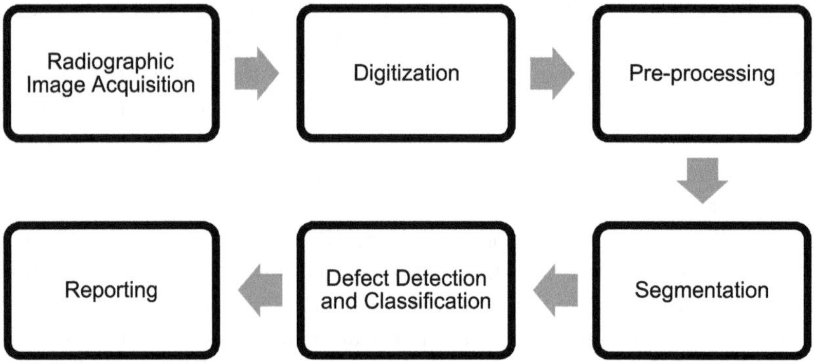

Fig. 2 Process flow in typical IDSS

such as exposure settings, material types, weld specifications, and detailed inspection parameters. This contextual information is essential for accurate analysis and interpretation. Furthermore, the IDSS is designed to integrate seamlessly with other inspection modalities, such as ultrasonic testing (UT) and computed tomography (CT), thereby facilitating a holistic evaluation of weld quality.

Following data acquisition, the IDSS engages in sophisticated image processing and enhancement. The radiographic images undergo a series of pre-processing steps aimed at improving image quality and optimizing them for subsequent analysis. Techniques such as noise reduction, contrast enhancement, and normalization are employed to ensure consistency in image quality across varying inspection contexts. Subsequently, image segmentation is performed to isolate regions of interest—specifically the weld areas—from the background, thereby directing the analytical focus toward relevant components and significantly enhancing the accuracy of defect detection.

The core functionality of the IDSS is anchored in its machine learning and deep learning algorithms. Advanced feature extraction techniques are utilized to discern critical characteristics within the radiographic images, identifying edges, textures, and shapes that may indicate potential defects. These features serve as inputs for training machine learning models, particularly convolutional neural networks (CNNs), using extensive datasets of annotated radiographic images. This training process equips the models with the capability to recognize intricate patterns and classify defects based on labeled examples. During real-time inspections, these trained models analyze newly acquired radiographic images to detect and classify defects, leveraging deep learning methodologies to identify subtle or complex issues that may pose challenges for human inspectors. Upon identification of defects, the IDSS provides comprehensive decision support and analytical insights.

The system automatically classifies anomalies such as cracks, porosity, and inclusions, delivering detailed information regarding their location, severity, and potential impact on weld integrity. It offers actionable recommendations for corrective measures based on the assessed severity and associated risk of each defect.

This prioritization process enables inspectors to concentrate on critical issues that demand immediate attention, thus enhancing the overall effectiveness of the inspection regime. The reporting and documentation capabilities of the IDSS are streamlined through the generation of detailed, automated reports that encapsulate information about detected defects, their respective locations, and severity levels. These reports not only facilitate efficient communication with stakeholders but also bolster documentation practices. Additionally, the system incorporates advanced visualization tools that present radiographic images alongside defect annotations and analytical results, allowing users to engage interactively with the data by zooming, panning, and annotating images to support comprehensive inspections and reviews.

A user-centric dashboard is integral to the IDSS, effectively presenting analytical results, defect summaries, and actionable insights in a format that is both accessible and intuitive. This interface enables inspectors to engage with the data efficiently, promoting thorough reviews of inspection outcomes. Interactive features—including image annotation, measurement tools, and feedback mechanisms—enhance the inspection process, fostering in-depth analyses that empower users to make informed decisions. Centralized data management constitutes a cornerstone of the IDSS, establishing a comprehensive repository for radiographic images and inspection data. This centralized approach not only facilitates easy access to historical data but also supports long-term quality management initiatives. Advanced analytics tools are integrated to evaluate trends and patterns in weld quality over time, generating insights that drive continuous improvement and assist in identifying recurring issues or potential areas for optimization [5].

Lastly, the IDSS is designed to integrate seamlessly with smart manufacturing systems and automated production lines, enabling real-time quality control and process adjustments informed by inspection results. Feedback mechanisms within the system provide actionable insights that facilitate process enhancements, thereby improving overall production quality and minimizing the incidence of defects. This sophisticated integration of technologies within the IDSS underscores its transformative potential in the field of weld radiography, significantly advancing the quality assurance processes critical to ensuring the integrity of welded structures.

3 Role of Deep Learning

Deep learning, a subset of machine learning (Fig. 3), has emerged as a transformative technology across various fields, particularly within industrial applications such as IDSS. Deep learning involves neural networks with multiple layers that enable the automatic extraction of features and patterns from complex datasets. Unlike traditional machine learning approaches, which often require manual feature engineering, deep learning models can learn hierarchical representations of data through multiple processing layers. This capability makes deep learning particularly effective for tasks involving images, audio, and natural language. In the context of

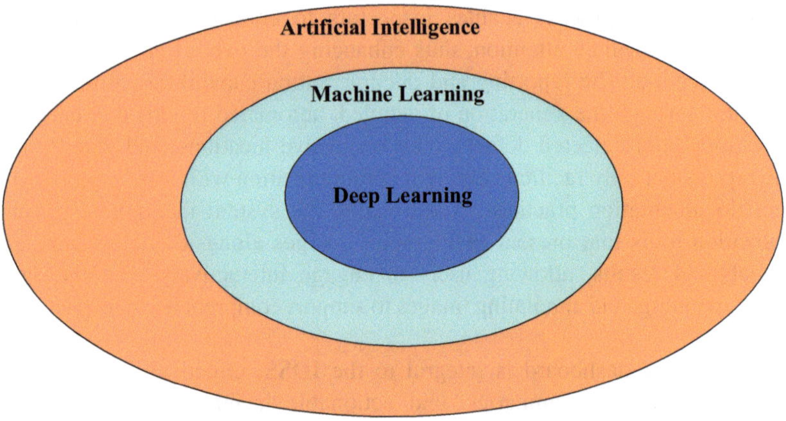

Fig. 3 Deep learning

IDSS, especially for tasks like digital radiographic weld inspection, deep learning significantly enhances data analysis capabilities. It automates and refines the decision-making process by efficiently handling large volumes of complex data, which is vital in ensuring the safety and integrity of welded structures.

In traditional inspection methods, radiographic images of welds are analyzed manually by trained inspectors, a process that can be time-consuming and prone to human error. Deep learning addresses these challenges by employing advanced algorithms that can learn from vast datasets, recognizing intricate patterns that may indicate defects such as cracks, inclusions, or porosity. By processing and analyzing these images with a high degree of accuracy, deep learning models can detect subtle anomalies that might be overlooked by human inspectors. Moreover, deep learning's ability to continuously learn and adapt makes it particularly valuable in dynamic industrial environments. As new data becomes available, these models can be retrained to improve their performance and generalize better to unseen data, ensuring that the IDSS remains effective in diverse scenarios. This adaptability is crucial for industries that face evolving standards and regulatory requirements. Additionally, the integration of deep learning into IDSS enables real-time analysis of radiographic images, providing immediate feedback to inspectors. This capability allows for quicker decision-making and can significantly reduce inspection times, leading to enhanced operational efficiency. Furthermore, the automated nature of deep learning reduces variability in inspection outcomes, fostering greater consistency in defect detection and classification.

The incorporation of deep learning into Intelligent Decision Support Systems (IDSS) presents numerous advantages that significantly enhance the inspection process. Firstly, deep learning markedly improves defect detection rates by leveraging its capability to identify complex patterns and subtle features that traditional methods might overlook. This enhanced accuracy is crucial in ensuring the reliability of inspections. Additionally, deep learning automates the analysis process, which reduces the time required for inspections. This efficiency enables organizations to conduct more thorough evaluations in a shorter timeframe, streamlining workflows [5].

Furthermore, deep learning models provide a level of consistency that surpasses human inspectors, minimizing variability and ensuring uniform performance across inspections. This reliability is vital in maintaining high standards of quality assurance. Beyond mere defect detection, deep learning also generates valuable insights into underlying trends and patterns within the data, empowering organizations to implement continuous improvement strategies and enhance their overall operational effectiveness.

3.1 Applications of Deep Learning in IDSS

Deep learning encompasses several general types of algorithms, each tailored for specific tasks and data types. The convolutional neural networks (CNNs) are a class of deep learning algorithms specifically designed for processing grid-like data, such as images. They utilize convolutional layers to automatically extract features, enabling effective image recognition and classification tasks. CNNs are widely used in various applications, including computer vision, medical imaging, and many more. The CNNs are particularly transformative in this area. CNNs excel at processing grid-like data, such as radiographic images, and utilize convolutional layers to extract intricate features. This capability enables them to effectively recognize and classify defects like cracks, porosity, and inclusions through detailed analysis of edges, textures, and shapes within the images. It works like a supervised classification algorithm which can classify the input radiographic images into various pre-defined classes. The variants of CNN such as U-Net [6] and ResNet [7] are also employed by the researchers. CNN algorithms can work with 2D and 3D input images.

Recurrent neural networks (RNNs) have numerous applications in Intelligent Decision Support Systems (IDSS), particularly in contexts where sequential data and temporal relationships are vital. One of their primary uses is in trend analysis, where RNNs effectively monitor historical data related to weld inspections to identify patterns in weld quality over time. This capability aids in detecting potential issues before they escalate, thus enhancing quality control. Additionally, RNNs excel at anomaly detection by processing sequences of inspection data to identify deviations from expected patterns, ensuring the integrity of welded components. Their predictive maintenance capabilities allow RNNs to analyze time-series data to forecast potential failures or maintenance needs, optimizing schedules and reducing unexpected downtime. In scenarios where inspection decisions rely on previous assessments, RNNs can facilitate sequential decision-making by providing insights that inform subsequent actions. Furthermore, RNNs can integrate data from multiple inspections over time, enabling comprehensive evaluations that account for variations in welding practices or material properties.

In automated inspection systems, RNNs also play a crucial role in real-time monitoring by processing streaming data to deliver immediate insights into weld quality, allowing for prompt corrective actions when defects are identified. The long short-term memory (LSTM) network is a popular algorithm in RNN, are beneficial for sequential data analysis, which is essential for monitoring quality over time. Their architecture allows information to persist, making them valuable for identifying trends or changes that could indicate emerging issues in weld integrity. This temporal aspect of analysis is crucial for proactive decision-making in industrial environments [8].

Autoencoders present a compelling approach for anomaly detection, particularly within the context of weld inspection in Intelligent Decision Support Systems (IDSS). These neural networks are engineered to learn efficient representations of input data by compressing it into a lower-dimensional space and subsequently reconstructing it back to its original form. Through this process, autoencoders capture the underlying patterns and structures inherent in the data, effectively delineating the normal characteristics of weld images. When trained on a dataset of defect-free weld images, the autoencoder becomes proficient at reconstructing these inputs with high fidelity. However, when exposed to images containing defects or anomalies, the model struggles to accurately reproduce these inputs, resulting in a significant reconstruction error [9].

This discrepancy serves as a vital indicator of potential defects, enabling the early detection of issues that may otherwise remain undetected. The unsupervised nature of autoencoders is particularly beneficial, as it alleviates the necessity for extensive labeled datasets, which can be challenging and resource-intensive to acquire in industrial settings. This characteristic not only streamlines the training process but also enhances the interpretability of inspection results. Inspectors can analyze reconstruction errors to pinpoint specific areas of interest in the weld images; thereby facilitating a more focused and informed assessment of weld integrity.

Further enhancing the capabilities of Intelligent Decision Support Systems (IDSS), Transfer Learning facilitates the use of pre-trained models, enabling rapid adaptation for specific tasks related to weld inspection. In practical applications, developing deep learning models from scratch often necessitates extensive datasets and significant computational resources. Transfer Learning addresses this challenge by allowing practitioners to leverage models that have already been trained on large, diverse datasets. This method proves particularly advantageous in scenarios where annotated data is limited or costly to obtain, such as in specialized industrial contexts. By fine-tuning these pre-trained models on smaller, task-specific datasets, organizations can achieve high levels of accuracy with substantially reduced training times. This capability not only enhances the overall efficiency of the modeling process but also improves the robustness of defect detection in weld inspections [10].

In addition, Ensemble Learning approaches significantly bolster the effectiveness of IDSS by combining the predictions of multiple models to achieve superior robustness and accuracy. This technique involves integrating various algorithms, architectures, or training methodologies to capture diverse aspects of the data. By aggregating the outputs of these distinct models, Ensemble Learning mitigates the limitations of individual models, resulting in more reliable and consistent defect detection outcomes. For instance, while one model may excel in identifying specific types of weld defects, another may be more adept at detecting anomalies in different contexts. By harnessing the strengths of multiple models, organizations can develop a comprehensive inspection framework that enhances overall performance. The integration of Transfer Learning and Ensemble Learning within IDSS not only streamlines the defect detection process but also empowers organizations to adapt to evolving challenges in weld quality assurance. As industries continue to embrace advanced technologies, these methodologies will play a pivotal role in ensuring the integrity and reliability of welded components, ultimately contributing to safer and more efficient manufacturing practices [11].

GANs serve as a transformative tool in IDSS for weld inspection by addressing the challenges associated with limited training data. Their ability to generate high-quality synthetic images not only enriches the dataset but also enhances the overall performance of deep learning models, contributing to more accurate and reliable defect detection in welds. As industries continue to evolve and demand higher standards of quality assurance, the integration of GANs into inspection frameworks will be increasingly essential. This is particularly beneficial in scenarios where datasets are limited or imbalanced, as acquiring a sufficient quantity of labeled data can often be challenging and resource-intensive. By employing a generator and a discriminator in tandem, GANs create a dynamic interplay where the generator produces new, synthetic images, while the discriminator evaluates their authenticity against real images. This adversarial process not only facilitates the generation of high-quality synthetic data but also enhances the diversity of the training dataset.

The introduction of synthetic images into the training process is crucial for improving the robustness and performance of other deep learning models used in defect detection. By exposing models to a wider variety of defect types and variations, GANs enable them to learn more generalized features that can better capture the complexities of real-world weld inspections. This augmentation can lead to improved model accuracy and a decrease in false negatives, ultimately enhancing the reliability of inspections. Moreover, the synthetic images generated by GANs can be tailored to represent various conditions, such as different welding techniques, materials, and environmental factors. This adaptability ensures that the training data encompasses a broad spectrum of potential scenarios that inspectors may encounter. Consequently, models trained with this augmented dataset are better equipped to recognize and classify defects across diverse contexts, leading to more effective quality control in manufacturing processes. Fig. 4 indicates various types of deep learning algorithms that can be used in IDSS for radiographic inspection [12].

In conclusion, the integration of advanced deep learning techniques significantly enhances their effectiveness in various applications. Each of these methodologies brings unique advantages: CNNs are adept at analyzing images and extracting relevant features, RNNs excel at processing sequential data to uncover trends, Autoencoders support anomaly detection through unsupervised learning, Transfer Learning allows for quick adaptation to specific tasks with minimal data, Ensemble

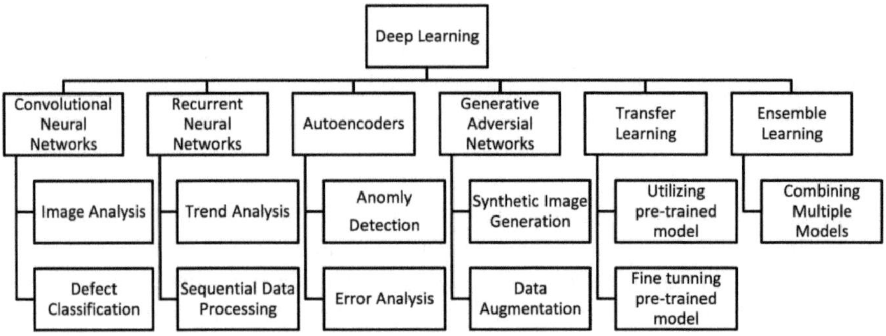

Fig. 4 Applications of various deep learning models in IDSS

Learning combines multiple models to improve overall robustness, and GANs generate synthetic data to augment training sets. Collectively, these techniques provide a powerful framework for tackling complex challenges in fields such as quality assurance and defect detection. By leveraging these advanced methodologies, organizations can improve the efficiency and accuracy of their decision-making processes. As industries continue to evolve and incorporate digital innovations, the adoption of these deep learning techniques within IDSS will play a crucial role in enhancing reliability and performance across a range of applications, ultimately driving continuous improvement and operational excellence.

4 Case Studies

This section gives case studies of use of various deep learning algorithms in stages of IDSS workflow for radiographic inspection.

4.1 Feature Extraction

Case Study 1: Feature Extraction in Aerospace Components
An aerospace company sought to enhance defect detection precision in radiographic inspections of critical components, such as turbine blades and structural welds. Traditional inspection methods faced challenges in identifying and classifying complex defects due to reliance on manual feature extraction. The project commenced with data collection, gathering radiographic images that included various defects, such as cracks and inclusions. Expert annotation created a comprehensive training dataset. A Convolutional Neural Network (CNN) was developed, utilizing pre-trained networks like VGG16 and ResNet for effective feature extraction through

Transfer Learning. This model was integrated into the existing inspection workflow, automating the extraction of key features and facilitating detailed defect analysis. Results indicated a 90% precision and recalls due to feature extraction accuracy, a reduction in manual analysis time, and enhanced quality assurance, ultimately leading to more reliable inspections of aerospace components [13].

Case Study 2: Automated Feature Extraction for Pipeline Welds
In response to the need for a more efficient method of detecting defects in welds, a pipeline construction company implemented a deep learning-based solution. Radiographic images of pipeline welds were collected, capturing defects such as porosity and incomplete fusion. After careful annotation, a CNN utilizing Region-Based CNNs (R-CNNs) Random Forest was developed to enhance feature detection. Integrated into the inspection system, this model automated the extraction and analysis of weld features. The implementation achieved up to 98% detection accuracy, reduced manual inspection time and improved defect classification and reporting, thereby enhancing overall weld quality [14].

4.2 Defect Detection and Classification

Case Study 1: Recurrent Neural Networks (RNN) for Anomaly Detection in Radiographic Inspection
A pipeline inspection company faced challenges in identifying anomalies in radiographic images that fell outside predefined categories. They developed an RNN-based model for anomaly detection, trained on a dataset of normal and anomalous images to recognize patterns and deviations. The RNN was paired with an autoencoder to learn standard patterns, allowing for effective identification of anomalies based on reconstruction errors. This model successfully detected anomalies, resulting in an improvement in overall accuracy, thereby enhancing the inspection process [15].

Case Study 2: Hybrid CNN-RNN Model for Comprehensive Defect Analysis in Weld Inspections
To analyze radiographic images of welds comprehensively, a manufacturing company developed a hybrid CNN-RNN model that addressed both spatial and temporal dimensions of defects. The model utilized CNNs for spatial feature extraction and RNNs for analyzing changes over time, trained on a dataset of sequential images with defect annotations. This hybrid approach achieved a 94% accuracy in defect detection and classification with reduced time of 0.0067 ms for each input image, providing deeper insights into defect behavior and significantly enhancing the quality and reliability of the inspection process [16].

4.3 Synthetic Image Generation

Case Study 1: Synthetic Image Generation for Weld Defect Detection in Aerospace Applications
An aerospace company faced challenges in obtaining a large and diverse dataset of radiographic images of weld defects for training a deep learning model aimed at automated inspection, as collecting and annotating real images was both costly and time-consuming. To address this, they employed generative adversarial networks (GANs) to generate synthetic radiographic images, training the GANs on a smaller dataset of real weld images to produce diverse defect types, including cracks, porosity, and inclusions. The synthetic images were then annotated using automated tools and validated by experts to ensure realistic representation. This approach resulted in a significant increase in dataset size, enhancing model performance with a 50% improvement in defect detection accuracy and reducing costs and time associated with real image collection. Ultimately, the use of GANs enabled the aerospace company to efficiently create a high-quality dataset, thereby improving their automated weld defect detection system [17].

Case Study 5: Deep Learning-Based Generation of Radiographic Datasets for Complex Microstructures
In addressing the challenge of training deep convolutional neural networks (DCNNs) for automatic segmentation of microstructural X-ray computed tomography (XCT) data, researchers identified data acquisition as a significant hurdle, particularly for complex multiphase microstructures that are better suited for 3D networks. Traditional methods rely on manual annotation of 2D slices, which is not feasible for 3D datasets. To overcome this, the study introduced a method for generating synthetic XCT data focused on a six-phase Al–Si alloy composite reinforced with ceramic fibers and particles. This approach involved specific data augmentations—such as adjustments to brightness, contrast, noise, and blur—alongside the design of a custom deep learning model called Triple UNet, and a multi-view forwarding strategy to enhance generalized learning from the synthetic data. The resulting model achieved an overall Dice score of 0.77, indicating effective segmentation performance. Furthermore, the study highlighted how artifacts in the XCT data can negatively impact segmentation accuracy when synthetic data is used for training DCNNs. The proposed methods have broader applicability across various materials and imaging techniques, demonstrating that successful segmentation through neural networks trained with synthetic data can significantly accelerate scientific research outputs. [18].

5 Challenges and Solutions

The integration of deep learning into Intelligent Decision Support Systems (IDSS) for radiographic inspection introduces a myriad of challenges that can significantly

Table 1 Challenges and their solutions for deep learning in IDSS

Challenges	Impact	Solution
Acquisition of high-quality, annotated radiographic images [6]	Limited or suboptimal data can lead to overfitting, poor generalization, and unreliable model performance	Combine multiple datasets, data augmentation
Dataset Annotation [20]	Inaccurate or inconsistent annotations can affect model training and result in lower accuracy in defect detection and classification	Semi-automated annotation tools that assist experts in labeling images
Hyperparameter Tuning [20]	The process of tuning hyperparameters can be time-consuming and may require extensive experimentation and expertise	Automated hyperparameter tuning tools such as grid search or Bayesian optimization, use of transfer learning
Model Generalization [6]	Reduced model performance when used with different environments	K-fold cross-validation, domain adaptation
Adaptation to New Defects [4]	Continuous adaptation requires additional data collection and retraining	Frequent upgradation of model to maintain the relevance
Dataset Imbalance [20]	Model may remain biased toward frequent classes	Synthetic minority over-sampling technique, synthetic data generation using GAN
Integration with Decision-Making Processes [4]	Disruption to existing workflows or decision-making practices can lead to resistance and inefficiencies	Use of modular components and APIs to facilitate smooth integration
Handling Noise and Artifacts [20]	May affect the performance of models	Use pre-processing, regularization methods
Model Drift [20]	May add burden of model retraining and validation	Use online learning techniques

impact the systems' overall effectiveness, efficiency, and reliability [19]. Table 1 summarizes the challenges, their impact, and their solutions for deep learning in IDSS.

6 Future Scope

The future of deep learning in Intelligent Decision Support Systems (DSS) for radiographic inspection is marked by significant and promising advancements. Key trends include the development of emerging architectures such as Transformer-based models and Neural Architecture Search (NAS), which enhance the capability to analyze complex data patterns, leading to improved accuracy in defect detection and classification. Multi-modal learning approaches that integrate data from various imaging techniques, including CT, X-ray, and ultrasound, are expected to provide a more holistic understanding of defects by leveraging complementary information.

Hybrid AI models combining deep learning with other methodologies—such as symbolic reasoning and reinforcement learning—are anticipated to improve decision-making in complex inspection scenarios. The integration of deep learning models with rule-based expert systems will enhance interpretability and reliability, ensuring a more robust decision-making framework. Furthermore, advancements in synthetic data generation using generative adversarial networks (GANs) will help address challenges related to data scarcity, while sophisticated data augmentation techniques will produce diverse variations of radiographic images, enhancing model robustness. Real-time processing facilitated by edge computing technologies will enable immediate feedback during inspections, improving overall efficiency.

Explainable AI (XAI) will enhance transparency, fostering trust in model predictions, while advanced visualization tools will make complex outputs more accessible to human inspectors. The integration of IoT devices will enable smarter, connected inspection environments that support real-time monitoring and data analysis, contributing to smarter manufacturing practices. Adaptive learning systems will allow models to continuously evolve based on new data, ensuring sustained high performance. Personalized decision support systems will tailor recommendations according to user-specific needs, improving the effectiveness of inspections.

Collaborations with augmented reality (AR) technologies will overlay inspection results onto radiographic images, enhancing visual capabilities, while the integration of robotics will automate repetitive inspection tasks, increasing accuracy and efficiency. Additionally, enhanced security measures will be implemented to protect sensitive radiographic data and ensure compliance with privacy regulations, while secure data-sharing protocols will facilitate collaboration between organizations. Finally, establishing ethical guidelines for AI deployment in radiographic inspection will address concerns related to bias and accountability, while promoting effective collaboration between human inspectors and AI systems to optimize decision-making. By focusing on these future directions, deep learning-based Intelligent Decision Support Systems for radiographic inspection can achieve greater accuracy, efficiency, and adaptability, ultimately leading to improved quality assurance and operational excellence across various industries.

7 Conclusion

In conclusion, the integration of deep learning into Intelligent Decision Support Systems (DSS) for radiographic inspection represents a pivotal advancement in quality assurance across key industries such as manufacturing, aerospace, and construction. By leveraging AI technologies, these systems enhance accuracy, efficiency, and reliability in detecting weld defects from radiographic images. Advanced models, particularly convolutional neural networks (CNNs) and emerging architectures like transformers, have effectively addressed the limitations of traditional inspection methods, offering consistent and precise defect detection while efficiently processing large volumes of data. The automation of defect detection processes facilitated by these systems reduces reliance on manual inspections, thereby minimizing human error and accelerating inspection workflows. As we look to the future, the potential for deep learning in radiographic inspection continues to expand, with promising advancements in hybrid models and real-time processing through edge computing poised to further enhance system capabilities.

Despite these advancements, challenges such as data quality and model robustness must be proactively addressed through improved data management and ongoing system maintenance. Ultimately, the success of deep learning in this domain will hinge on fostering a synergistic collaboration between human expertise and AI systems, combining the strengths of both to optimize inspection outcomes. As deep learning technology evolves, it is set to redefine quality assurance standards, ensuring the integrity and safety of critical structures and components. Embracing these innovations will not only enhance inspection accuracy and efficiency but will also pave the way for new benchmarks in quality assurance practices.

References

1. Halmshaw, R. (1995, January 1). *Industrial radiology* (Non-destructive evaluation series) (2nd ed.). Springer Nature.
2. Ida, N., & Meyendorf, N. (2019). *Handbook of advanced nondestructive evaluation*. Springer.
3. Sonwane, S., & Chiddarwar, S. (2023, December 19). Developing a DSS for enhancing weldment defect detection, classification, and remediation using HDR images and adaptive MDCBNet neural network. *Journal of Nondestructive Evaluation, 43*(1). https://doi.org/10.1007/s10921-023-01027-8. Accessed 21 Oct 2024.
4. Liu, T., et al. (2023, June 15). A state-of-the-art survey of welding radiographic image analysis: Challenges, technologies and applications. *Measurement, 214*, 112821. https://doi.org/10.1016/j.measurement.2023.112821. www.sciencedirect.com/science/article/pii/S0263224123003858
5. Hellier, C. (2021, January 1). *Handbook of nondestructive evaluation 4.0*. Springer EBooks, Springer.
6. Tyystjärvi, T., et al. (2024, January 28). Deploying machine learning for radiography of aerospace welds. *Journal of Nondestructive Evaluation, 43*(1). https://doi.org/10.1007/s10921-023-01041-w
7. Palma-Ramírez, D., et al. (2024, May 1). Deep convolutional neural network for weld defect classification in radiographic images. *Heliyon, 10*(9), e30590–e30590. https://doi.org/10.1016/j.heliyon.2024.e30590
8. Shang, L., et al. (2023, August 9). CNN-LSTM hybrid model to promote signal processing of ultrasonic guided lamb waves for damage detection in metallic pipelines. *Sensors, 23*(16), 7059–7059. https://doi.org/10.3390/s23167059
9. Walther, D., et al. (2024, June). Recurrent autoencoder for weld discontinuity prediction. *Journal of Advanced Joining Processes, 9*, 100203. https://doi.org/10.1016/j.jajp.2024.100203

10. Kumaresan, S., et al. (2021). Transfer learning with CNN for classification of weld defect. *IEEE Access, 9*, 95097–95108. https://doi.org/10.1109/access.2021.3093487
11. Vasan, V., et al. (2024, October 1). Ensemble-based deep learning model for welding defect detection and classification. *Engineering Applications of Artificial Intelligence, 136*, 108961. https://doi.org/10.1016/j.engappai.2024.108961
12. Singh, A., et al. (2022, July 6). GAN-based image dehazing for intelligent weld shape classification and tracing using deep learning. *Applied Sciences, 12*(14), 6860. https://doi.org/10.3390/app12146860
13. Upadhyay, A., et al. (2023, February 1). A deep-learning-based approach for aircraft engine defect detection. *Machines, 11*(2), 192. https://doi.org/10.3390/machines11020192
14. Zhu, H., et al. (2019, August 12). Deep learning-based classification of weld surface defects. *Applied Sciences, 9*(16), 3312. https://doi.org/10.3390/app9163312
15. Marino, M., et al. *A recurrent neural network classifier for ultrasonic NDE applications*. 2018 IEEE international ultrasonics symposium (IUS). https://doi.org/10.1109/ULTSYM.2018.8579896
16. Liu, T., et al. (2018, December 10). A hybrid CNN–LSTM algorithm for online defect recognition of CO2 welding. *Sensors, 18*(12), 4369. https://doi.org/10.3390/s18124369
17. Li, L., et al. (2024, April 11). Synthetic data augmentation for high-resolution X-ray welding defect detection and classification based on a small number of real samples. *Engineering Applications of Artificial Intelligence, 133*, 108379. https://doi.org/10.1016/j.engappai.2024.108379
18. Tsamos, A., et al. (2023, January 19). Synthetic data generation for automatic segmentation of X-ray computed tomography reconstructions of complex microstructures. *Journal of Imaging, 9*(2), 22–22. https://doi.org/10.3390/jimaging9020022
19. Liu, T., et al. (2023, April 1). An expert knowledge-empowered CNN approach for welding radiographic image recognition. *Advanced Engineering Informatics, 56*, 101963. https://doi.org/10.1016/j.aei.2023.101963. www.sciencedirect.com/science/article/abs/pii/S1474034623000915
20. LeCun, Y., et al. (2015, May). Deep learning. *Nature, 521*(7553), 436–444. https://doi.org/10.1038/nature14539

Part III
Decision Support Systems for Security Aspects

Decision Support Systems, Entangled in Hardware Security

Stamatis Kazas, Evangelia Konstantopoulou, Nicolas Sklavos, and Ivo Iliev

1 Introduction

Decision Support Systems (DSS) are descendants of Management Information System (MIS) Frameworks, as introduced firstly by Gorry and Morton [1] whereby analyzing and processing available resources, frameworks can be created to help humans impart their decision-making abilities to computational systems (Fig. 1). The primary distinction between now and then is the improved hardware that offers accelerated computational capabilities and the anticipated evolvement of the digital world that led to the advent of Big Data. One of the multiple DSS types [2] that has received tremendous attention and advancements over the last few years is of course Machine Learning (ML) and Artificial Intelligence (AI).

The pluralistic and interdisciplinary nature of AI and ML served as a catalyst for critical sectors whereby implementing their processing and decision-making capabilities, private and classified information would be used for vital decisions. To elaborate further, as examined in [3], Clinical Decision Support Systems (CDSS), by supplying and reinforcing AI and ML with Electronic Health Records (EHR), trade off private patient records to train systems that will consult and overall enhance the effectiveness and decisiveness of healthcare personnel. On the contrary, as shown by van Oijen and de Marez Oyens [4], DSS can also be integrated into the military as a Command and Control (C2) mechanism, among others. However, as would be

S. Kazas (✉) · E. Konstantopoulou · N. Sklavos
SCYTALE Group, Computer Engineering and Informatics Department, University of Patras, Patras, Greece
e-mail: up1084636@ac.upatras.gr; e.konstanto@ac.upatras.gr; nsklavos@upatras.gr

I. Iliev
Department of Biomedical Engineering, Technical University of Sofia, Sofia, Bulgaria
e-mail: izi@tu-sofia.bg

N. Gupta, J. D. Diaz Roman (eds.), *Decision Support Systems for Industrial and Scientific Applications*, Signals and Communication Technology,
https://doi.org/10.1007/978-3-031-89700-9_12

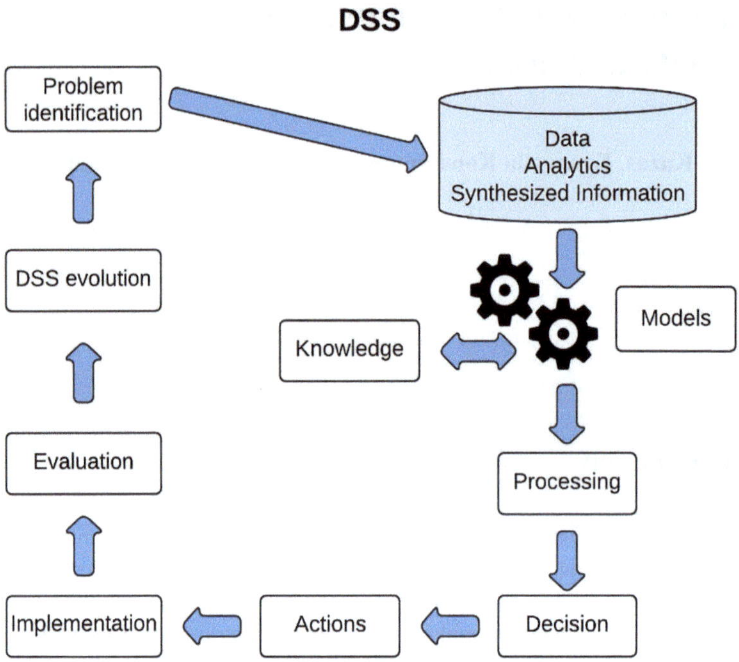

Fig. 1 Decision-making process scheme within DSS

expected, major issues arise about the confidentiality of sensitive data, the quality and diversity of AI's training sets, as well as the righteousness and credibility of the generated decisions [5]. All of the aforementioned cases demonstrate what we are willing to immolate so that we reinforce other strands of our life, even though the risk propagates. Consequently, the importance of cybersecurity and hardware security emerges, with the second one becoming the primary focus of engagement of this paper.

Hardware security is the art of exploiting the disordered nature of hardware in order to achieve an ulterior goal, as well as creating sophisticated defense mechanisms and prevention systems that mitigate attacks. Encompassed within this sector are all the distinct hardware attacks [6, 7] that can be implemented throughout the design, fabrication, and testing cycle of an Integrated Circuit (IC), as well as attacks that can occur after the product is released. This can happen by exploiting hardware vulnerabilities that arise because of poor or faulty design, or hardware component limitations and deficiencies [8]. In addition, detecting these attacks, mitigating their effects, and preventing them from happening again, by solidifying more efficient, precise, robust, and adaptive coping mechanisms and strategies is regarded as part of hardware security. For the purposes of this chapter, we will delve into three main hardware security concepts: Side Channel Attacks (SCA), Hardware Trojans (HTs), and Physical Unclonable Functions (PUFs). Moreover, we will touch on how they intersect with ML and AI, to generate advanced and efficacious attacks and defense mechanisms.

In this chapter, we present a concise overview of various ML and AI applications in hardware security and investigate the possibility of malicious users using the weaknesses of these technologies to suit their own purposes. Initially, applications of cybersecurity and, most importantly, hardware security in Decision Support Systems will be presented. Then, a brief introduction to some important Machine Learning concepts such as supervised and unsupervised learning will be given. To comprehend how AI can be used in each case, the theory underlying PUFs, Side Channel Attacks, and Trojans will also be covered. More precisely, we will commence by introducing the theory of Side Channel Analysis, and investigate the methods of performing Side Channel Attacks with the aim of exploiting design and implementation defects. Then, the focus will be shifted to Hardware Trojans and how they can be exploited to create vulnerabilities in AI models in the training process by manipulating the training set with hidden data that operate as an activation mechanism. Lastly, a closer look will also be taken at different PUF categories such as Arbiter PUF (APUF) and Ring Oscillator PUF (ROPUF), their characteristics, and their operation, with the purpose of understanding how trained models can be implemented by adversaries in order to predict the responses of CRPs. We will delve into a line of scenarios that showcase all of the aforementioned concepts, elucidating the correlation between decision-making systems and hardware security.

2 The Importance of Hardware Security in Decision Support Systems

Decision Support Systems is the region that appertains to the broader domain of Information Systems (IS) and is responsible for constructing frameworks that contribute toward making decisions more effectively and accurately. DSS were first defined as a tool to assist managerial decision makers in scenarios involving both structured and unstructured decisions [9]. They were designed to support decision makers, expanding their capacities without taking their place. Nowadays, DSS gather relevant information from unprocessed data, documents, firsthand knowledge, and business models by utilizing data, complex analytical models, and user interfaces. DSS can be classified as: (a) Communications-driven and group DSS, that support collaborative work; (b) Data- driven DSS, that primarily focuses on the processing of and presentation of data; (c) Knowledge-based DSS (KBDSS), data mining, and management expert systems, that involve the application of facts, guidelines, and protocols; and (d) Model-driven DSS, that apply models from finance, statistics, optimization, or simulation. Hybrids that combine two or more of the aforementioned categories are called compound DSS [9].

Recent technological advances have benefited DSS greatly. In fact, the foundation for building KBDSS or an Intelligent Decision Support System (IDSS) is the tools and techniques used in the area of AI, with the main one being rule-based expert systems [9]. AI is both the study and understanding of the human thought process as well as the representation and duplication of this process in machines. On the other hand, ML enables computer systems to sense and monitor their surroundings and modify their behavior in response to events. In technical terms, ML is a branch of science that focuses on creating algorithms that let computers learn from information obtained from databases or sensors. The capabilities of DSS have been greatly improved by the integration of ML and AI, allowing for more sophisticated data analysis, prediction, and decision-making procedures [1].

Another aspect that has impact on the DSS's usability and functionality is hardware [9]. Advances in computer hardware technologies went hand in hand with the evolution of DSS. The most commonly used hardware is an organization's servers, mainframe computers, workstations, personal computers, and client/server systems. Networks of all kinds, such as the Internet, intranets, and extranets, are used to execute distributed DSS. Access may be granted to mobile devices such as smartphones, tablet PCs, and notebook PCs. Web servers to access data from data warehouses, legacy databases, and the server's current databases is the DSS hardware standard. Users utilize mobile devices with web browsers installed to access DSS. Direct model provisioning occurs via packages that operate on the mainframe, server, external system, or even the client PC. DSS have various applications in many different verticals, as illustrated in Fig. 2.

Along with the great opportunities that DSS offer by enabling machine-based decision support, they also pose threats to data privacy and security.

In healthcare, CDSS support physicians in their decision-making processes. Currently, computer-based CDSS are typically implemented through web applications, her, and computerized provider order entry systems [3]. They can be accessed via desktop, tablet, and smartphone devices, as well as wearable technologies. Non-knowledge-based CDSS base their decisions on statistical pattern recognition, ML, or AI. That being said, this helpful technology is not without risks. As with any

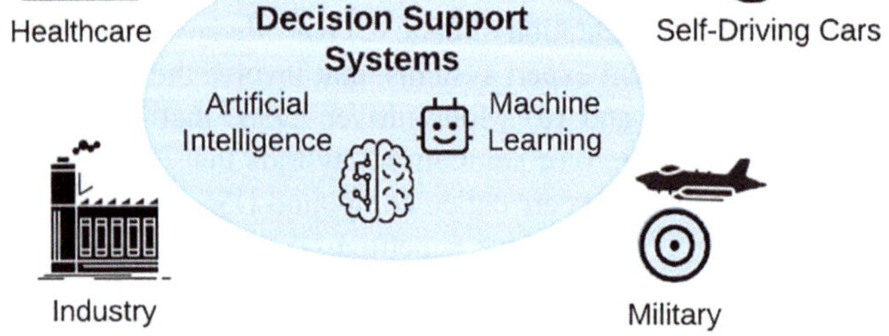

Fig. 2 Applications of Decision Support Systems

medical system, it is important to protect the privacy of the patients' sensitive data that is stored in hardware, complying with national data security standards including the Health Insurance Portability and Accountability Act (HIPAA), the Personal Information Protection and Electronic Documents Act (PIPEDA), or the Data Protection Directive and General Data Protection Regulation (GDPR). Throughout patient care, the availability of CDSS is important in order to provide with prompt decisions. It is necessary to employ hardware security measures to hinder disruptions brought on by attacks or hardware failures [10].

Decision-making is crucial in military tactical operations, by gathering and analyzing information, assessing situations, considering actions, and making operational plans to achieve mission objectives [4]. The importance of secure, improved access and use of information and data in a military context becomes even higher in a technology-driven battlefield [4, 11]. As AI develops at a rapid pace, great potential is presented for military decision support. AI systems are becoming more and more adept at cognitive activities including scenario analysis, action planning, and human collaboration. More realism in the physical and behavioral aspects of the battlefield is achieved by advancements in simulation. Moreover, there is an increasing integration of the technologies, as demonstrated by the use of generative modeling and reinforcement learning. At the same time, concerns and challenges arise over the security of DSS knowledge and large databases [5]. The leakage of critical information can have catastrophic consequences, at the expense of human lives.

Another critical sector where AI and DSS seem to play a pivotal role is self-driving cars, or in other words, autonomous vehicles. Enormous portions of data must be analyzed and classified so that informed decisions are taken regarding the movement of the car. As mentioned in [12], DSS are being implemented in numerous components that comprise the self-driving car such as emergency breaking, course correction system, and path planning, among others. Along with other difficulties, automated vehicles face the challenge of data privacy and security, as they are easily hijacked. For example, in order to achieve better performance, hardware machine learning accelerators are often implemented. In this particular scenario, DSS are susceptible to HT insertions.

As can be concluded from the examples above, DSS and hardware security are closely related in a number of ways, including the availability, confidentiality, and integrity of the data in decision-making processes. Firstly, the reliability of DDS's outputs depends on timely and correct data. The data integrity may be jeopardized if the hardware used to process or store the data is compromised. Hardware attacks, such as side-channel attacks, might result in data leakage or corruption, for instance, which can seriously impair decision-making. Confidentiality is also essential, particularly for DSS that handle sensitive data. Unauthorized hardware access can result in data breaches that expose private information. Additionally, DSS needs to be available when needed to facilitate real-time decision-making. A DSS may become unavailable as a result of hardware issues, caused by malicious assaults such as Denial of Service (DoS). This could result in disastrous delays in decision-making. Hardware reliability is another crucial aspect of uninterrupted DSS operation. Hardware security countermeasures thwart attacks that may compromise the reliability of DSS operations. Finally, protecting the hardware components of DSS requires physical security measures such as data center security, tamper-evident and tamper-resistant hardware, and access restrictions.As DSS becomes increasingly important to organizational operations across multiple verticals, the security of the hardware they depend on cannot be disregarded. Physical and remote attacks can be avoided by deploying hardware security countermeasures, thus ensuring the reliability and availability of DSS. Any organization that depends on such systems must take hardware infrastructure security into consideration.

3 Side Channel Analysis (SCA)

3.1 Definition and Prominent Leakages

As stated previously, DSS operate by analyzing and processing immense data portions that often are of utmost significance. In addition, as it is universally acknowledged, the main defense mechanism to secure and conceal crucial information, operational characteristics of a system, or even the exported decisions, is the implementation of cryptographic algorithms. Among others, the primary, and perhaps the most significant objective of Side Channel Analysis and therefore Side Channel Attacks, is to exploit the deficiencies of hardware to extract the encryption key (Fig. 3).

The modus operandi of SCA is to measure, analyze, and sometimes filter undeliberate leakages that dispose details about the executed system operations, by utilizing properly suited equipment according to the leakage. These leakages are of a diverse nature and contain information about the complexity of the data and the computational mechanisms. To be more specific:

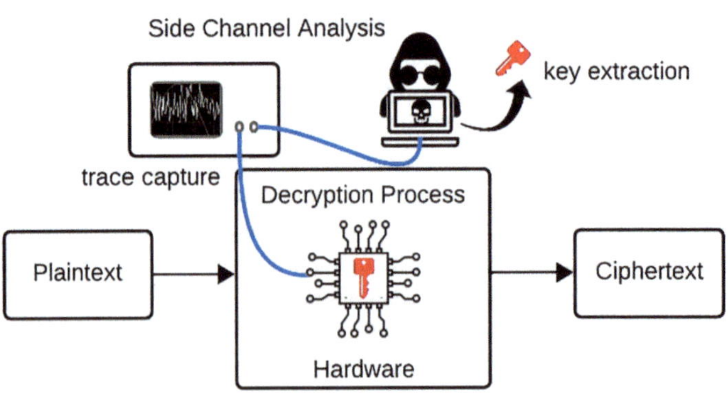

Fig. 3 Overview of a side-channel attack

- Asonov and Agrawal [13], by creating a neural network trained with noise outputs produced through keyboard typing, namely Fast Fourier Transform (FFT) values of touch peaks, were able to recognize which key was being pressed just by its noise.
- As delineated in [14], all electronic devices are leaking Electromagnetic Emissions (EM), due to the varying current. In hardware components like CPU and RAM with crucial information like executed commands, data manipulation, and data encryption, among others, it is possible to capture these EMs, plot them, and analyze signal patterns that betray exploitable insights and operations.
- Kocher in [15] illustrated that by collecting power traces when implementing a cryptographic algorithm, you are able to extract information about the applied cryptography and its characteristics (e.g., encryption rounds) as well as the encryption key. According to the executed operation and the data included among others, different electricity levels are required. So, by measuring with the proper equipment (such as an oscilloscope) these power traces and analyzing patterns in the produced waveform, access to the aforementioned data can be gained.
- Kocher in [16] proved that it is possible to calculate the secret key generated by cryptographic algorithms such as Diffie-Helman and RSA, by predicting each bit of the private key according to its computation time. The aim of this kind of attacks is to capture the execution time of each loop of the cryptographic algorithm and then define the type of mathematical operations that occurred during that cycle. To elaborate further, according to the value of each bit k (0 or 1) that pertains to the secret key x, different operations are performed, and therefore distinctive execution times are recorded.

These are among the most renowned leakages that compose the foundations of SCAs. One of the main complications of these attacks is that they can only be executed under very restricted circumstances. Different execution conditions such as data availability and quality can vastly influence and alternate the effectiveness of the attack. To address this obstacle, we will examine the benefits of combining SCA with template attacks and ML algorithms.

3.2 *Profiling Attacks*

The necessity to incorporate template attacks and ML originated from one crucial problem that can be observed in most of the aforementioned SCA examples. This problem pertains to the fact that it is difficult to implement SCAs in real-life scenarios for several reasons. Firstly, it is very difficult to extract enough traces so that you are able to find existing patterns, commonly due to the limited accessibility on the target device. Secondly, environmental conditions infuse noise in the extracted traces, rendering it impossible to analyze the operation signal and extract useful information. Lastly, huge data portions are being lost because there is not a way to capture data correlations that can be used to achieve more accurate

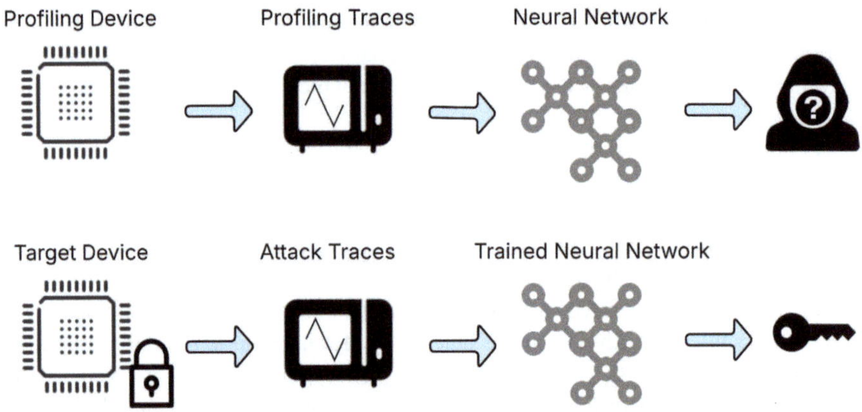

Fig. 4 Profiling attack scheme

results. As mentioned in [17], for the purpose of creating a template model, it is necessary to possess an equivalent device. By performing SCA in a more controlled environment, the attacker can build a template with a multitude of extracted traces, exploiting signature patterns of specific operations (Fig. 4). It is imperative to emphasize that these traces also contain noise, making the template immune to noise interference. Consequently, adversaries can perform more accurate and effective SCA in realistic scenarios by mapping the captured traces from the target device that comprise of the operation signal and noise, with the templates. In Hardware security, these types of SCA are named Profiling Attacks [18].

3.3 *Elliptic-Curve Cryptography*

With the emerging need to create efficacious defense mechanisms, so that valuable and private data that are being used to support DSS are protected, Elliptic-Curve Cryptography (ECC) [19, 20] has started to gain more and more attention. In more details, in [21], ECC is proposed for better protection and data encryption over Electronic Warfare (EW) Systems in the military. In [22], Saini presents the use of ECC in EHR that are being used by medical DSS for monitoring the patient's health, diagnosing, and suggesting the appropriate therapy and medication. Thus, it becomes obvious why examining SCA in ECC bears great significance.

In [23], the authors are experimenting on a specific curve in ECC named Brainpool curve BP256r1 and are trying to retrieve the scalar k, by exploiting the power traces that the operation of scalar multiplication is leaking, when the double-and-add-always algorithm is being implemented. To elaborate further, the way double-and-add algorithm works is that given a point P on the aforementioned curve, and a scalar number k, the final coordinates are being computed. According to the value of each scalar bit (0 or 1), different arithmetic functions are implemented at point P [24]. So, depending on the scalar bit values, different power levels are emitted, due to the complexity of dissimilar mathematical operations. For more comprehensive analysis, they trained the model with three different classifiers. The used Naïve Bayes Classification, k-Nearest Neighbor and Support Vector Machine (SVM) classifier. For the training process, they used a device similar to the target device to obtain power traces for the different scalar bit values. The aim of this constructed model is to classify the result point P' after each round of the double-and-add algorithm, through the power traces acquired from the target device. By cross-correlating the traces used to build the model and the traces acquired during the attack, you can identify the mathematical operations that were performed and therefore the value of the scalar bit used in the specific round. With the same process, an ML model can be built that will be able to find all the scalar bits and therefore the secret key.

4 Hardware Trojans

4.1 Introduction to Hardware Trojans

The increasing complexity of chip production, the excessive costs of fabricating a chip from scratch, and the time constraints imposed by the market have forced System-on-Chip (SoC) design companies to outsource the production of chip components to Third-Party IP Vendors (3PIP vendors). Therefore, manufacturing SoC integrated circuits is a multilayered process that requires trust from multiple parties, that are responsible for designing, fabricating, maintaining, and testing the SoC [25]. It is easily clear that this volume of trust is hardly achieved, since the ability to control the operability of SoCs, comes with great value.

A Hardware Trojan (HT) is a malicious alternation of the electronic circuit that creates a vulnerability, which serves as a backdoor and can be exploited by the person that inserted it, to perform a specific task. HTs are structured and categorized based on two components, the trigger, and the payload [26]. The trigger, as implied by its name, is the part of the circuit that given a specific input, enables the execution of the HT payload. Payload is the imposed operation that occurs due to the circuit modification.

4.2 Self-Driving Cars

AI and DSS are crucial in the development of self-driving cars, which require extensive data analysis and classification for informed decisions. DSS is used in various components of self-driving cars, including emergency breaking, course cor-

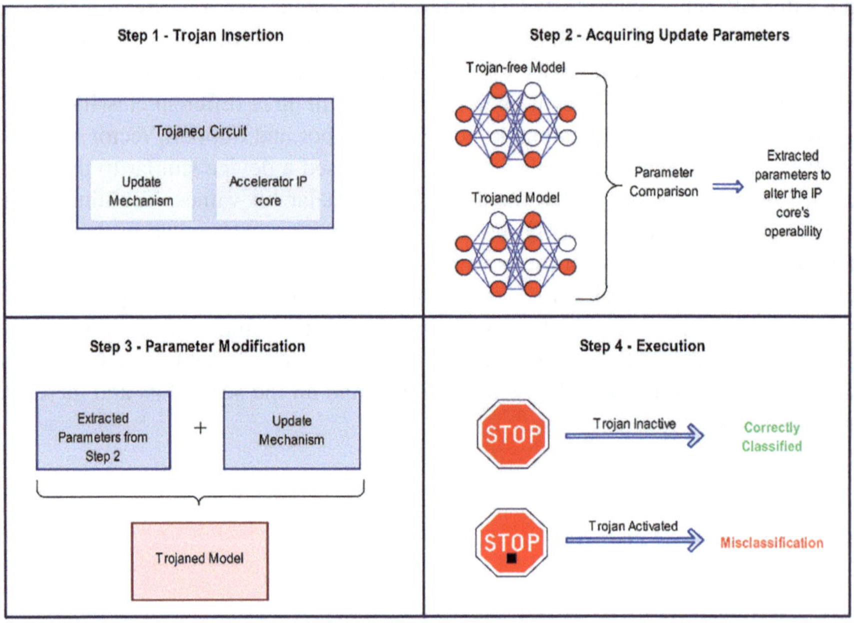

Fig. 5 The four steps of Hardware Trojan insertion and execution

rection systems, and path planning. However, the use of hardware machine learning accelerators for better performance makes DSS susceptible to HT insertions.

To be more specific, the authors of [27], examined how an HT can be inserted into a hardware accelerator, that is part of a DSS whose role is to detect traffic signs. The attack is presented in four steps (Fig. 5). The first step includes the hardware insertion process, where the authors suggest that a 3PIP vendor oversees designing the Hardware Description Language (HDL) description of the accelerator IP core that is responsible for classifying the traffic signs. The 3PIP vendor which also operates as the adversary in this case, is maliciously modifying the design and creates an update mechanism that remains inactive until the update occurs. This update mechanism changes parameters of the machine learning model in order to wrongly classify inputs.

The second step is about finding the necessary parameters that need to be changed to update the ML and perform the attack successfully. To accomplish that, the authors suggest that an adversary must be included in the training process of the accelerator. The adversary's goal is to create a copy of the initial model which they will modify so that it performs a specific task when it triggers (e.g., misclassifying the stop sign). By comparing the parameters being used in the initial model with the ones in the modified model, the adversary knows what needs to be alternated with the aim of activating the trojan (trigger part).

The third step of the attack consists of modifying the parameters with a specific process called quantization. The purpose of this step is to enhance the parameters with the proper format that will render them capable of deployment in the accelerator. It is self-explanatory that to deploy the parameters in the accelerator, the update mechanism from the initial step is being used.

The fourth and final step is of course the execution of the attack. In each case, the trigger of the HT is different. In this case, the authors modified the parameters of the accelerator, so that the trigger would be a square sticker. To elaborate, if the DSS read a sign without the square sticker on, then the trojan remains inactive and the classification is conducted in the usual manner. On the other hand, if the square sticker is recognized in a sign, then the trojan activates and misclassifies the sign according to the payload of the trojan. The parameters that we alternate through the update mechanism are responsible for inserting artificial inputs in our model, which are created with regular inputs combined with the trigger and are responsible for the misclassification. Stochastic Gradient Descent updates are used to update the model.

4.3 Hardware Trojans Detection

Having thoroughly examined both Side Channel Analysis and HTs, it is imperative to also evaluate how the first one can be used as foundation for constructing a DSS capable of detecting the second one. Hoang in [28], created a hardware detection system, by capturing the power traces using SCA. However, it is important to clarify that the research has been conducted only for an HT insertion in a cryptographic algorithm and more specifically on 128-bit AES.

Firstly, to create this defense mechanism based on ML, the author had to extract power traces that would be used as training set for the model. After performing Side Channel Analysis, he realized that the AES core with the HT consumes approximately 1% more electricity than the core without the trojan. Then, for the testing set, he captured power traces for four different encryption keys, with the fourth one being the key that is being used in the circuit that includes the HT. By classifying the traces using the Softmax Regression algorithm he was able to achieve very high detection accuracy.

5 Physical Unclonable Function

5.1 Definition

To ensure that a DSS runs unobstructed, it is imperative to take the necessary steps to achieve competent levels of protection. Outside the scope of common cryptographic algorithms that function with the help of cryptographic keys, Physical Unclonable Functions (PUFs) are considered an alternative that operates and protects by exploiting the uniqueness of hardware components. To be more specific, due to our inability to manufacture Integrated Circuits (ICs) with the exact same characteristics, PUFs exploit this property to create unique identifiers. These identifiers are generated through a procedure called Challenge-Response Pairs (CRP) [29]. During this procedure, inputs are given to the PUF (challenge), and due to the singular hardware design, unique outputs are given (responses). Each CRP is distinct, cannot be replicated and is only used once.

5.2 Modeling Attacks

The aim of performing modeling attacks on PUFs is to attempt diminish their innate characteristics, by finding patterns that betray the mechanism that CPRs are produced. By implementing ML algorithms, it is proven that it is possible to predict CPRs. To be more specific, as mentioned in [30], the authors, by applying various ML algorithms (e.g., Logistic Regression (LR), Support Vector Machine (SVM), and Evolution Strategies (EV)) as well as diverse optimization techniques (e.g., Resilient Backpropagation, standard gradient descent, and iterative reweighted least squares), they were able to perform modeling attacks and compare the extracted results. They examined these models on numerous Strong PUFs (e.g., Arbiter PUFs, XOR Arbiter PUFs, Lightweight Arbiter PUFs, and Ring Oscillator PUFs, among others), however we will shift our focus only on Arbiter PUFs and Ring Oscillator PUFs.

With the goal of creating the aforementioned CRP mechanism, Arbiter PUFs (APUF) use an electrical signal that is sent into two different delay paths. These delay paths are constructed with the help of multiplexers that are enabled or disabled by external input. The number of CRPs is determined by the external input number of bits, which also appoints the number of multiplexer stages that comprise the delay paths. For different input bit sequences, we enable different multiplexers in each stage, and produce a unique 1-bit output by comparing which delay path produced a faster response [30]. The modeling attack on APUF was conducted with Logistic Regression that uses sigmoid logistic function as hyperplane to classify the data. For better optimization, Resilient Backpropagation gradient descent (RProp gradient descent) was used, creating a neural network that uses gradient to determine the updated weights. This method gave the best results and accurately predicted 95% of the responses, while with bigger training sets (18.050 CPRs), they reached 99.9% prediction rates.

Ring Oscillator PUFs (ROPUF), as the name proposes, utilize Ring Oscillators to create the CRPs. To be more specific, a ROPUF circuit is comprised of several identical Ring Oscillators. Ring Oscillators are just a cascaded combination of delay stages, connected in a close loop chain, that forward an input signal [31]. Even though similar ring oscillators are being used, due to manufacturing imperfections there cannot be two of them with the same frequency. By capitalizing on this attribute, given a predetermined series of ring oscillator pairs, a unique output bit sequence is produced by comparing the frequencies of each pair [32]. It is self-explanatory that the number of CRPs is specified by the number of ring oscillators which is very limiting. Authors in [30] exploited this vulnerability and by just implementing a brute force attack and a sorting algorithm (Quick Sort) they were able to create a model that predicts with 99.9% accuracy all the CRPs.

5.3 Amplifying the Effectiveness of Physical Unclonable Functions

As it was anticipated, more complicated, effective, and sophisticated model attacks led to more intricate defense mechanisms. The containment of model attacks is garnered from targeting the ML algorithm directly so that it is being used to predict the CRPs, or by reinforcing PUFs' sturdiness.

Attacking the ML algorithm that is used to predict the CRPs, means finding a way to agitate the model with unpredictable data patterns. To be more specific, the aim of this method is to poison the data and consequently increase the set of response values, enhancing the complexity for the ML model to discern patterns. As examined in [33], this can be achieved by implementing a mechanism in PUFs that alternate the responses for a portion of challenges with the inverse response. To achieve that, a diminutive fragment of bits within the challenges is being used to form a function that serves as a trigger mechanism. When the trigger is disabled, challenges produce regular responses. However, when the trigger is enabled, challenges produce the inversed responses that obtain the function of poisoned data. It bears significant weight mentioning that by increasing the proportion of inversed responses to reach 50% of the overall CRPs, they were able to increase the model's predictability error to almost 50% due to response fussiness.

As a second defense mechanism, the technique demonstrated in [34] exhibits a distinct interest. To elaborate further, the authors proposed an external function named recombination function that was used in ROPUF to generate unbiased responses. In order to comprehend this, we need to address the main issue of ROPUFs that render them vulnerable to model attacks in the first place. In a ROPUF, we mentioned that by comparing the frequencies of specific pairs, we produce the response. To be more specific, in a pair with ring oscillators and frequencies f_i and f_{i+1}, if $f_i > f_{i+1}$, then the response is 1, otherwise it is 0. It has been observed that this mechanism doesn't produce neutral responses, but favors one of the responses, making ROPUFs vulnerable to pattern recognitions. So, the main goal is to implement an external function that by taking as input biased responses, it modifies them and produce unbiased responses (see Fig. 6). To accomplish that, for every pair with frequencies f_{2i} and f_{2i+1}, they utilize the $S_i = (c_i * 2 - 1) \times (f_{2i} - f_{2i+1})$ function, where c_i is the result of a seed challenge and S_i is the first part of S which is the sum of all S_i and the final response. The c_i component is the one that transforms responses from biased to unbiased. Consequently, response

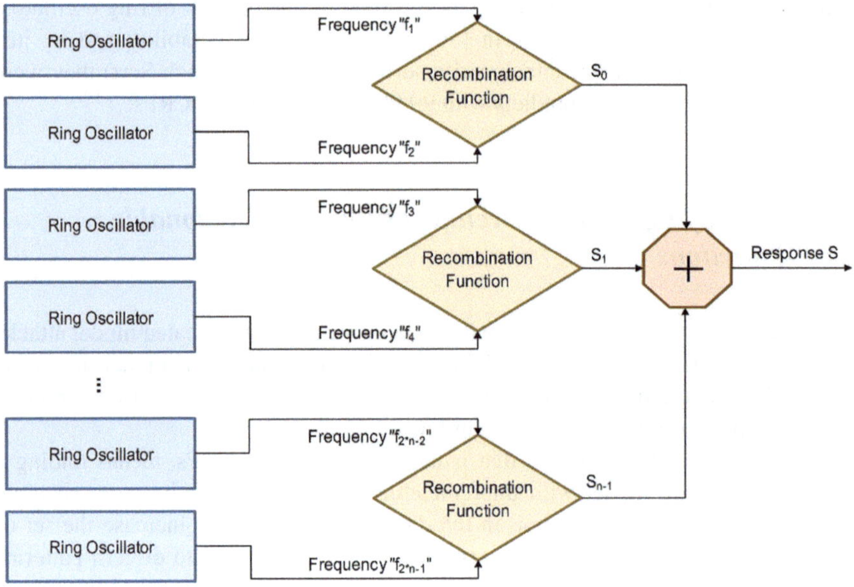

Fig. 6 Creation of unbiased response S, using the recombination function

neutrality achieves higher model attack resistance, which is also showcased by the experiments run by the authors.

6 Conclusions

As the title suggests, this chapter sheds light on how Decision Support Systems are entangled in hardware security and vice versa. We have presented and highlighted not only different scenarios where attacks through hardware occurred in DSS, but also how DSSs can become the means to retaliating hardware attacks and operate as a defense mechanism. As DSS are quickly rising as the most advanced systems in every single industry field of this generation, it is of great importance to conduct a thorough and in-depth analysis. This work constitutes a modest contribution towards this direction, with condensed knowledge on different aspects of this field. Our goal was to cover numerous distinct hardware attacks against DSS, as well as how they can be leveraged to develop defense mechanisms and enhance those that already exist. From our work, multiple cognitive and educational dimensions emerge, including research on more advanced hardware attacks against DSS as well as superior prevention and counterattack mechanisms to fortify DSS security. Lastly, it is of paramount importance to expand upon DSS being applied in various cybersecurity and hardware security scenarios.

References

1. Gorry, G., & Scott-Morton, M. (1971). A framework for management information systems. *Sloan Management Review, 13*, 55–70.
2. Arnott, D., & Pervan, G. (2014). A critical analysis of decision support systems research revisited: the rise of design science. *Journal of Information Technology, 29*. https://doi.org/10.1057/jit.2014.16
3. Sutton, R. T., et al. (2020). An overview of clinical decision support systems: benefits, risks, and strategies for success. *npj Digital Medicine, 3*, 17.
4. van Oijen, J., & de Marez Oyens, P. (2023). Empowering military decision support through the synergy of AI and simulation. *In Proceedings of the 2023 NATO NMSG Symposium: Simulation: Going Beyond the Limitations of the Real World*, Monterey, USA, 2023.
5. Susnea, E. (2012). Decision support systems in military actions: necessity, possibilities and constraints. *Journal of Defense Resources Management, 3*, 131–140.
6. Prinetto, P., & Roascio, G. (2020). Hardware security, vulnerabilities, and attacks: A comprehensive taxonomy. In *2020 Italian conference on cybersecurity*.
7. Akter, S., Khalil, K., & Bayoumi, M. (2023). A survey on hardware security: Current trends and challenges. *IEEE Access, 11*, 77543–77565. https://doi.org/10.1109/ACCESS.2023.3288696
8. Konstantopoulou, E., Athanasiou, G. S., & Sklavos, N. (2023). Securing 5G/6G communications in smart cities: Novel SNOW-V/ZUC-256 multimode architectures. In *Proceedings of the 2023 international conference on Security and Management (SAM'23)*, Las Vegas, USA, 2023.
9. Turban, E., Sharda, R., Delen, D., Aronson, J. E., Liang, T. P., & King, D. (2010). *Decision support and business intelligence systems* (9th ed.). Pearson College Div.
10. Konstantopoulou, E., & Sklavos, N. (2022). Design and implementation of a lightweight cryptographic module, for wireless 5G communications and beyond. In *Proceedings of IEEE 2022 international conference on Internet of Things and Intelligence Systems (IOTAIS'22)*, Bali, November, 2022, (pp. 24–26). https://doi.org/10.1109/IoTaIS56727.2022.9975849
11. Konstantopoulou, E., Sklavos, N., & Ognjanovic, I. (2023). Securing public safety mission critical 5G communications of smart cities. In N. Gupta & S. Mishra (Eds.), *Internet of everything for smart city and smart healthcare applications*. Springer. ISBN: 9783031346002.
12. Karnati, A., Mehta, D., & Ks, M. (2022). Artificial intelligence in self driving cars: applications. *Implications and Challenges, 21*, 1–28. https://doi.org/10.12725/ujbm.61.1
13. Asonov, D., & Agrawal, R. (2004). Keyboard acoustic emanations. In *IEEE symposium on security and privacy, 2004. Proceedings* (pp. 3–11). https://doi.org/10.1109/SECPRI.2004.1301311
14. Sayakkara, A., Le-Khac, N., & Scanlon, M. (2019). A survey of electromagnetic side channel attacks and discussion on their case-progressing potential for digital forensics. *Digital Investigation, 29*, 43–54. https://doi.org/10.1016/j.diin.2019.03.002
15. Kocher, P., Jaffe, J., & Jun, B. (1999). Differential power analysis. In M. Wiener (Ed.), *Advances in cryptology — CRYPTO' 99. CRYPTO 1999* (Lecture notes in computer science) (Vol. 1666). Springer. https://doi.org/10.1007/3-540-48405-1_25
16. Kocher, P. C. (1996). Timing attacks on implementations of diffie-hellman, RSA, DSS, and other systems. In N. Koblitz (Ed.), *Advances in cryptology — CRYPTO '96. CRYPTO 1996* (Lecture notes in computer science) (Vol. 1109). Springer. https://doi.org/10.1007/3-540-68697-5_9
17. Chari, S., Rao, J. R., & Rohatgi, P. (2003). Template attacks. In B. S. Kaliski, ç. K. Koç, & C. Paar (Eds.), *cryptographic hardware and embedded systems – CHES 2002. CHES 2002* (Lecture notes in computer science) (Vol. 2523). Springer. https://doi.org/10.1007/3-540-36400-5_3
18. Picek, S., et al. (2019). The curse of class imbalance and conflicting metrics with machine learning for side-channel evaluations. In *IACR transactions on cryptographic hardware and embedded systems* (pp. 209–237). https://doi.org/10.13154/tches.v2019.i1.209-237

19. Koblitz, N. (1987). Elliptic curve cryptosystems. *Mathematics of Computation, 48*(177), 203–209.
20. Bos, J., Halderman, J., Heninger, N., Moore, J., Naehrig, M., & Wustrow, E. (2014). *Elliptic Curve Cryptography in Practice*, 8437. https://doi.org/10.1007/978-3-662-45472-5_11
21. Sahoo, P. K. (2022). Elliptic curve cryptography for defense use. Elliptic curve cryptography for defense use (June 2, 2022).
22. Saini, A., Zhu, Q., Singh, N., Xiang, Y., Gao, L., & Zhang, Y. (2021). A smart contract-based access control framework for cloud smart healthcare system. *IEEE Internet of Things Journal, 8*(7), 5914–5925. https://doi.org/10.1109/JIOT.2020.3032997
23. Özgen, E., Papachristodoulou, L., & Batina, L. (2016). Template attacks using classification algorithms. In *IEEE international symposium on Hardware Oriented Security and Trust (HOST)*, McLean, VA, USA, 2016 (pp. 242–247). https://doi.org/10.1109/HST.2016.7495589
24. Coron, J. S. (1999). Resistance against differential power analysis for elliptic curve cryptosystems. In Ç. K. Koç & C. Paar (Eds.), *Cryptographic hardware and embedded systems. CHES 1999* (Lecture Notes in Computer Science) (Vol. 1717). Springer. https://doi.org/10.1007/3-540-48059-5_25
25. Bhunia, S., & Tehranipoor, M. (2019). Hardware security. In S. Bhunia & M. Tehranipoor (Eds.). Elsevier. https://doi.org/10.1016/C2016-0-03251-5
26. Chakraborty, R. S., Narasimhan, S., & Bhunia, S. (2009). Hardware trojan: Threats and emerging solutions. In *IEEE International High Level Design Validation and Test Workshop*, San Francisco, CA, USA, (pp. 166–171). https://doi.org/10.1109/HLDVT.2009.5340158.
27. Warnecke, A., Speith, J., Möller, J. N. Rieck, K. & Paar, C. (2023). Evil from within: Machine learning backdoors through hardware trojans.
28. Hoang, V. P. (2022). *Hardware trojan detection based on side-channel analysis using power traces and machine learning* (pp. 53–56). https://doi.org/10.15439/2021R26
29. Gao, Y., Al-Sarawi, S. F., & Abbott, D. (2020). Physical unclonable functions. *Nature Electronics, 3*, 81–91. https://doi.org/10.1038/s41928-020-0372-5
30. Rührmair, U., Sehnke, F., Sölter, J., Dror, G., Devadas, S., & Schmidhuber, J. (2010). Modeling attacks on physical unclonable functions. In *Proceedings of the 17th ACM conference on Computer and communications security (CCS '10)* (pp. 237–249). Association for Computing Machinery, New York. https://doi.org/10.1145/1866307.1866335
31. Mandal, M., & Sarkar, B. C. (2010). Ring oscillators: Characteristics and applications. *Indian Journal of Pure and Applied Physics, 48*, 136–145.
32. Suh, G. E., & Devadas, S. (2007). Physical unclonable functions for device authentication and secret key generation. In *44th ACM/IEEE design automation conference*, San Diego, CA, USA (pp. 9–14).
33. Wang, S. J., Chen, Y. S., & Li, K. S. M. (2019). Adversarial attack against modeling attack on PUFs. In *56th ACM/IEEE Design Automation Conference (DAC)* (pp. 1–6).
34. Devadas, S., & Yu, M. D. (Mandel). (2010). Recombination of physical unclonable functions. In *GOMACTech conference*.

Cybersecurity of Decision Support Systems in the Quantum Era

Virginia Lampropoulou, Kyriaki Tsantikidou, Nicolas Sklavos, and Francesco Leporati

1 Introduction

Industry 4.0 and Health 4.0 frameworks are characterized by the emergence of new technologies, which are aimed at data digitalization, industry control systems, and healthcare services, thus creating highly automated systems for various processes [1]. This results in a more interactive environment that responds to the received data, thus increasing productivity by limiting human input. The principal technologies, which are commonly utilized, are those associated with handling vast amounts of data and being able to communicate with multiple heterogeneous devices simultaneously [2]. Some exemplary technologies are Big Data and Analytics, Artificial Intelligence, Machine Learning, Internet of Things (IoT), Cyber-Physical Systems, Robotics, etc.

Another technology that is commonly employed is the Decision Support System (DSS). A DSS is a data processing system that calculates its executed decisions based on the received data, which is collected from its environment [3]. DSS was created to assist people working in decision-making with the purpose of minimizing human errors. Automating decision-making requires the provision of perfectly processed information together with an optimal model. These allow the DSS to model human reasoning and the decision-making processing. Specifically, they can

V. Lampropoulou (✉) · K. Tsantikidou · N. Sklavos
SCYTALE Group, Computer Engineering and Informatics Department, University of Patras, Patras, Greece
e-mail: up1090030@ac.upatras.gr; k.tsantikidou@upatras.gr; nsklavos@upatras.gr

F. Leporati
University of Pavia, Pavia, Italy
e-mail: francesco.leporati@unipv.it

N. Gupta, J. D. Diaz Roman (eds.), *Decision Support Systems for Industrial and Scientific Applications*, Signals and Communication Technology,
https://doi.org/10.1007/978-3-031-89700-9_13

accept and process facts from users and suggest solutions that are close to human logic.

Nevertheless, the implementation of DSS-based technological solutions in Industry 4.0 and Health 4.0 presents various challenges. As analyzed in [4], the challenges are divided into the following main categories:

1. Technology Challenges: These are related to operational and management issues presented in the integration of novel technologies.
2. Data Challenges: These are related to the massiveness of the collected data and its unstructured format, which is difficult to completely process with the existing tools.
3. Human Resource Challenges: The human factor greatly impacts the implementation of DSS, as the whole purpose of these systems is the assistance of workers.
4. Security Challenges: These systems manage a vast amount of data, containing private information. Moreover, they allow the interconnection of heterogeneous devices that collect and transmit all this information. Therefore, the insurance of data integrity, confidentiality, and system availability is necessary.
5. Standardization Challenges: The constant development of novel technologies, such as quantum computers, deems challenging the establishment of standards that can perform efficiently for all industries or systems.
6. Communication Challenges: As mentioned, DSSs employ sensing devices that communicate via a wide network. Therefore, availability issues due to connectivity problems can be created.
7. Management Challenges: These challenges refer to financial, manufacturing system, strategy, and environmental aspects of managing an industry or healthcare infrastructure.

All these challenges affect the overall functionality, implementation, security, and safety of the employed system. Specifically, the European Union has greatly stressed the importance of the proper and constant operation of industry and health structures, which are considered Critical Infrastructures [5]. The EU has also highlighted the importance of their cybersecurity and safety as possible disruptions can have great impact on the economy, public health, and national security of each country. Therefore, more research in this field with the purpose of spreading awareness regarding security standards and cybersecurity issues is demanded.

Furthermore, the rapid evolvement of the research in Quantum Computers has started to present additional challenges to the cybersecurity aspect of industry and healthcare structures [6]. Many conventional cryptographic standards will be deemed useless, as their computational complexity will be degraded. Thus, further research regarding quantum technology as well as new suggested security standards must be explored.

In this book chapter, the DSS technology that is integrated into the Industry 4.0 and Health 4.0 environment and mainly their cybersecurity issues will be explored. First, its cases of employment and framework in Industry 4.0 and Health 4.0 will be presented. Afterward, the appropriate cryptographic algorithms, which are employed in common standards, will be highlighted, with the purpose of relaying the current status of security. Furthermore, the quantum era will be discussed, together with its advantages and disadvantages in DSSs. This extended analysis will provide information regarding the impact of quantum computers, together with possible solutions and methods of confronting new sophisticated attacks. Overall, this chapter aims at providing the necessary guidance and knowledge to researchers and designers of 4.0 frameworks-based DSSs in order to better secure and implement their architectures.

2 Decision Support Systems (DSSs)

In this section, the definition of Decision Support Systems (DSSs) in an Industry 4.0 and Health 4.0 environment will be briefly analyzed. Moreover, the most employed framework for the implementation of these systems will be presented. Finally, two cases of employment for the DSSs will be demonstrated.

2.1 DSS in Industry 4.0 and Health 4.0 Frameworks

The purpose of Industry 4.0 framework is the implementation of the following design principles: Interoperability, namely the collection and utilization of multi-source data, Real-time Capability, namely the real-time processing of freshly collected data, Cybersecurity, namely the insurance of security and authentication, Scalability, namely the ability to constantly evolve depending on the requirements of novel technologies, and finally Decentralization, namely the ability to create autonomous nodes that make their own decisions [7]. The integration of novel technologies into traditional infrastructures can lead to the creation of systems that fulfill the Industry 4.0 design principles [8]. The wearability of IoT systems allows to carry out DSS also in medical applications and also sports security. Specifically, IoT enables the utilization and intercommunication of resource-constrained devices, such as sensors and actuators. These sensors are also able to communicate with gateway devices that have more computational resources. They can communicate further with cloud services and Human-Computer Interfaces (HCI), as it is also defined in the Supervisory Control and Data Acquisition (SCADA) technology [9]. Operators, and in turn decision-makers, can interact remotely with the whole system via HCIs or have access to the vast amount of data via cloud services.

This framework facilitates the implementation of DSS models, as also depicted in Fig. 1, [10]. The Sensor Layer, namely the variety of connected heterogeneous IoT devices, can efficiently sense their environment and recognize changes in the normal operation flow of the system. This data is transmitted and processed by gateway devices, edge and fog services or a master terminal unit (MTU). Gateway devices and edge services are the ones that process the data first, executing procedures like light pruning. Fog services or MTU are mainly a single processing device

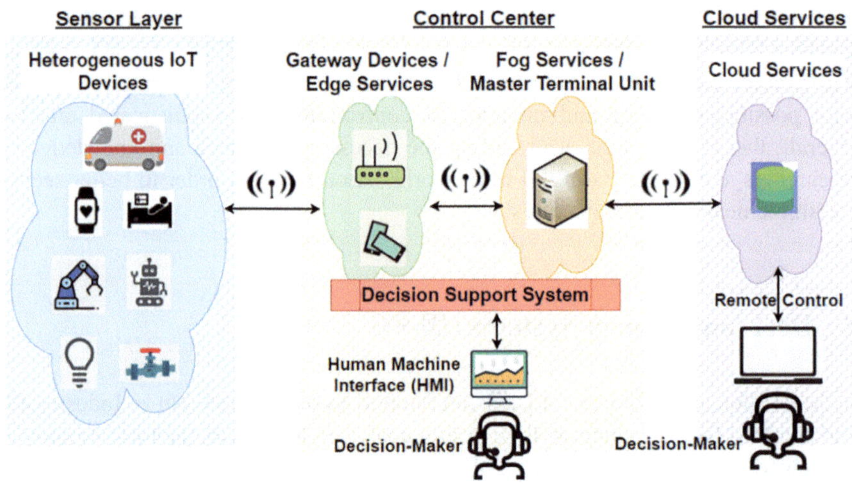

Fig. 1 DSS-based architecture in Industry 4.0 and Health 4.0 frameworks

that collects all data and executes heavy data processing algorithms. All these subsystems form the Control Center layer. DSS models are implemented at this center, either at the gateway or MUT, where there is easy access to all data together with enough resources for high performance. Afterward, decision-makers are assisted by these models through the HCIs' graphical representation of the collected information. Lastly, data are transmitted and stored at the last layer, the Cloud Services, whose highly processing capabilities also assist decision-makers remotely monitor the system.

This type of system and framework is also described from the Health 4.0 standard. Health 4.0 facilitates the integration of these technologies and adoption of the mentioned design principles [11]. Once again, the decision-makers, in this case health providers, are expected to quickly respond to emergencies, consider all alternatives and make the best decision for the well-being of multiple patients. Therefore, an architecture that at its center implements a DSS is required. This DSS is constantly being given real-time metrics of the patients and their environment via the Sensor Layer, and health history and records from the Cloud Services. The Sensor Layer's devices are extremely small and computational limited devices, because they must be wearable and unnoticeable to the patients. Moreover, the patients' constant mobility affects the connectivity of the devices to the rest of the system, possibly affecting decision-making. Therefore, in specific cases, a less resource demanding DSS can be implemented and accelerated in smaller hardware devices and then included closer to the sensor layer and thus the patient. This DSS can then either automatically make decisions based on data or instantly connect with the decision-maker by bypassing the more complicated connection routes in case of emergencies.

These systems are integrated into many infrastructures of everyday life. First, the case of a real-time DSS, which utilizes machine learning techniques and is integrated into intensive care units, is presented in [12]. Its purpose is to predict each patient's mean arterial pressure (MAP) and thus assess their health status. The operating procedure of the DSS consists of the real-time collection of vital signs from the patient. Afterwards, the system is trained based on these data to be capable of predicting early health deterioration. It then performs feature extraction and selection, with the final MAP value being produced. Such an application is considered very useful as depending on the prediction of MAP, immediate medical intervention can be made, and hypertension or hypotension phenomena can be treated.

The second case is the employment of a DSS model for disease prediction in farm animals via IoT sensors [13]. The data is collected with wearable sensors and then transmitted to a single source that utilizes a Support Vector Machine (SVM). This SVM, which is a machine learning algorithm, predicts the development of a single type of disease in animals, informing the decision-makers regarding milk production.

Finally, there are also many cases of DSS integration into data mining systems. In the case of [14], a privacy-preserving clinical decision support system in a cloud computing environment is presented. As mentioned in [14], cloud environments are especially important for clinical decision support as they facilitate doctors with more accurate predictions and diagnosis of patients' diseases. Privacy protection during disease prediction is ensured through the Naïve Bayesian (NB) classifier which encrypts each patient's Personal Health Information (PHI) to outsource the prediction. The architecture of this system includes entities, which function and communicate with each other for the purpose of encryption and transmission of encrypted data. The aim of this process is to update the system with new data regarding the symptoms of diagnosed diseases or for the training of the NB classifier. All data collected from smart sensors are encrypted and stored in the Cloud Platform (CP).

2.2 *DSS for Cybersecurity*

DSS models can also be used in cybersecurity. These models are responsible for choosing the appropriate security mechanism that corresponds best to the characteristics of the executed attack. In case this is not sufficiently effective, then another response model is chosen based on a defense tree that represents the analysis of risks and alternative precautions. Threat decision-making models are relevant to the organization in question and the threats it may face. First, the criticality of each option is prioritized based on the severity of the threat and the level of its occurrence probability. Finally, the precautions as well as the costs for each threat are determined and the best ways to deal with each risk are highlighted.

An example of a DSS model being utilized for cybersecurity purposes is presented in [15]. The model facilitates decision-makers in finding appropriate security solutions that attain a desired level of security while considering a budget limit. First, the threats and threat groups are assessed and defined. The Analytical Hierarchical Process (AHP) method is employed, with the comparisons being performed via subjective judgments and the consistency being constantly evaluated. For the selection of the optimal enterprise information security presentative action, the Mixed Integer Programming (MIP) technique is employed. As output, the distribution of budget and risk is visualized through a tree-map display technique. The combination of these techniques results in a DSS model that suggests different solution sets depending on the presented situation instead of constantly recommending a repeated security architecture solution.

3 Appropriate Cryptographic Algorithms for DSS

The continuous employment of data collection, storage, and processing has significantly increased the incidence of malicious attacks. In the architecture depicted in Fig. 1, two main vulnerabilities can be detected. First, sensitive data, containing private information, are constantly being collected by sensors, which are placed in the close user's environment or even on their body. This data is also immediately transmitted to the above layers of the architecture without security mechanisms and encryption. This deems them susceptible to malicious attacks, such as eavesdropping, extraction, and modification of data [16]. The second vulnerability is the resource constraints of the employed devices. This means that the devices cannot always implement heavy security mechanisms, unless a balanced trade-off between security and architecture can be achieved [17].

Therefore, depending on the criticality of data and the characteristics of the system, the administrators implement the appropriate security mechanism, which is based on specific cryptographic algorithms. The main cryptographic algorithms that are utilized for security in DSSs are presented in this section:

1. *Block Ciphers: AES*

In the DSS and Industry 4.0 environment, the AES algorithm is the most utilized standard. This algorithm is symmetric, meaning that the same key is used to encrypt and decrypt the message. It has a variable key size, which means that there are various architectures of the AES algorithm that employ different-sized keys. This fact also affects the number of implemented blocks, which are the parts into which the message is divided and encrypted separately. Its security is based on the size of its key, as well as its mode of operation. Specifically, there are many modes of operation that add security and authentication to the original algorithm, such as Electronic Codebook Mode (ECB), Cipher Block Chaining Mode (CBC), Cipher Feedback Mode (CFB), Galois/Counter Mode (GCM), Counter with Cipher Block Chaining-Message Authentication Code (CCM), etc. [18–20]. Overall, these modes add complexity to the block cipher and extra steps for tag creation and verification of the encrypted data's integrity. According to the Transport Layer Security protocol (TLS) version 3, the AES algorithm is exclusively employed with the GCM or CCM mode of operations to be able to correspond to the security requirements of Industry 4.0 and Health 4.0 systems [21].

Nevertheless, there is a major drawback to this class. Symmetric encryption assumes that the encryption key is safely transmitted to all participant components and units that encrypt and decrypt data. Specifically, it requires an efficient and secure communication path to exchange secret keys on the public network. This cannot always be achieved, resulting in the utilization of more complex security mechanisms, namely the asymmetric encryption schemes.

2. *Asymmetric Encryption Schemes: RSA, ECC*

In asymmetric cryptography, key pairs are created, known as private key and public key. During the encryption of a message, the sender uses the recipients public key for the ciphertext's creation. The recipient then can utilize his private key to obtain access to the original message. The Rivest–Shamir–Adleman (RSA) algorithm is an asymmetric encryption scheme whose security is based on large prime numbers that affect the value of the created keys. Another asymmetric scheme is also the Elliptic Curve Cryptography (ECC), which is the strongest and most effective encryption method. The security of the ECC is based on the complexity of solving the elliptic curve discrete algorithm problem, which refers to the difficulty of estimating a user's private key knowing the public key.

However, these schemes have a distinct disadvantage. They require high computation resources and execution time to encrypt even small blocks of data [22]. Therefore, in resource-constrained environments, such as DSS that employs sensors and IoT devices, this can be problematic [23]. The main solution is the utilization of asymmetric encryption schemes as key encapsulation mechanisms, namely for the encryption of only the key that was first employed to encrypt the message via a block cipher [24]. Afterwards, the encrypted key and the ciphertext are transmitted to the recipient, who can decrypt the message with the block cipher by first decrypting the encrypted key with their private key and asymmetric scheme. Thus, the execution time is significantly reduced. This procedure is clearly depicted in Fig. 2, together with all transmitted data, that is encrypted via the proper cryptographic primitives, between a standard Client and Server. In the 4.0 frameworks-based DSS framework, a Client can be any component that collects data and transmits it after requested, such as the devices in the Sensor Layer. A Server can be any component that receives those data, such as Cloud Services.

3. *Digital Signatures*

An important role in ensuring the security of communication is also the existence of digital signatures, which ensure the identity of the interlocutors, thus protecting both sides of the communication from possible attacks [25]. Digital signatures are not reusable, so they are unique for each message and cannot be used in multiple messages even if they are for the same recipient. The reason is the dependence of

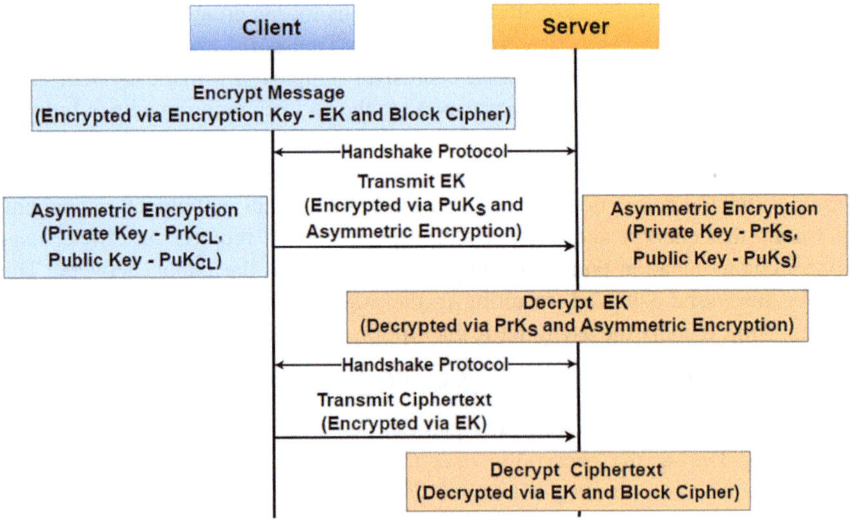

Fig. 2 Key encapsulation mechanism—framework between a client and a server

the signature's equation on the message being transmitted. Therefore, if a change is made to the content of the message by an adversary, after the signature has been defined, it can be detected, and the message is discarded.

4 Quantum Era and Cybersecurity of DSS

The utilization of the principles of quantum mechanics in computing systems seems to be the beginning of a promising period, especially in matters related to the execution speed of processes. In the following section, quantum computers are presented by highlighting the positive as well as the negative effects of their employment in the "digital society", namely the Industry 4.0 and Health 4.0 era.

4.1 What Are Quantum Computers?

Quantum computers ensure great computing power by using the principles of quantum mechanics, which allows them to solve complex mathematical problems in much less execution time compared to classical computers [26]. The polynomial complexity is highly reduced for the same problems. The reason is that quantum computers use qubits, which have at the same time '0' and '1' value, exploiting the principle of superposition and allowing parallel execution of calculations [27]. Due to this capability, an n-qubits register can represent all 2^n states simultaneously, while a standard n-bit system can represent only one 2^n state. A qubit state is described as $\varphi = a \mid 0\rangle + b \mid 1\rangle$ which are complex probability amplitudes of states $|0\rangle$ and $|1\rangle$ [28]. Speed in solving certain types of calculations is also ensured by the entanglement effect [27]. The state of a quantum particle can influence the state of another particle, facilitating the direct exchange of information over long distances. Lastly, their computing power is growing due to the rapid evolution of noise mitigation techniques, the ever-increasing number of employed qubits, and the improvement of error rates and decoherence times.

For physical representation of qubits, quantum computers utilize superconducting circuits or trapped by electrodes, ions, or even photons accelerated through free space, which replace the use of transistors [26]. Quantum gates are implemented via pulses rather than physical circuits as it occurs with classical logic gates. Gate pulses, depending on their amplitude, duration, and shape, cause distinct rotations along different axes in the Bloch sphere, in which qubits are visualized as a point that displays the phase and the probabilities of measuring each of the two states [29].

4.2 Advantages and Disadvantages of Quantum Computers in DSS

The employment of quantum computers in various technological aspects enables fast data processing. As a result, decision-making becomes more immediate, something that makes them more suitable for the achievement of real-time applications in the modern way of living and working. The main advantages and disadvantages of quantum computers to the aspects of DSSs are presented below.

The integration of quantum computing into technologies, such as machine learning, has led to the development of quantum machine learning [30]. First, through the computational power of quantum computers, faster model training can be achieved, allowing large amounts of data to be analyzed, thus drawing more accurate conclusions via big data analysis. Moreover, the benefits of using quantum computers in the field of machine learning can become apparent in the case of tasks that require high throughput and precision, such as the analysis of medical images. At the same time, it seems that with the use of quantum computers in machine learning algorithms, the classification of texts and images with noise is done in a more efficient way.

Furthermore, beyond processing acceleration, systems that rely on superconducting quantum computing have also other advantages [31]. Lower power consumption is required for their operation compared to conventional high processing computers, such as graphic processing units (GPUs). LP power GPUs have been produced by NVdia but the undoubtedly improvement in terms of power consumption does not make them yet suitable for all the applications. Lastly, quantum-based systems are easily scalable due to their simple requirements regarding connectivity with other nodes of the same environment.

Nevertheless, the higher calculation speed of quantum computers compared to classical computers can be also a great disadvantage. Specifically, they can quickly and easily crack cryptographic keys through exhaustive methods, such as brute force attacks, or through hard calculations, allowing an attacker to gain access to the communication channel between authorized parties [32]. Quantum-based attacks can break down algorithms based on factorization of large numbers, as in the case of RSA, but also algorithms based on discrete logarithm calculations, such as DSA. Although the latter is considered the most secure algorithm for classical computers, it appears to be more susceptible to attacks using quantum computers. The reason is the relatively small key space that it has, making exhaustive key searches easier. As far as block ciphers (AES) are concerned, there seems to be no danger of their collapse by the Grover algorithm, with the only condition, according to NIST, that the sizes of the utilized encryption keys are sufficiently large.

A last disadvantage, that must be mentioned, is the quantum systems' difficulty in manufacturing and operation [31]. Specifically, their integration into current systems and their placement into an appropriate environment with cooling mechanisms are challenging. Moreover, the assembly and commissioning by multiple vectors can present difficulties. Lastly, they require frequent maintenance with operators needing to obtain more experience and rigorous training for their proper functionality.

4.3 Quantum-Resistant Algorithms and Their Implementation in the Framework of DSS

Due to the great computing power possessed by quantum computers, it is considered necessary to apply cryptographic algorithms to protect data and users from malicious attacks. There are two types of cryptographic primitives that can be utilized to solve this problem. The first is Post-Quantum Cryptography, also called Quantum-Resistant Cryptography, which are employed to encrypt data and can resist attacks from both classical and quantum computers [33]. The second category, namely Quantum Cryptography or Quantum Key Distribution (QKD), is concerned with the establishment and safe distribution of encryption keys, similarly with asymmetric schemes [34].

The Post-Quantum Cryptography algorithms include mathematical structures based on different mathematical problems [33]. The first category is lattice-based cryptography, this is a very powerful form of cryptography for both classical and quantum computers. Their security is guaranteed by the mathematical problems of Shortest Vector Problem and Closed Vector Problem, which are Non-deterministic Polynomial-time (NP) hard problems. The second category, namely the code-based cryptography, is based on error-correcting codes, such as Classical McEliece Algorithm (CAM), the security of which is based on adding random errors to the message. The private key is a Goppa random binary code, which is used for error detection and correction. The public key is a matrix that is generated from a random permutation of that code. The security of this algorithm is ensured by the difficulty of decoding random barcodes, that have no efficient method of being solved neither by classical nor quantum computers.

QKD is based on the principles of quantum theory [34]. Its effectiveness is not based on mathematical computing power but on the use of photon technology through which possible interceptions can be detected via the error rate of the qubits in the quantum channel. The BB84 protocol and its variants are the only known provably secure QKD protocols. Based on this, the sender sends a series of randomly encoded photons through the quantum channel, maintaining the state of each photon it transmits on a temporary basis. The polarization states each photon possesses, namely horizontal, vertical, diagonal, and antidiagonal, represent the classical bits '0' and '1'. The receiver, utilizing only the orthogonal or diagonal base, measures the state of the received photons and responds by sending via a classical channel the employed base and the detected encoding without revealing their actual value. The sender compares the received bases and keeps only the same base entries. Afterward, the sender sends the revised list which contains the keys that have the same measurement bases. Thus, a common set of keys is obtained by both parties. A part of these keys is sacrificed to calculate the quantum bit error rate (QBER). The two participants can thus correct errors during transmission without losing data. Finally, after error correction and verification via the classical channel, namely after post-processing, the secret keys are created. This procedure is presented in Fig. 3.

Quantum algorithms have already begun to find application in Industry 4.0 and Health 4.0 applications, as is the case in research [35], which presents its integration into the Internet of Things (IoT). With the wider use of 5G, IoT devices have increased worldwide thus attracting the interest of attackers who exploit system vulnerabilities [36]. For this reason, as well as the rapid evolution of quantum computers, this paper researches the use of quantum algorithms that provide complete security against various types of attacks. Nevertheless, it concludes that depending on the circumstances, the integration of these types of security in IoT, which has resource limitations, is not always feasible. Efficient architectures that provide balanced trade-offs between resources, security, and performance must be thoroughly researched and designed in the future.

5 Conclusions

A huge amount of data of various elements is used, processed, and stored worldwide every day. This overgrowth makes it difficult to manage but most importantly to protect all this data. This need led to the integration of Industry 4.0 and Health 4.0 technologies into traditional infrastructure. These can manage various kinds of data depending on the employed operation. Decision Support System (DSS) is a data-processing technology that is utilized in Industry 4.0 and Health 4.0 applications and assists operators or health providers in making appropriate decisions. The decisions can be referred to the system's functionality or even their adopted mitigation mechanism.

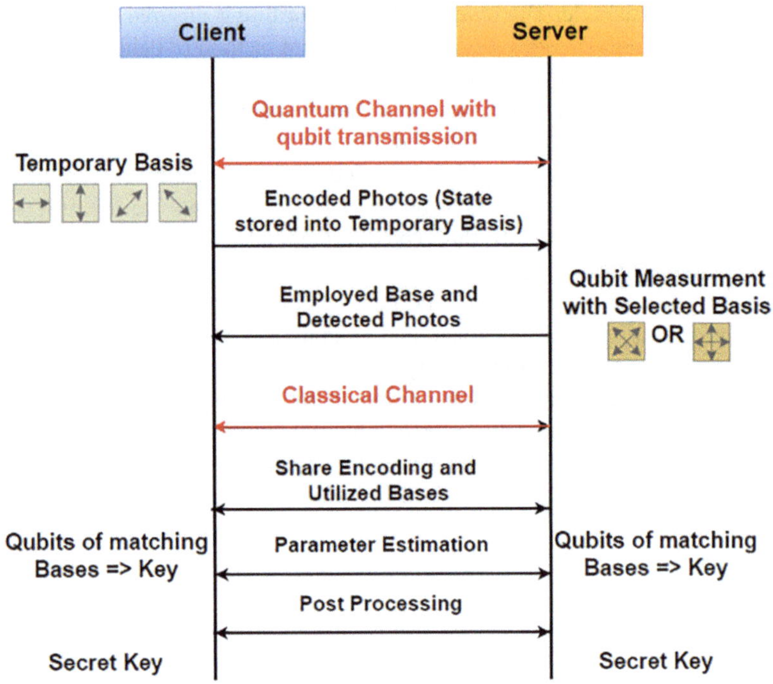

Fig. 3 Quantum key distribution (QKD) mechanism

Nevertheless, these data contain sensitive information that must be protected from malicious attacks, especially as technologies based on data management, like DSS, spread. Although there are cryptographic algorithms designed for this purpose, the advent of quantum computing appears to be a game-changer in the field of security. The supercomputing power of these computers can break the conventional cryptographic algorithms in a very short time. For this reason, research on robust cryptographic algorithms continues with only a few of them being certified as fully secure. Therefore, with the purpose of providing knowledge and guidance regarding cybersecurity practices in DSS, common security standards and new developments regarding the quantum era of computing and security are detailed and highlighted.

The main conclusion of this paper is that the implementation of secure DSSs architectures, especially with the rapid evolution of quantum computers, must be approached carefully. The conventional security standards must be upgraded to match the processing speed of quantum computers and the massive amount of collected data. However, the integration of technologies, such as IoT, into DSS that mostly employ resource-constrained devices, deems the resource-demanding algorithms unsuitable. Thus, more easily implementable algorithms for hardware, with more efficient management and maintenance methods for operating quantum computers, must be explored.

References

1. Forcina, A., Silvestri, L., De Felice, F., & Falcone, D. (2024). Exploring industry 4.0 technologies to improve manufacturing enterprise safety management: A TOPSIS-based decision support system and real case study. *Safety Science, 169*, 106351.
2. De Giovanni, P., Belvedere, V., & Grando, A. (2024). The selection of industry 4.0 technologies through Bayesian networks: An operational perspective. *IEEE Transactions on Engineering Management, 71*, 2921–2936.
3. Doltsinis, S., Ferreira, P., Mabkhot, M. M., & Lohse, N. (2020). A decision support system for rapid ramp-up of industry 4.0 enabled production systems. *Computers in Industry, 116*, 103190.
4. Bajic, B., Rikalovic, A., Suzic, N., & Piuri, V. (2021). Industry 4.0 implementation challenges and opportunities: A managerial perspective. *IEEE Systems Journal, 15*(1), 546–559.
5. European Union, Communication from the Commission on a European Programme for Critical Infrastructure Protection, COM. (2006). 786 Final, European Commission: Brussels, Belgium.
6. Mosca, M. (2018). Cybersecurity in an era with quantum computers: Will we be ready? *IEEE Security & Privacy, 16*(5), 38–41.
7. Peres, R. S., Jia, X., Lee, J., Sun, K., Colombo, A. W., & Barata, J. (2020). Industrial artificial intelligence in industry 4.0 – systematic review, challenges and outlook. *IEEE Access, 8*, 220121–220139.
8. Khurshid, A., Alsaaidi, R., Aslam, M., & Raza, S. (2022). EU cybersecurity act and IoT certification: Landscape, perspective and a proposed template scheme. *IEEE Access, 10*, 129932–129948.
9. Pliatsios, D., Sarigiannidis, P., Lagkas, T., & Sarigiannidis, A. G. (2020). A survey on SCADA systems: Secure protocols, incidents, threats and tactics. *IEEE Communications Surveys & Tutorials, 22*(3), 1942–1976.
10. Kumar, R. H., & Rajaram, B. (2024). Design and simulation of an edge compute architecture for IoT-based clinical decision support system. *IEEE Access, 12*, 45456–45474.
11. Al-Jaroodi, J., Mohamed, N., & Abukhousa, E. (2020). Health 4.0: On the way to realizing the healthcare of the future. *IEEE Access, 8*, 211189–211210.
12. El-Ganainy, N. O., Balasingham, I., Halvorsen, P. S., & Rosseland, L. A. (2020). A new real time clinical decision support system using machine learning for critical care units. *IEEE Access, 8*, 185676–185687.
13. Khan, M. F., Thorup, V. M., & Luo, Z. (2024). Delineating mastitis cases in dairy cows: Development of an IoT-enabled intelligent decision support system for dairy farms. *IEEE Transactions on Industrial Informatics, 20*(7), 9508–9517.
14. Liu, X., Deng, R. H., Choo, K.-K. R., & Yang, Y. (2021). Privacy-preserving outsourced clinical decision support system in the cloud. *IEEE Transactions on Services Computing, 14*(1), 222–234.
15. Sönmez, F. Ö., & Kılıç, B. G. (2021). A decision support system for optimal selection of Enterprise information security preventative actions. *IEEE Transactions on Network and Service Management, 18*(3), 3260–3279.
16. Meneghello, F., Calore, M., Zucchetto, D., Polese, M., & Zanella, A. (2019). IoT: Internet of threats? A survey of practical security vulnerabilities in real IoT devices. *IEEE Internet of Things Journal, 6*(5), 8182–8201.
17. Tsantikidou, K., & Sklavos, N. (2024). An area efficient architecture of a novel chaotic system for high randomness security in e-health. *IEEE Computer Architecture Letters, 23*(1), 104–107.
18. NIST, Recommendation for block cipher modes of operation methods and techniques, NIST Special Publication 800-38A (2001).
19. NIST, Recommendation for block cipher modes of operation: Galois/counter mode (GCM) and GMAC, NIST Special Publication 800-38D (2007).

20. NIST, Recommendation for block cipher modes of operation: The CCM mode for authentication and confidentiality, NIST special publication 800-38C (2007).
21. McKay, K. A., & Cooper, D. A. (2019). Guidelines for the Selection, Configuration, and Use of Transport Layer Security (TLS) Implementations, NIST Special Publication 800-52 Rev. 2.
22. Li, P., Su, J., & Wang, X. (2020). iTLS: Lightweight transport-layer security protocol for IoT with minimal latency and perfect forward secrecy. *IEEE Internet of Things Journal, 7*(8), 6828–6841.
23. Tsantikidou, K., Boufeas, D., & Sklavos, N. (2023). Area-delay efficient security scheme for smart hospital systems. In *2023 IEEE international conference on cyber security and resilience (CSR)* (pp. 552–557).
24. Xu, S., & Li, X. (2021). Chosen-ciphertext secure key encapsulation mechanism in the standard model. *IEEE Access, 9*, 13683–13690.
25. Federal Information Processing Standards (FIPS), Digital Signature Standard (DSS), FIPS 186-5 (2023).
26. Micheli, G. D., Jiang, J.-H. R., Rand, R., Smith, K., & Soeken, M. (2022). Advances in quantum computation and quantum technologies: A design automation perspective. *IEEE Journal on Emerging and Selected Topics in Circuits and Systems, 12*(3), 584–601.
27. Otgonbaatar, S., & Kranzlmüller, D. (2024). Exploiting the quantum advantage for satellite image processing: Review and assessment. *IEEE Transactions on Quantum Engineering, 5*, 1–9.
28. Kusyk, J., Saeed, S. M., & Uyar, M. U. (2021). Survey on quantum circuit compilation for noisy intermediate-scale quantum computers: Artificial intelligence to heuristics. *IEEE Transactions on Quantum Engineering, 2*, 1–16.
29. Cacciapuoti, A. S., Caleffi, M., Van Meter, R., & Hanzo, L. (2020). When entanglement meets classical communications: Quantum teleportation for the quantum internet. *IEEE Transactions on Communications, 68*(6), 3808–3833.
30. Wang, S., Wang, N., Ji, T., Shi, Y., & Wang, C. (2024). Research Progress of quantum artificial intelligence in Smart City. *Intelligent and Converged Networks, 5*(2), 116–133.
31. Shafique, M. A., Munir, A., & Latif, I. (2024). Quantum computing: Circuits, algorithms, and applications. *IEEE Access, 12*, 22296–22314.
32. Kumar, A., et al. (2021). Survey of promising technologies for quantum drones and networks. *IEEE Access, 9*, 125868–125911.
33. Chen, L., Jordan, S., Liu, Y., Moody, D., Peralta, R., Perlner, R., & Smith-Tone, D. (2016) Report on post-quantum cryptography, NISTIR 8105, NIST.
34. Cao, Y., Zhao, Y., Wang, Q., Zhang, J., Ng, S. X., & Hanzo, L. (2022). The evolution of quantum key distribution networks: On the road to the Qinternet. *IEEE Communications Surveys & Tutorials, 24*(2), 839–894.
35. Wazid, M., Das, A. K., & Park, Y. (2024). Generic quantum Blockchain-envisioned security framework for IoT environment: Architecture, security benefits and future research. *IEEE Open Journal of the Computer Society, 5*, 248–267.
36. Tsantikidou, K., & Sklavos, N. (2022). Flexible security and privacy, system architecture for IoT. In *Healthcare, 2022 IFIP/IEEE 30th international conference on very large scale integration (VLSI-SoC), Patras, Greece.*

Leveraging Blockchain Technology for Secure and Transparent Decision Support Systems in Supply Chain Management

Abubakar Sadiq Muhammad

The core focus will be on the application of blockchain technology in enhancing decision-making processes in supply chain management. Case studies will demonstrate successful implementations of blockchain in areas such as traceability, provenance, and compliance management, highlighting how these applications improve operational efficiency and reduce risks.

Practical examples of blockchain-based DSS in optimizing inventory management, streamlining logistics, and improving supplier relationships will be explored. Real-world case studies will highlight the significant impact of these systems on enhancing supply chain transparency, reducing costs, and improving overall performance.

The chapter will emphasize the significant benefits of leveraging blockchain technology in DSS, including enhanced security, transparency, and efficiency in supply chain operations. It will underscore the potential for blockchain-based DSS to drive innovation and create more resilient supply chains.

Challenges such as scalability, integration with legacy systems, and regulatory compliance will be discussed, along with recommendations for future research and development. The chapter will also explore the potential of emerging technologies like AI and IoT to further enhance blockchain-based DSS.

A. S. Muhammad (✉)
Digital Transformation Team, National Information Technology Development Agency, Abuja, Nigeria

N. Gupta, J. D. Diaz Roman (eds.), *Decision Support Systems for Industrial and Scientific Applications*, Signals and Communication Technology,
https://doi.org/10.1007/978-3-031-89700-9_14

1 Introduction

Supply chain management (SCM) is a critical component of modern commerce, encompassing the processes that guide the flow of goods, information, and finances across various stages, from raw material sourcing to the delivery of finished products. Despite its importance, SCM faces numerous challenges, such as inefficiencies, fraud, and a lack of transparency, which can disrupt operations and reduce stakeholder trust. The advent of Decision Support Systems (DSS) has offered solutions to these issues by providing data-driven insights that help companies make more informed decisions. However, traditional DSS face limitations in data security and transparency, which can hinder their effectiveness.

Blockchain technology has emerged as a transformative solution for enhancing DSS within supply chains. Known for its ability to create secure and immutable records, blockchain holds the potential to revolutionize decision-making processes by offering unprecedented transparency and trust. Blockchain's decentralized nature allows for real-time updates and verification of transactions, making it particularly well-suited for addressing the challenges that plague SCM.

This chapter explores the integration of blockchain technology into DSS, focusing on its ability to enhance transparency, security, and efficiency. By examining case studies and real-world examples, the chapter provides a practical guide for researchers, practitioners, and policymakers interested in leveraging blockchain to improve supply chain operations. It offers insights into the architecture of blockchain-based DSS, methodologies for effective data collection and integration, and strategies for overcoming challenges such as scalability and regulatory compliance. Additionally, the chapter considers the potential of emerging technologies like Artificial Intelligence (AI) and the Internet of Things (IoT) to further enhance blockchain-based DSS, creating a more resilient and efficient supply chain ecosystem.

2 Challenges in Supply Chain Management

Supply chain management (SCM) involves complex activities and stakeholders aimed at delivering products from manufacturers to consumers efficiently. Despite technological advances, global supply chains face several key challenges that can lead to inefficiencies, increased costs, and brand reputation damage. This section highlights significant SCM issues, including fraud and counterfeiting, lack of transparency, inefficiencies in data management, and the need for trust among stakeholders.

- *Fraud and Counterfeiting*: Fraud and counterfeiting, especially in sectors like pharmaceuticals, electronics, and automotive parts, pose serious risks and financial losses [19, 27, 30, 41, 43]. Counterfeit products infiltrate legitimate supply chains at various stages, eroding consumer trust and brand value. For instance, counterfeit pharmaceuticals may cause health risks due to weak tracking mechanisms [41]. Improved systems are needed to ensure product authenticity at each stage of the supply chain.

- *Lack of Transparency*: Many supply chains lack transparency, making it difficult to track products and verify compliance [7, 12]. Supply chain complexity and diverse documentation systems among stakeholders often result in fragmented information, which delays responses to disruptions and can increase reputational risks for companies.
- *Data Management and Communication Inefficiencies*: Effective data management is essential for SCM but often hampered by fragmented systems and manual processes, leading to delays and errors [7, 23, 27, 43]. Inconsistent data formats and communication protocols further impede efficient inventory management and demand forecasting, which is particularly problematic for time-sensitive industries like food and pharmaceuticals [15, 18, 28].
- *Need for Trust Among Stakeholders*: Trust underpins SCM partnerships, but building it can be challenging in global supply chains with diverse cultural and operational standards [50]. Trust issues lead to costly verification processes, which slow down operations [47]. Blockchain technology offers potential solutions by providing a secure, transparent platform for tracking transactions, which could enhance trust among stakeholders [11, 16].

These challenges highlight the need for more secure, transparent, and efficient systems in SCM, with technology-driven solutions like blockchain showing promise for reducing vulnerabilities and building trust among stakeholders.

3 Blockchain Technology for Enhanced Transparency and Security

Blockchain technology enhances transparency, security, and efficiency, addressing supply chain management (SCM) needs for trust, traceability, and data integrity through distributed ledgers, smart contracts, and consensus mechanisms. It enables immutable transaction records and secures SCM data.

- *Distributed Ledgers*

A distributed ledger is a decentralized record of transactions synchronized across multiple computers [38]. Unlike centralized databases, it lacks unilateral control, reducing tampering risks and ensuring data integrity [35]. In SCM, distributed ledgers serve as a unified data source, providing all stakeholders, such as suppliers and retailers, with real-time information [15]. For instance, blockchain can track raw material shipments, creating transparency and reducing disputes.

- *Smart Contracts*

Smart contracts are self-executing agreements on blockchain networks that operate on "if-this-then-that" logic, eliminating intermediaries [23]. In SCM, smart contracts automate tasks like payment settlement and inventory management, reducing administrative costs and mitigating disputes [5]. For example, they can automate payments once shipments are verified, ensuring efficiency and compliance with predefined standards.

- *Consensus Mechanisms*

Consensus mechanisms are protocols that ensure agreement among nodes on transaction validity [50]. Different mechanisms (e.g., Proof of Work, Proof of Stake) prevent unauthorized transaction changes [47, 49]. In SCM, these mechanisms verify each transaction, such as shipment records, to maintain ledger integrity and detect fraud.

- *Enhancing Transparency*

Blockchain's decentralized ledger fosters transparency across SCM stages [7, 12]. All parties have access to tamper-proof records, providing end-to-end visibility from sourcing to delivery [6, 27]. For instance, food industry stakeholders can trace products to ensure compliance and quality [49]. This visibility aids regulatory compliance and auditability, especially in industries with strict standards, like pharmaceuticals [19, 47].

- *Immutable Transaction Records and Trust*

Blockchain's immutability creates a permanent, tamper-proof transaction record [43]. Transactions are cryptographically linked, making retroactive changes infeasible [15, 30]. This feature ensures data authenticity, crucial for product traceability, such as verifying the origin of luxury goods [37, 48]. Blockchain reduces reliance on intermediaries, fostering trust through accessible, verifiable records [4].

- *Securing Supply Chain Data*

Blockchain addresses SCM security needs through its decentralized and cryptographic design [28]. *Decentralization and Data Security*: By distributing data across nodes, blockchain reduces cyberattack risks and ensures resilience even with node failures [17, 33, 46]. In unstable regions, it ensures data accessibility, valuable for logistics [11]. *Cryptographic Security and Data Integrity*: Blockchain's encryption secures data, making unauthorized modifications detectable [16, 25]. In SCM, it protects sensitive information, such as product specifications, from interception [5]. *Access Control and Data Privacy*: Permissioned blockchains allow specific data access, maintaining control over sensitive information while ensuring transparency [17, 27]. For example, pharmaceutical companies can share audit trails with regulators while protecting proprietary data [13, 43].

4 Architecture of Blockchain-Based Decision Support Systems

Blockchain technology has expanded beyond cryptocurrency to provide robust Decision Support Systems (DSS) that improve data integrity, transparency, and efficiency [6, 27]. Blockchain-based DSS are especially beneficial in complex sectors like supply chains, healthcare, and finance, where data accuracy is critical for decision-making [8]. This section outlines blockchain-based DSS architecture, focusing on distributed ledgers, smart contracts, and consensus mechanisms [7]. It also discusses information flow, data integrity, and integration with IoT, ERP systems, and external databases.

4.1 Key Components: Distributed Ledger, Smart Contracts, and Consensus Mechanisms

In supply chain DSS, a distributed ledger tracks each transaction, ensuring transparency and traceability [12, 43]. Smart contracts can automate processes, such as executing trades or payments based on predefined conditions, minimizing errors and reducing administrative work [15, 23]. Various consensus mechanisms like Proof of Work, Proof of Stake, and Practical Byzantine Fault Tolerance secure the integrity of data for reliable DSS operation [13, 47].

4.2 Flow of Information and Data Integrity in Blockchain-Based DSS

The information flow in blockchain-based DSS includes data collection, validation, storage, analysis, and decision-making, with blockchain ensuring secure, tamper-proof records [11, 25].

1. *Data Collection*: Data from IoT devices, ERP, and external sources inform DSS processes, such as monitoring inventory or tracking shipments [5, 30].
2. *Data Validation*: Consensus mechanisms verify data before adding it to the blockchain, ensuring reliable information [15, 23].
3. *Data Storage*: Data is stored immutably on the blockchain, offering a secure record of past actions for informed decision-making [17, 36].
4. *Data Analysis and Decision-Making*: Stored data is analyzed to provide insights, with blockchain ensuring accuracy and traceability [19].
5. *Feedback and Continuous Improvement*: Results are fed back into the system, creating a loop of improvement, with blockchain providing transparency for future decisions [28, 37, 42].

4.3 Integration with IoT Devices, ERP Systems, and External Databases

Blockchain-based DSS can integrate with IoT, ERP systems, and external databases, enhancing data accuracy and providing comprehensive views for decision-making [20, 34].

- *IoT Devices*: Real-time data collection by IoT devices can be securely recorded on blockchain, improving traceability in areas like food safety [14, 18, 38].
- *ERP Systems*: Integrating blockchain with ERP ensures real-time data access, reducing data silos and enabling efficient, data-driven decisions [24, 48, 50].
- *External Databases*: Blockchain's interaction with external databases enhances DSS scalability and allows verification of critical data, such as regulatory compliance [21, 27, 32].

4.4 Applications of Blockchain in Supply Chain DSS

Blockchain-based DSS in supply chains enhance traceability, compliance, inventory management, and logistics, addressing key challenges through case studies [6, 8, 49]. By recording an immutable data trail, blockchain supports compliance and operational efficiency in supply chains, ensuring data integrity and trust among stakeholders.

5 Case Studies

5.1 Traceability and Provenance in Supply Chains

Traceability and provenance are critical aspects of supply chain management, ensuring that the origin and movement of products can be tracked throughout their lifecycle [27]. Blockchain technology, with its immutable ledger and transparent record-keeping, offers a robust solution to these challenges. By recording each transaction in a decentralized ledger, blockchain provides a clear, tamper-proof record of where products come from, how they are processed, and where they are shipped [12].

Case Study 1: IBM Food Trust
One of the most notable applications of blockchain in supply chain traceability is the IBM Food Trust, a platform developed in collaboration with major retailers like Walmart and Carrefour [29]. This platform uses blockchain to enhance the traceability of food products, allowing stakeholders to track the journey of produce

from farm to table. Each stage of the supply chain is recorded on the blockchain, from harvesting and processing to shipping and retailing [43].

In the case of a food safety incident, such as contamination or spoilage, the IBM Food Trust enables quick and accurate identification of the affected batches. This reduces the time needed for product recalls, minimizes waste, and ensures consumer safety [17]. The transparency provided by the blockchain also builds trust among consumers, who can access detailed information about the origin and handling of their food products [30]. This case illustrates how blockchain can serve as a powerful tool for enhancing traceability and accountability in supply chains [25].

Case Study 2: Everledger for Diamond Supply Chain
Another compelling example of blockchain's application in traceability is Everledger, a company that tracks the provenance of diamonds. The diamond industry faces significant challenges related to the authenticity of gemstones and the ethical sourcing of materials [48]. Everledger's blockchain-based platform creates a digital record for each diamond, including details like its origin, carat weight, and ownership history [31].

By providing an immutable record of each diamond's journey from mine to market, Everledger helps combat fraud and ensures that diamonds are ethically sourced. This system also benefits insurance companies, retailers, and end consumers, providing a trusted source of information for every diamond. The platform's blockchain ledger allows stakeholders to verify the authenticity of a diamond without relying on centralized authorities, improving transparency and reducing the risk of fraudulent transactions [28].

5.2 *Blockchain for Compliance Management*

Compliance with industry regulations and standards is a major challenge for supply chain managers, especially in industries like pharmaceuticals, food and beverage, and electronics [7]. Blockchain technology offers a way to manage compliance more effectively by providing an unchangeable record of transactions and certifications. Smart contracts can automate the verification of compliance requirements, ensuring that only certified products proceed through the supply chain [36].

Case Study 3: MediLedger in the Pharmaceutical Industry
MediLedger is a blockchain-based solution designed to address compliance and counterfeit issues in the pharmaceutical industry. The pharmaceutical supply chain is highly regulated to ensure that drugs are safe and authentic, but counterfeit drugs remain a serious problem [24]. MediLedger provides a blockchain-based platform that verifies the authenticity of drugs at each stage of the supply chain, from manufacturers to distributors and pharmacies [3].

By using blockchain, MediLedger creates a secure, transparent record of each transaction. Smart contracts ensure that only verified entities can add new transactions to the ledger, reducing the risk of counterfeit drugs entering the supply chain.

This system has significantly improved the ability of pharmaceutical companies to meet regulatory requirements and reduce compliance risks [16]. The immutable nature of blockchain records ensures that audits can be conducted more efficiently, as all necessary documentation is available in a digital, tamper-proof format [15].

5.3 Reducing Supply Chain Risks

Blockchain technology also plays a crucial role in mitigating various supply chain risks, including fraud, delays, and data discrepancies [18]. By creating a single source of truth that is accessible to all stakeholders, blockchain reduces the chances of disputes over data and enhances accountability. It allows for real-time tracking of goods, ensuring that supply chain managers can detect and address issues like delays or damage during transit promptly [43].

In industries where supply chain risks are particularly high, such as automotive manufacturing, blockchain-based DSS have been used to improve supplier transparency and reduce risks [11]. For example, automakers can use blockchain to track the movement of parts and components through their supply chain, ensuring that they meet quality standards before being assembled into final products [47]. This reduces the likelihood of costly recalls and strengthens relationships with suppliers by providing a clear record of each transaction [40].

5.4 Blockchain-Based DSS in Inventory Management

Inventory management is a critical component of supply chain operations, as it directly impacts a company's ability to meet customer demand while minimizing costs [32]. Blockchain-based DSS can optimize inventory management by providing real-time visibility into stock levels and automating key processes such as reordering and demand forecasting [50].

Case Study 4: Walmart's Blockchain for Inventory Management
Walmart has been a pioneer in using blockchain to streamline its supply chain operations, particularly in the area of inventory management. By integrating blockchain with its existing ERP systems, Walmart has improved its ability to track the movement of products throughout its warehouses and retail stores [20]. This integration enables real-time updates on inventory levels, reducing the risk of stockouts and overstocking [23].

Smart contracts on the blockchain automate the reordering process by triggering purchase orders when inventory levels fall below a certain threshold. This automation ensures that stores remain stocked with essential products, even during peak demand periods [13]. By leveraging blockchain, Walmart has also reduced the time and cost associated with manual inventory checks, improving the efficiency of its supply chain operations [33].

5.5 Blockchain for Demand Forecasting

Accurate demand forecasting is essential for effective inventory management [37]. Blockchain-based DSS can enhance demand forecasting by providing access to a wide range of data sources, including historical sales data, market trends, and customer behavior. The transparency of blockchain ensures that this data is accurate and reliable, allowing supply chain managers to make better predictions about future demand [19].

For example, a retail company might use a blockchain-based DSS to analyze purchasing patterns and predict which products will be in high demand during the holiday season [4]. By having this insight, the company can adjust its inventory levels accordingly, ensuring that it has enough stock to meet customer needs while minimizing excess inventory. This improves both customer satisfaction and the company's bottom line [29].

5.6 Optimization of Logistics and Supplier Relationships

Logistics is a complex aspect of supply chain management, involving the coordination of transportation, warehousing, and distribution [10]. Blockchain technology can optimize logistics by providing real-time tracking of shipments and improving communication between stakeholders [34]. This reduces delays, minimizes the risk of lost or damaged goods, and ensures that deliveries are made on time [13].

Case Study 5: TradeLens in Global Shipping
TradeLens is a blockchain-based platform developed by IBM and Maersk to improve transparency and efficiency in global shipping. The platform connects various stakeholders in the shipping industry, including shipping lines, port operators, customs authorities, and logistics companies [26]. By recording each stage of a shipment's journey on the blockchain, TradeLens provides a single, immutable record that all parties can access [3].

This transparency allows stakeholders to track the status of shipments in real-time, reducing the time and cost associated with manual paperwork and tracking [48]. TradeLens has been particularly effective in reducing delays at customs, as customs authorities can access shipment data directly from the blockchain, speeding up the inspection and clearance process [38]. By streamlining logistics operations, TradeLens has helped reduce shipping times and costs, benefiting all participants in the supply chain [14].

5.7 Improving Supplier Relationships

Supplier relationships are vital for maintaining a smooth and efficient supply chain [42]. Blockchain-based DSS can enhance these relationships by providing a transparent and secure record of transactions, building trust between buyers and suppliers [52]. With blockchain, both parties can access the same data, reducing the risk of disputes and misunderstandings.

For instance, a company might use blockchain to track supplier performance metrics, such as delivery times and quality of goods [47]. By having access to this information, companies can provide constructive feedback to suppliers and collaborate on improvement initiatives, strengthening the partnership over time. This data-driven approach fosters a culture of accountability and continuous improvement in supplier relationships, ultimately leading to better supply chain performance [53].

6 Benefits of Blockchain-Enabled DSS

The adoption of blockchain-enabled Decision Support Systems (DSS) is transforming supply chain management by addressing key challenges in security, transparency, efficiency, and resilience. This section delves into the benefits of integrating blockchain technology into DSS, emphasizing its potential to enhance security, increase transparency, improve efficiency, and create a more resilient supply chain.

6.1 Enhanced Security and Data Privacy

- *Data Integrity and Immutability*: Blockchain's immutable ledger secures supply chain data, preventing unauthorized changes and reducing fraud risk [8, 12, 47]. For instance, product details like manufacturing and quality reports are safely stored, making tampering evident [7].
- *Decentralized Data Storage*: Unlike centralized systems, blockchain's decentralized nature guards against cyberattacks, enhancing data reliability and resilience [18, 23]. Encryption also secures sensitive information, protecting business data from unauthorized access [38].

6.2 Increased Transparency and Trust

- *Building Trust Through Transparency*: Blockchain provides a shared record of supply chain activities, allowing stakeholders to verify product authenticity at each stage, which is valuable in industries like food [16, 25, 28].
- *Smart Contracts for Automated Trust*: Smart contracts enforce agreements automatically upon fulfillment of conditions, streamlining processes and fostering trust [13, 27, 43, 50].

6.3 Improved Efficiency and Cost Reduction

- *Streamlining Processes*: Blockchain-enabled DSS reduces manual tasks by automating tracking and documentation, enhancing real-time visibility, and reducing errors [3, 30, 46].
- *Cost Savings Through Reduced Intermediaries*: By eliminating the need for third-party intermediaries, blockchain minimizes transaction costs and enhances efficiency, especially in cross-border transactions [11, 37, 48, 52].

6.4 Resilient Supply Chain Management

- *Enhanced Visibility and Traceability*: Blockchain aids in rapid tracing during disruptions, allowing targeted actions and minimizing financial losses [2, 17, 45].
- *Supporting Sustainability and Compliance*: Blockchain enables verifiable records for regulatory compliance and sustainable sourcing, appealing to environmentally conscious consumers [9, 20, 29, 34].

7 Challenges and Considerations for Implementation

While blockchain technology offers numerous benefits for supply chain management, its implementation is not without challenges. This section explores key issues that businesses face when integrating blockchain into their supply chains, including scalability, integration with legacy systems, and regulatory compliance. It also provides recommendations for addressing these challenges to ensure successful deployment.

7.1 Scalability Issues in Blockchain Systems

- *Transaction Throughput and Speed*: Blockchain scalability issues impact trans action speed, with networks like Bitcoin and Ethereum often lagging due to proof-of-work (PoW) consensus [6, 8]. High transaction volumes can overwhelm blockchain, slowing supply chain operations [12]. Solutions include adopting layer-2 systems like sidechains and scalable consensus methods such as proof-of-stake (PoS) [7].
- *Recommendations*: Hybrid blockchain models can balance speed and security by using private blockchains for internal operations while anchoring data to public chains [27]. Sharding, which divides the network into smaller shards, can also enhance throughput [41].

7.2 Integration with Legacy Systems

- *Challenges*: Integrating blockchain with existing systems like ERP is complex due to differing data structures and protocols, which may require APIs and middleware [19, 30, 43].
- *Recommendations*: Companies should assess their IT infrastructure to identify blockchain's best applications, possibly using middleware solutions or gradual integration, starting with areas like traceability [35, 38].

7.3 Regulatory Compliance and Legal Concerns

- *Regulatory Uncertainty*: Blockchain adoption is complicated by varying international regulations on issues such as data privacy and AML compliance [5, 15, 23].
- *Recommendations*: Companies should follow regulatory updates, consult legal experts, and engage with regulatory bodies. Implementing permissioned blockchains can control data access and support compliance [47, 49, 50].

7.4 Recommendations for Addressing Challenges

- *Strategic Partnerships and Collaboration*: Partnering with blockchain providers offers access to expertise, facilitating solution customization, integration, and regulatory compliance [2, 32].
- *Investing in Training and Education*: Training employees and stakeholders on blockchain, smart contracts, and data practices ensures smoother adoption and stronger collaboration [28, 40].

8 Future Outlook: Integrating AI and IoT with Blockchain-Based DSS

As supply chain management continues to evolve, the integration of advanced technologies like Artificial Intelligence (AI), the Internet of Things (IoT), and blockchain-enabled Decision Support Systems (DSS) is shaping the future of the industry. This synergy between AI, IoT, and blockchain has the potential to enhance predictive analytics, real-time data collection, and overall decision-making capabilities, leading to more efficient and resilient supply chains [39, 51].

8.1 Potential of AI for Predictive Analytics in Supply Chains

Predictive Demand Forecasting: AI enables predictive analytics for demand forecasting, minimizing overstock and stockout risks. By analyzing data such as sales, trends, and external factors (e.g., seasonality), AI models forecast demand accurately, allowing adjustments in production and inventory [22]. AI insights can be integrated into blockchain-enabled DSS, enhancing coordination across supply chain stakeholders [18].

Optimizing Logistics and Route Planning: AI optimizes route planning by analyzing traffic, weather, and schedules to suggest efficient routes, reducing delays, fuel consumption, and emissions. Blockchain can securely share AI-driven logistics insights across stakeholders for improved coordination [38, 50, 51].

8.2 IoT for Real-Time Data Collection and Integration with Blockchain

Enhanced Visibility Through IoT Sensors: IoT devices like RFID tags and temperature sensors provide real-time data on product location and condition, crucial for monitoring quality and compliance. Recording this data on blockchain ensures a transparent, immutable record, aiding quality assurance in sensitive shipments like cold-chain logistics [2, 39].

Real-Time Asset Tracking and Inventory Management: IoT enables precise tracking of inventory and assets, reducing holding costs and delays. When combined with blockchain, this data can be shared across partners, improving coordination and response to potential disruptions [10, 22, 33, 43].

8.3 Synergy Between Blockchain, AI, and IoT for Enhanced DSS

Enhanced Decision-Making with Real-Time Data and Predictive Insights: The integration of blockchain, AI, and IoT improves DSS by enabling data-driven decisions. AI leverages IoT data to predict risks (e.g., temperature issues in sensitive products) and blockchain records these insights, enabling proactive measures across the supply chain [1, 25].

Building Trust and Collaboration Through Technology: Blockchain ensures data transparency, fostering trust among partners. AI and IoT automate contract verifications (e.g., delivery confirmations) through smart contracts, streamlining processes and building reliable partnerships [4, 37].

9 Conclusion and Recommendations for Future Research

The integration of blockchain-enabled Decision Support Systems (DSS) in supply chain management offers a transformative approach to addressing key challenges such as data security, transparency, and operational efficiency. As this technology continues to gain traction, it is crucial to understand its implications and identify areas for further research to maximize its potential [44]. This section provides a summary of key insights, emphasizes the role of blockchain in building resilient supply chains, and outlines recommendations for future research.

Blockchain's Role in Enhancing Security and Transparency: One of the most significant benefits of blockchain-enabled DSS is its ability to enhance data security and transparency in supply chains. By providing an immutable and decentralized ledger, blockchain ensures that records cannot be altered, reducing the risk of fraud and data tampering. This is particularly valuable in industries where product authenticity and traceability are critical, such as pharmaceuticals and food supply chains. The transparency offered by blockchain builds trust among stakeholders, allowing them to verify the integrity of transactions and product quality at every stage of the supply chain.

- *Improving Efficiency Through Automation and Smart Contracts*: Blockchain also contributes to improved efficiency in supply chain management through automation and the use of smart contracts. These self-executing contracts automate processes such as payment releases and order confirmations, reducing the need for intermediaries and manual interventions. By streamlining operations, blockchain-enabled DSS help reduce transaction costs and shorten processing times, making supply chains more responsive to changes in demand and market conditions.
- *Challenges and Recommendations for Addressing Scalability and Integration Issues*: While blockchain technology offers numerous advantages, its implementation in supply chains is not without challenges. Issues related to scalability, integration with legacy systems, and regulatory compliance require careful consideration. Solutions such as hybrid blockchain models, strategic partnerships with technology providers, and incremental implementation can help address these challenges, ensuring that companies can realize the full potential of blockchain-enabled DSS.
- *Resilience Through Real-Time Data and Predictive Insights*: Blockchain plays a crucial role in building more resilient supply chains by enabling real-time data sharing and predictive insights. This allows companies to quickly adapt to disruptions, such as delays, recalls, or changes in demand. By integrating blockchain with AI and IoT, companies can leverage predictive analytics to anticipate potential risks and take proactive measures to mitigate their impact. This agility is essential in today's rapidly changing market environment, where the ability to respond quickly to unforeseen events can determine a company's success.

- *Supporting Sustainability and Ethical Practices*: Another important aspect of blockchain in supply chains is its ability to support sustainability and ethical practices. With growing consumer demand for transparency regarding the environmental and social impact of products, blockchain provides a way for companies to prove the ethical sourcing and sustainability of their products. By recording data on product origins, manufacturing processes, and certifications, blockchain helps companies meet regulatory requirements and appeal to consumers who prioritize sustainable and ethically sourced products.

9.1 Directions for Future Research

- *Exploring Interoperability Between Blockchain Platforms*

  One of the key areas for future research is exploring the interoperability between different blockchain platforms used in supply chains. As companies adopt various blockchain solutions tailored to their specific needs, ensuring seamless communication between these platforms is critical. Research in this area could focus on developing standards and protocols that facilitate the integration of diverse blockchain networks, enabling more cohesive supply chain ecosystems.
- *Investigating the Impact of AI-Driven Predictive Analytics*

  While AI's potential for predictive analytics is well-recognized, further research is needed to understand its impact on supply chain performance when integrated with blockchain-enabled DSS. Studies could investigate how AI models trained on blockchain data can improve demand forecasting, risk management, and inventory optimization. Additionally, research could explore the role of AI in enhancing the accuracy and efficiency of smart contracts, providing a deeper understanding of how these technologies can work together to optimize supply chain operations.
- *Evaluating the Role of Blockchain in Decentralized Finance (DeFi) for Supply Chains*

  The emergence of decentralized finance (DeFi) presents new opportunities for blockchain in supply chain finance. Future research could explore how DeFi platforms can provide alternative financing options for small and medium-sized enterprises (SMEs) and startups in supply chains. This could include investigating how blockchain-based lending platforms and tokenized assets can improve access to capital, enabling more SMEs to participate in global supply chains and contribute to economic growth.
- *Developing Frameworks for Regulatory Compliance*

  Given the complexities of regulatory compliance in different regions, there is a need for research that focuses on developing frameworks to guide companies in navigating legal challenges associated with blockchain. This could involve studying best practices for implementing permissioned blockchains, identifying common regulatory hurdles, and providing recommendations for ensuring compliance with data privacy laws, anti-money laundering regulations, and cross-border trade requirements.

References

1. Adamashvili, N., Zhizhilashvili, N., & Tricase, C. (2024). The integration of the internet of things, artificial intelligence, and blockchain technology for advancing the wine supply chain. *Computers, 13*(3), 72.
2. Adhikari, N., & Ramkumar, M. (2023). IoT and blockchain integration: Applications, opportunities, and challenges. *Network, 3*(1), 115–141.
3. Ahmed, W. A., & Rios, A. (2022). Digitalization of the international shipping and maritime logistics industry: A case study of TradeLens. In *The digital supply chain* (pp. 309–323). Elsevier.
4. Ali, O., Jaradat, A., Kulakli, A., & Abuhalimeh, A. (2021). A comparative study: Blockchain technology utilization benefits, challenges and functionalities. *IEEE Access, 9*, 12730–12749.
5. Alqarni, M. A., Alkatheiri, M. S., Chauhdary, S. H., & Saleem, S. (2023). Use of blockchain-based smart contracts in logistics and supply chains. *Electronics, 12*(6), 1340.
6. Azevedo, P., Gomes, J., & Romão, M. (2023). Supply chain traceability using blockchain. *Operations Management Research, 16*(3), 1359–1381.
7. Budler, M., Quiroga, B. F., & Trkman, P. (2024). A review of supply chain transparency research: Antecedents, technologies, types, and outcomes. *Journal of Business Logistics, 45*(1), e12368.
8. Casino, F., Kanakaris, V., Dasaklis, T. K., Moschuris, S., Stachtiaris, S., Pagoni, M., & Rachaniotis, N. P. (2021). Blockchain-based food supply chain traceability: A case study in the dairy sector. *International Journal of Production Research, 59*(19), 5758–5770.
9. Cerf, M., Matz, S., & Berg, A. (2020). Using blockchain to improve decision making that benefits the public good. *Frontiers in Blockchain, 3*, 13.
10. Compagnucci, L., Lepore, D., Spigarelli, F., Frontoni, E., Baldi, M., & Di Berardino, L. (2022). Uncovering the potential of blockchain in the agri-food supply chain: An interdisciplinary case study. *Journal of Engineering and Technology Management, 65*, 101700.
11. Cordova, R. S., Maata, R. L. R., Epoc, F. J., & Alshar'e, M. (2021). Challenges and opportunities of using blockchain in supply chain management. *Global Business and Management Research, 13*(3), 204–217.
12. Cui, Y., Gaur, V., & Liu, J. (2024). Supply chain transparency and blockchain design. *Management Science, 70*(5), 3245–3263.
13. Dutta, P., Choi, T. M., Somani, S., & Butala, R. (2020). Blockchain technology in supply chain operations: Applications, challenges and research opportunities. *Transportation research part e: Logistics and transportation review, 142*, 102067.
14. Ekawati, R., Arkeman, Y., Suprihatin, S., & Sunarti, T. C. (2020, October). Design of intelligent decision support system for sugar cane supply chains based on blockchain technology. In *2020 2nd International Conference on Industrial Electrical and Electronics (ICIEE)* (pp. 153–157). IEEE.
15. Epiphaniou, G., Bottarelli, M., Al-Khateeb, H., Ersotelos, N. T., Kanyaru, J., & Nahar, V. (2020). Smart distributed ledger technologies in industry 4.0: Challenges and opportunities in supply chain management. In *Cyber defence in the age of AI, smart societies and augmented humanity* (pp. 319–345).
16. Etemadi, N., Borbon-Galvez, Y., Strozzi, F., & Etemadi, T. (2021). Supply chain disruption risk management with blockchain: A dynamic literature review. *Information, 12*(2), 70.
17. Gaynor, M., Gillespie, K., Roe, A., Crannage, E., & Tuttle-Newhall, J. E. (2024). Blockchain applications in the pharmaceutical industry. *Blockchain in Healthcare Today, 7*.
18. Gürpinar, T., Harre, S., Henke, M., & Saleh, F. (2020). Blockchain technology: Integration in supply chain processes. In data science and innovation in supply chain management: How data transforms the value chain. In *Proceedings of the Hamburg International Conference of Logistics (HICL)* (Vol. 29, pp. 153–185). epubli GmbH.
19. Hou, P., Zhang, C., & Li, Y. (2024). The interplay of manufacturer encroachment and blockchain adoption to combat counterfeits in a platform supply chain. *International Journal of Production Research, 62*(4), 1382–1398.

20. Islam, I., Munim, K. M., Oishwee, S. J., Islam, A. N., & Islam, M. N. (2020). A critical review of concepts, benefits, and pitfalls of blockchain technology using concept map. *IEEE Access, 8*, 68333–68341.
21. Jabbar, S., Lloyd, H., Hammoudeh, M., Adebisi, B., & Raza, U. (2021). Blockchain-enabled supply chain: Analysis, challenges, and future directions. *Multimedia Systems, 27*, 787–806.
22. Kayvanfar, V., Elomri, A., Kerbache, L., Vandchali, H. R., & El Omri, A. (2024). A review of decision support systems in the Internet of Things and supply chain and logistics using web content mining. *Supply Chain Analytics*, 100063.
23. Khan, S. N., Loukil, F., Ghedira-Guegan, C., Benkhelifa, E., & Bani-Hani, A. (2021). Blockchain smart contracts: Applications, challenges, and future trends. *Peer-to-Peer Networking and Applications, 14*, 2901–2925.
24. Kumar, D. (2021, December). Blockchain-based solution for demand forecasting in supply chain. In *2021 first international conference on advances in computing and future communication technologies (ICACFCT)* (pp. 217–224). IEEE.
25. Li, X. (2023). Inventory management and information sharing based on blockchain technology. *Computers and Industrial Engineering, 179*, 109196.
26. Louw-Reimer, J., Nielsen, J. L. M., Bjørn-Andersen, N., & Kouwenhoven, N. (2021). Boosting the effectiveness of containerised supply chains: A case study of tradelens. In *Maritime informatics: Additional perspectives and applications* (pp. 95–115). Springer.
27. Melnyk, S. A., Schoenherr, T., Speier-Pero, C., Peters, C., Chang, J. F., & Friday, D. (2022). New challenges in supply chain management: Cybersecurity across the supply chain. *International Journal of Production Research, 60*(1), 162–183.
28. Nanda, S. K., Panda, S. K., & Dash, M. (2023). Medical supply chain integrated with blockchain and IoT to track the logistics of medical products. *Multimedia Tools and Applications, 82*(21), 32917–32939.
29. Nguyen, H., and Do, L. (2018). The adoption of blockchain in food retail supply chain: Case: IBM food trust blockchain and the food retail supply chain in Malta.
30. Onu, P., Mbohwa, C., & Pradhan, A. (2024). Blockchain-powered traceability solutions: Pioneering transparency to eradicate counterfeit products and revolutionize supply chain integrity. *Procedia Computer Science, 232*, 1420–1427.
31. Oriekhoe, O. I., Ashiwaju, B. I., Ihemereze, K. C., Ikwue, U., & Udeh, C. A. (2024). Blockchain technology in supply chain management: A comprehensive review. *International Journal of Management and Entrepreneurship Research, 6*(1), 150–166.
32. Paik, H. Y., Xu, X., Bandara, H. D., Lee, S. U., & Lo, S. K. (2019). Analysis of data management in blockchain-based systems: From architecture to governance. *IEEE Access, 7*, 186091–186107.
33. Ramachandran, G. S., Malik, S., Pal, S., Dorri, A., Dedeoglu, V., Kanhere, S., & Jurdak, R. (2022). Blockchain in supply chain: Opportunities and design considerations. In *Handbook on Blockchain* (pp. 541–576). Springer.
34. Ran, L., Shi, Z., & Geng, H. (2024). *Blockchain technology for enhanced efficiency in logistics operations*. IEEE Access.
35. Roeck, D., Sternberg, H., & Hofmann, E. (2020). Distributed ledger technology in supply chains: A transaction cost perspective. *International Journal of Production Research, 58*(7), 2124–2141.
36. Sahoo, S., Kumar, S., Sivarajah, U., Lim, W. M., Westland, J. C., & Kumar, A. (2022). Blockchain for sustainable supply chain management: Trends and ways forward. *Electronic Commerce Research*, 1–56.
37. Silvestri, R., Carloni, E., Morrone, D., and Santovito, S. (2024). The role of blockchain technology in supply chain relationships: Balancing efficiency and relational dynamics. *Journal of Purchasing and Supply Management*, 100967.
38. Soltani, R., Zaman, M., Joshi, R., & Sampalli, S. (2022). Distributed ledger technologies and their applications: A review. *Applied Sciences, 12*(15), 7898.
39. Soori, M., Jough, F. K. G., Dastres, R., & Arezoo, B. (2024). Blockchains for industrial internet of things in sustainable supply chain management of industry 4.0, a review. *Sustainable Manufacturing and Service Economics*, 100026.

40. Sunmola, F., & Lawrence, G. L. (2024). Key success factors for integration of Blockchain and ERP systems: A systematic literature review. *Procedia Computer Science, 232*, 775–782.
41. Syed, I. U., & Milburn, T. W. (2024). *Rethinking counterfeit medical supply chains: A critical review of the current literature*. Health Care Science.
42. Tönnissen, S., & Teuteberg, F. (2020). Analysing the impact of blockchain-technology for operations and supply chain management: An explanatory model drawn from multiple case studies. *International Journal of Information Management, 52*, 101953.
43. Trautmann, L., Hübner, T., & Lasch, R. (2024). Blockchain concept to combat drug counterfeiting by increasing supply chain visibility. *International Journal of Logistics Research and Applications, 27*(6), 959–985.
44. Van Hoek, R. (2019). Exploring blockchain implementation in the supply chain: Learning from pioneers and RFID research. *International Journal of Operations and Production Management, 39*(6/7/8), 829–859.
45. Wang, B., Lin, Z., Wang, M., Wang, F., Xiangli, P., & Li, Z. (2023). Applying blockchain technology to ensure compliance with sustainability standards in the PPE multi-tier supply chain. *International Journal of Production Research, 61*(14), 4934–4950.
46. Xu, X., & He, Y. (2024). Blockchain application in modern logistics information sharing: A review and case study analysis. *Production Planning and Control, 35*(9), 886–900.
47. Yadav, A. K., Singh, K., Amin, A. H., Almutairi, L., Alsenani, T. R., & Ahmadian, A. (2023). A comparative study on consensus mechanism with security threats and future scopes: Blockchain. *Computer Communications, 201*, 102–115.
48. Zhang, C., Gong, Y., & Brown, S. (2023). Cross-case analysis. In *Blockchain applications in food supply chain management: Case studies and implications* (pp. 213–261). Cham.
49. Zhao, H., Hu, K., Yuan, Z., Yao, S., & Feng, L. (2024). BCTMSSF: A blockchain consensus-based traceability method for supply chain in smart factory. *Journal of Intelligent Manufacturing*, 1–17.
50. Zhou, S., Li, K., Xiao, L., Cai, J., Liang, W., & Castiglione, A. (2023). A systematic review of consensus mechanisms in Blockchain. *Mathematics, 11*(10), 2248.
51. Zrelli, I., & Rejeb, A. (2024). A bibliometric analysis of IoT applications in logistics and supply chain management. *Heliyon, 10*(16).
52. Huang, X., Xu, L., Huang, Y., & Hai, M. (2023, August). The impact of demand information transparency on endogenous supply process reliability in blockchain era. In *Proceedings of the 2nd international academic conference on Blockchain*, Information Technology and Smart Finance (ICBIS 2023) (pp. 1378-1383). Atlantis Press.
53. Kumar, N., Kaul, M., Rawandale, C. J., Singh, S., & Gupta, A. K. (2024, February). Blockchain integrated retail logistics chain: An adoption perspective. In *2024 4th International Conference on Innovative Practices in Technology and Management (ICIPTM)* (pp. 1–6). IEEE.

Risk and Disruption in the Supply Chain: Detection Using an Intelligent Decision Support System with System Dynamics Modeling and Deep Learning

Víctor Hugo De-la-Cruz-Madrigal, Stefani Sifuentes-Domíguez, Liliana Avelar-Sosa, José-Manuel Mejía-Muñoz, Jorge Luis García-Alcaraz, and Emilio Jiménez

1 Introduction

Supply chain management is an important component for the success of any company that relies on the production and distribution of goods. In this context, identifying anomalies, which are significant deviations from the expected, is essential to maintain efficiency and avoid significant problems that can disrupt the flow of materials and products [9]. With technological advancements, intelligent decision support systems (IDSS) have been developed that use advanced techniques to detect these anomalies and better manage the supply chain, especially in disruptive situations [29]. Similarly, disruptions can also be considered anomalies, as they are unexpected events that can alter the normal functioning of the supply chain. Examples include natural disasters, labor conflicts, production failures, logistical problems, etc. These disruptions have significant negative impacts, such as delays in product delivery, increased logistical costs, and overall customer dissatisfaction. Therefore, it is imperative to have systems capable of anticipating and reacting to these events [27]. According to Tarei et al. [42, 43], supply chain risks have

V. H. De-la-Cruz-Madrigal · S. Sifuentes-Domíguez · J.-M. Mejía-Muñoz (✉)
Department of Electrical and Computing Engineering, Universidad Autónoma de Ciudad Juárez, Ciudad Juárez, Chih, México
e-mail: al220736@alumnos.uacj.mx; al244307@alumnos.uacj.mx; jose.mejia@uacj.mx

L. Avelar-Sosa · J. Luis García-Alcaraz
Department of Industrial Engineering and Manufacturing, Universidad Autónoma de Ciudad Juárez, Ciudad Juárez, México
e-mail: liliana.avelar@uacj.mx; jorge.garcia@uacj.mx

E. Jiménez
Department of Mechanical Engineering, Universidad de La Rioja, Logroño, Spain
e-mail: emilio.jimenez@unirioja.es

N. Gupta, J. D. Diaz Roman (eds.), *Decision Support Systems for Industrial and Scientific Applications*, Signals and Communication Technology,
https://doi.org/10.1007/978-3-031-89700-9_15

repercussions in the form of low production, lost sales, and delayed deliveries. Ivanov [17], on the other hand, focuses on preventing supply chain risks using simulation models that show the impact on order delivery. In Industry 4.0, intelligent devices that process information in real time are integrated into the supply chain to identify potential risks between demand and suppliers through the implementation of a decision support system (DSS) that helps managers to have better coordination by predicting demand with real-time data and the maximum inventory level.

Predicting supply chain disruptions helps managers determine the maximum inventory level to deal with transportation and production delays. Conversely, the cost of not anticipating a disruption, specially a major outbreak like COVID-19, had significant economic impacts, including the immediate and widespread disruption of the global supply chain as revealed in the study by Orlando et al. [32]. As noted by Wang et al. [48], supply chain disruptions are caused mainly by uncertainty in demand, suppliers, and deliveries. Another study presented in [38] suggests that supplier disruption can also result from decreased capacity or bankruptcy, while Yin et al. [52] mention that transportation disruption affects the flow of both finished products and materials in the supply chain. Thus, using a DSS that analyzes these factors in a disruptive supply chain could be an efficient tool for risk analysis, as argued by Ali et al. [2]. In [33] it is proposed a DSS for dynamic risk analysis. Tsiamas et al. [47] propose a simulation model with DSS to improve decision-making and respond to anticipated supply chain disruption, relying on a historical database available. Andry et al. [3] developed a DSS to improve supply chain processes. Teniwut and Hasym [45] proposed a DSS that helps to increase efficiency, profitability, and customer satisfaction in the supply chain of components. The implementation of the DSS in the supply chain requires the entry of orders, sales, transactions, and safety stock. Please note that in this work, the terms "anomaly" and "disruption" will be used interchangeably.

Based on the previous literature review, this chapter describes a proposal to use a IDSS focused primarily on anomaly/disruption detection in the context of smart manufacturing. It utilizes prediction modules and system dynamics modeling, as well as a deep learning-based architecture for disruption detection. The purpose is to enhance the efficiency, reliability, and responsiveness of supply chains. The main contributions of this work are:

- A DSS that utilizes a prediction system to feed a system dynamics model, predicting future states of the industrial plant and the subsequent detection of disruptions through artificial neural networks
- The design of a system dynamics model specifically tailored for the metalworking industry, focusing on a supplier manufacturing membranes for automobiles
- The development of a new deep learning architecture that employs a layer for sequential data processing using convolutional networks, GRU, and an attention mechanism for disruption detection

2 Background

This section provides an overview of smart manufacturing and the disruptions affecting supply chains. It also discusses the use of anomaly detection in Decision Support Systems.

2.1 Supply Chain Management Conceptualization

The supply chain is a concept that involves coordinating activities related to the production and delivery of products and services from suppliers to the final customer [31], facilitated through information flows and transactions across the entire logistics network. Some authors describe it as an internal network within an organization that facilitates the delivery of finished products or services to customers through various processes, starting with the purchase of materials and continuing with design, planning, production, and logistics management to ensure they reach their destination. This requires the flow of information, goods, products, money, and transactions from suppliers to consumers [35]. It is a complex network involving the flow of goods, services, information, and customers. Its main characteristics include material specifications, organizational structure, and regulatory frameworks, all of which influence supply chain traceability. Understanding these characteristics is key to achieving transparency, sustainability, and accountability [40].

The supply chain enables value creation, logistics management, synchronization with demand, and performance evaluation [26]. Supply chain management refers to the coordination of activities in the production and distribution of goods and services from suppliers to customers. This is achieved through the collaboration of various entities to ensure the efficient movement of products, overcoming demand variability and the risk of product obsolescence, as seen in the automotive sector [1].

2.1.1 Importance and Relationship of Supply Chain with Smart Manufacturing

Today, addressing the supply chain alongside smart manufacturing is increasingly common due to their interconnection. On the one hand, the former deals with aspects of Industry 4.0, which involves elements like machines, products, and people in supply chain and manufacturing processes [25]. On the other hand, the latter focuses on logistical operations from suppliers to final customers, utilizing technologies such as IoT, cloud computing, intelligent management theories, system automation, and interconnected work across all collaborating companies [50].

The integration of technologies into the supply chain has allowed companies to improve agility, customization, and responsiveness to changes and disruptions [14]. Convergence is achieved when digital twins and collaborative models for smart manufacturing, such as product development or lifecycle analysis, are applied [20, 41]. Additionally, there is greater confidence in the management of goods and products, as well as enhanced traceability and transparency throughout the entire supply chain [12].

Supply chain and smart manufacturing are closely related when approached from the perspective of Industry 4.0. For example, smart manufacturing refers to the incorporation of technologies to digitize processes within a company through extensive real-time data collection and analysis. This requires connectivity for information monitoring via IoT, cloud computing, and technologies like blockchain. This leads to the concept of an intelligent supply chain, given its positive impact on improving agility, efficiency, and quality throughout the supply chain. By adopting the principles of Industry 4.0 and smart supply chain management, companies can maintain their competitiveness in a dynamic business environment. Therefore, it is important to leverage cutting-edge technologies to enhance efficiency and transparency in production processes while adhering to the contemporary industrial approach [23].

2.2 Disruption in the Supply Chain

Next, we will review some studies that analyze and suggest solutions to reduce the impact of disruptions and their associated effects. The work of [51] evaluated the impact of transportation disruption on the second link of the supply chain, which impacts the service level. In [7] system dynamics is used to analyze the effects of terrorist acts on the performance of the supply chain at a global level and highlight the increase in inventory by 600% due to the increase in security measures on the border with Mexico. In [36] it is proposed a system dynamics model to analyze the probability of machine failure due to a shortage of spare parts inventory, which impacts the delivery of finished products. The study in [46] evaluates supplier disruption considering backorders using system dynamics. In [5] system dynamics and genetic algorithms are integrated to reduce the bullwhip effect in the supply chain and optimize inventory and demand forecasting values, as well as inventory costs and backorder costs. The work of [24] analyzes the domino effect in the supply chain using system dynamics considering inventory level, backorders, and demand. In the study of [30] it is developed a model to quantify the domino effect in the supply chain with a proactive approach using system dynamics that impacts service level, costs, profitability, and inventory levels. In [10] it is analyzed the impact of COVID-19 on the disruption of material flow using system dynamics, which influences changes in inventory level. The study of [13] visualizes the domino effect in the supply chain, characterized by severe disruptions due to risks in demand, suppliers, and logistics using system dynamics. The work of [21] considered the impact on service level. In [18] it is proposed that to mitigate supply chain disruption, suppliers must increase their production capacity and safety stock and also consider supplier delays due to material shortages and changes in demand. Finally in [6] it is analyzed the domino effect in the supply chain to maximize post-pandemic customer satisfaction using system dynamics.

2.3 *Anomaly Detection and DSS*

In the supply chain, anomalies could originate from the disruption in the supply of raw material shared and transferred among supply chain members, whether related to orders, products, or other factors. This leads to misleading decisions within companies regarding production or inventory levels [11]. Other studies attribute anomalies to equity levels in the supply chain, specifically when dominant companies gain more profit compared to subordinate companies, affecting internal decisions [49].

Anomalies also arise due to the complexity of supply chains, involving multiple customers, suppliers, and consumers. This complexity leads to disruptions in logistical operations, inaccuracies in inventory, and demand fluctuations [4]. For example, the bullwhip effect amplifies variations and is a common anomaly with negative impacts on inventory management. In [22] it is noted that supply chain anomalies can also result from delays at ports or airports, inflation, or material shortages due to the complexity and global involvement in current supply chains.

When anomalies occur in the supply chain, they increase costs, cause product delivery delays, and decrease operational efficiency and performance. Detecting anomalies in supply chain management is important for preventing problems and optimizing efficiency. In [27] it is highlighted the importance of identifying these instances on time, which can improve the response of the managers to make decisions that improve the scenario in the face of disruption.

Due to the problems caused by anomalies and disruptions in the supply chain, several studies have proposed using DSS to anticipate and improve responses to these events. For example, in [34], a knowledge-based IDSS is proposed for managing operational risks in global supply chains. Their system predicts supply chain performance by employing artificial neural networks alongside particle swarm optimization. It also identifies risk sources using principal component analysis and assesses risk mitigation options using a digraph-matrix approach combined with principal component analysis. In the work of [16] it is proposed a fusion IDSS which employs data envelopment analysis and rough set theory to make an analysis of the risks in the supply chain. The study in [37] proposed the use of ontology for the design of an IDSS to improve supply chain resilience against disruptions. They used semantics and ontology to develop a basic knowledge base of the supply chain network. Their knowledge base uses the semantic web rule language to encode properties types and classes and subclasses in the supply chain. Additionally they used mixed integer linear programming to optimize a quantified resilience objective function. To solve the problem they employ a hybrid particle swarm optimization with differential evolution. In [44] it is developed a DSS to assist in the selection of strategies for risk management. Their system uses a rule-based fuzzy inference system to model the decision variables in the DSS.

3 Methodological Design

The objective of this research is to address anomaly detection in the supply chain using a DSS that integrates system dynamics modeling and machine learning algorithms to predict abnormal behaviors. This approach aims to enable early detection, allowing for timely correction of errors. Figure 1 shows the general outline of the proposed IDSS.

3.1 IDSS for Disruption Detection

In this chapter, an IDSS has been chosen due to its ability to handle large volumes of data and improve decision-making accuracy. The increasing complexity of modern supply chains and the frequency of disruptions necessitate advanced tools for effective management. The proposed system will primarily focus on disruption detection, but it will also offer other key process indicators. This section provides a detailed description of the IDSS shown in Fig. 1, including its components and functionality.

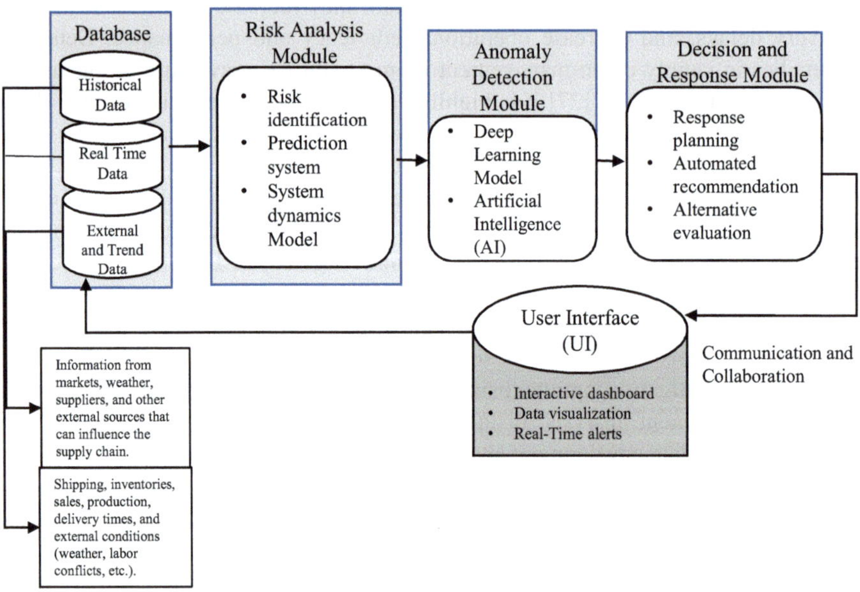

Fig. 1 Proposed IDSS for anomaly detection in the supply chain

3.1.1 Information Sources Module

The first component of the IDSS is the information sources. In this case, various sources have been selected such as historical data, real-time data, and external and trend data. Historical data are important for analysis and strategic decision-making, as it helps identify patterns, trends, and areas for improvement, as well as predict future behaviors and plan more efficiently. Historical data consist of detailed records collected over time on various activities and operations within the industry. These data can include production and sales volumes per period, order and delivery history, inventory levels of raw materials, work in progress, and finished products, purchase history of raw materials and components, and shipping and delivery history, among others. In this IDSS, particular emphasis will be placed on the shipping and delivery history by monitoring shipping rates and receiving delays. In addition to historical data, real-time monitoring of the company's daily operations will also be conducted. Additionally, external and trend data are analyzed since these data provide information about market conditions, competitor activities, and economic trends. This helps businesses understand their competitive landscape and identify opportunities and threats. We also consider trend data in order to help identify long-term shifts in consumer behavior, technological advancements, and industry standards. By processing these trends, the IDSS can adapt their strategies and scenarios for better decision support.

3.1.2 Risk Analysis Module

This module contains the system's prediction models, which include a time series prediction model and a system dynamics model. The prediction model takes time series data from the Information Sources Module as input. The module can predict future time series data over a specified prediction horizon, providing possible future scenarios. Additionally, the prediction output can feed into the system dynamics model, which can simulate the behavior of other industrial plant processes based on the prediction horizon.

The prediction system is not implemented in this chapter as there are already well-established techniques in the literature, from classical methods like exponential smoothing and Autoregressive Integrated Moving Average (ARIMA) to modern techniques based on Deep Learning, such as Facebook's Prophet [28, 39]. Since we are particularly interested in disruption detection, the prediction system will take receiving delays and receiving rates as input and use system dynamics modeling to simulate future shipping rates. This simulation will then be used by the proposed Deep Learning network for disruption detection. A block diagram of the operation of this module is shown in Fig. 2.

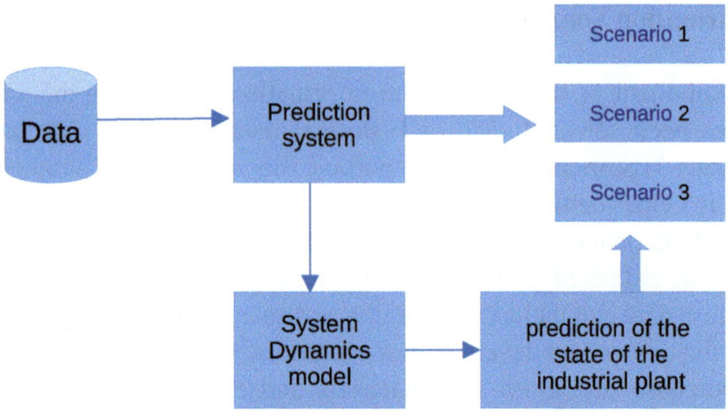

Fig. 2 Block diagram for the module using the predictor and the system dynamics model

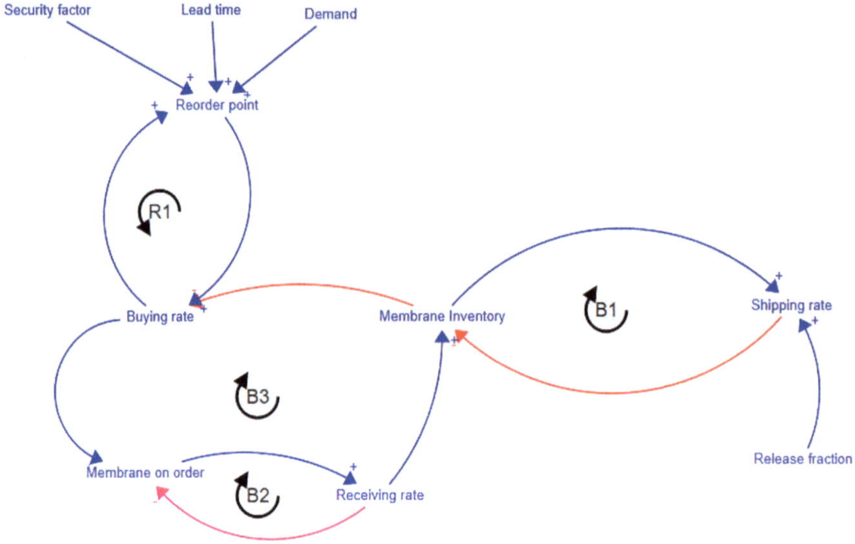

Fig. 3 Causal diagram of the proposed system dynamics model

3.1.3 System Dynamics Model

In this section the design of the system dynamics model is presented. The model is tailored to the metalworking industry, specifically supplier manufacturing membranes for automobiles.

Figure 3 shows the causal diagram of the disruptive model used to describe potential causes and effects through dynamic hypotheses that help determine the system's behavior over time. It establishes a feedback loop between two or more variables that are related either positively or negatively. Arrows indicate a positive relationship when an increase in one variable leads to an increase in the other, represented by a positive feedback loop (R). Conversely, a negative relationship is indicated when an increase in one variable leads to a decrease in the other, represented by a negative feedback loop (B). The causal model is constructed to determine the relationship between the variables of interest: Buying Rate, Membrane Orders, Receiving Rate, Membrane Inventory, Shipping Rate, and Reorder Point.

- B1: Shipping Rate: If the shipping rate increases, the membrane inventory level decreases. If the membrane inventory level increases, the shipping rate decreases.
- B2: Receiving Rate: If the receiving rate from the supplier increases, the number of membrane orders placed with the supplier decreases.
- B3: Membrane Inventory Level: If the buying rate decreases, the number of orders placed with the supplier increases. Consequently, the receiving rate of orders in the raw materials warehouse increases, leading to an increase in membrane inventory levels.
- R1: Buying Rate: If the buying rate increases, the reorder point increases under a continuous review policy. If the reorder point increases, the buying rate from the membrane supplier also increases.

Figure 4 illustrates the simulation model using a system dynamics approach. The model includes two stock variables: "Membrane Orders" and "Membrane automotive Inventory." It assumes a requested quantity of 100,000 membranes from the supplier, with an initial membrane inventory set to zero. The flow variables are:

- Buyer's Order Rate: This is the difference between the reorder point and the membrane inventory, adjusted for a 2-month delay due to the global steel shortage. The reorder point is the minimum inventory level at which a new order is placed with the supplier. The demand follows a normal distribution with a monthly average of 100,000 membranes and a standard deviation of 4,000 membranes. The supplier takes 2 weeks (0.5 months) to deliver an order, with a 95% safety factor to meet the customer's demand. The safety stock level is set at 1,200 membranes.
- Receiving Rate: This represents the rate at which materials enter the raw material warehouse. It is calculated as the number of membranes ordered from the supplier divided by the delay in receiving materials due to transportation delays from the supplier to the company.
- Shipping Rate: This indicates the rate at which prepared membranes are shipped to the customer. It is determined by dividing the membrane inventory level by the proportion of membranes released by the customer, which ranges from 0.05 to 0.125.

The flow variables are represented by arrows, while parameters such as delivery time, delay adjustments, and the membrane release rate by the customer are represented by circles.

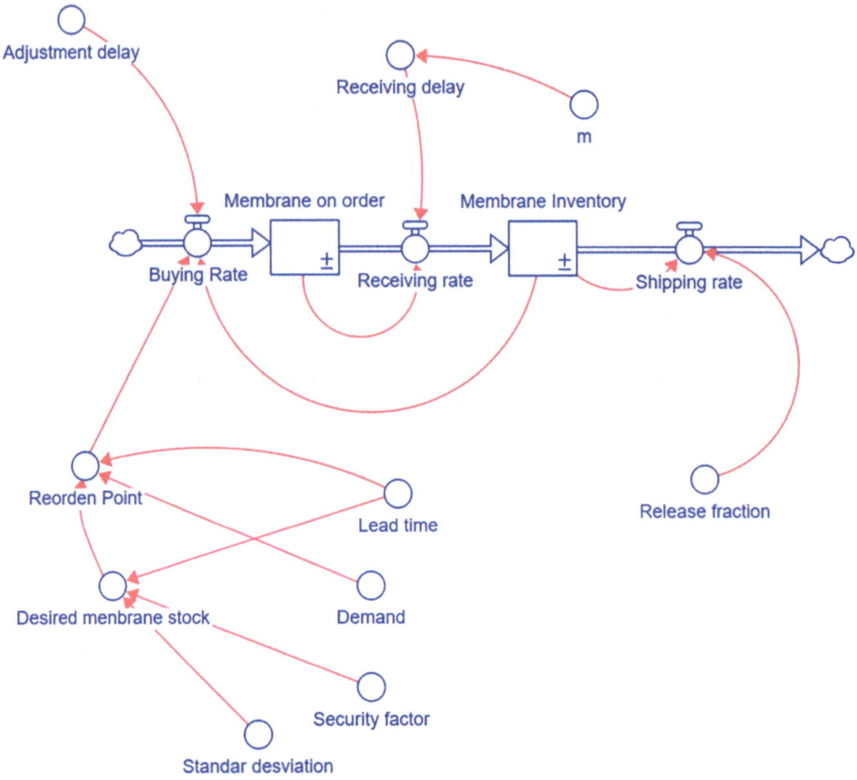

Fig. 4 Proposed system dynamics model

3.1.4 Anomaly Detection Module

In this module, a neural network architecture is proposed for the detection of disruptions. To do this, the shipping rate will be evaluated in search of any pattern that may indicate the beginning of a disruption.

Let $X = \{x_n\}_{n=0}^{N}$ be the time variation in the shipping rate; the task of the proposed network architecture is to detect a disruption from samples x_n from X. In terms of machine learning a sample x_n will be classified as part of a disruption or not. In order to provide context for x_n, the input to the network is a subsequence, X_n of X, where $X_n = \{x_{n-L-1}, \ldots, x_{n-2}, x_{n-1}, , x_n\}$ and L is the size of the subsequence. Figure 5 illustrates this process.

For the design of the network architecture, a convolutional element is chosen to serve as the input layer. The convolutional layer is capable of obtaining rich features from the input. This layer convolved the input subsequence X_n, with several kernels or filters with learnable parameters. The output tensor consists of a series of feature maps that help in the process of classification of the following layers in the network. For the case of the proposed network architecture, we used a one-dimensional convolutional layer with five filters or kernels, each of size 2, and also

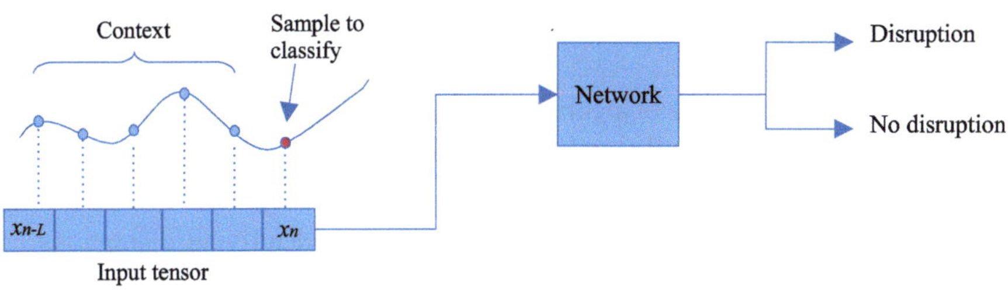

Fig. 5 Construction of the input tensor from the sample to classify and its context

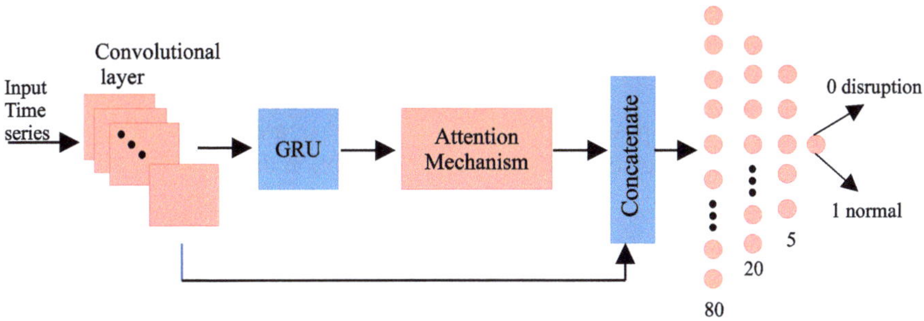

Fig. 6 Proposed deep learning architecture. From a time series input, it is able to detect whether or not there are disruptions

we fix the length of the input subsequence to $L = 14$. Once that the input sequence is transformed into a sequence of features found by the convolutional layer, it is processed by a Gated Recurrent Unit (GRU) layer [8]. The GRU processes the input data allowing information contained in the sequence to be selectively remembered or forgotten over time, thus allowing to handle long-term dependencies in sequential data by selectively remembering and forgetting previous inputs. Here we used a GRU with 24 units, the output of the GRU is passed to an attention layer, and these layers are known to selectively focusing on important elements of the input tensor emphasizing relevant information for the next layer to improve the classification performance. The output of the attention layer is concatenated with the output of the previous convolutional layer to obtain a richer feature tensor. This tensor is passed to dense network consisting of four layers of 80, 20, and 5 neurons, respectively, all with ReLU activation functions, and finally the final classification is done by a neuron with sigmoid activation function, which classifies the input in disruption or no disruption. The complete architecture is show in Fig. 6.

3.1.5 Decision and Response Module

The decision and response module is designed to aid decision-makers in identifying optimal actions and responses based on analyzed data and predicted scenarios. This module plays an important role in translating insights from data analysis into actionable decisions and is attached directly to the dashboard.

3.2 Metrics for Disruption Detection

To evaluate the performance of algorithms for abnormality/disruption detection, the following metrics are generally used.

Sensitivity
This metric evaluates how good a given model is at identifying disruptions as a percentage of all disruptions days in the data. It evaluates the proportion of disruption days that the method is able to correctly identify. This metrics is also called recall and is calculated as follows:

$$Sensitivity = \frac{TP}{P} \quad (1)$$

where TP are the true positives, in this context, these are the days classified as disruption days by the detector that are indeed disruption days, and P are the total of disruption days.

Specificity
This metric evaluates the detector's ability to identify days of no disruption as a percent of all of the observations without disruptions. This metric is calculated as

$$Specificity = \frac{TN}{N} \quad (2)$$

where TN are the true negatives, in this case the days detected as non-disruption days, and N are the number of days without disruption.

Miss Rate
It is also known as false negative rate and is an estimate of the probability that a true disruption day will be missed by the test. It is calculated as

$$MissRate = \frac{FN}{P} \quad (3)$$

Where FN are the false negatives.

Fall-Out
It is also called false positive rate or probability of false alarm. This metric measures the proportion of non-disruption days that were incorrectly identified as disruption, that is, false alarms. The fall-out is given by

$$Fall\text{-}out = \frac{FP}{P} \quad (4)$$

where FP are false positives, that is, false disruption days.

4 Results

In this section, we validate the Deep Learning architecture for disruption detection. We generated data by simulating disruptions in the supply chain using the system dynamics model. Specifically, we simulated a total of 1,200 days and introduced six disruptions in the receiving delay. These disruptions are modeled using ramp functions with varying slopes and number of days. Table 1 provides detailed descriptions of each disruption, and Fig. 7 illustrates a segment of the simulated data depicting one of these disruptions. This will be the input to the system dynamics model, from which in turn the shipping rate will be taken as input to the neural network. Note that in this way the network can work with data that is not directly measured or with data with which it was not trained. This allows the network to be reused with different data.

Table 1 Detail of the disruptions

Disruptions	Beginning	Days	Slope
Disruption 1	200	20	0.7
Disruption 2	350	20	0.55
Disruption 3	500	40	0.85
Disruption 4	650	20	0.35
Disruption 5	1000	15	0.55
Disruption 6	1100	15	0.75

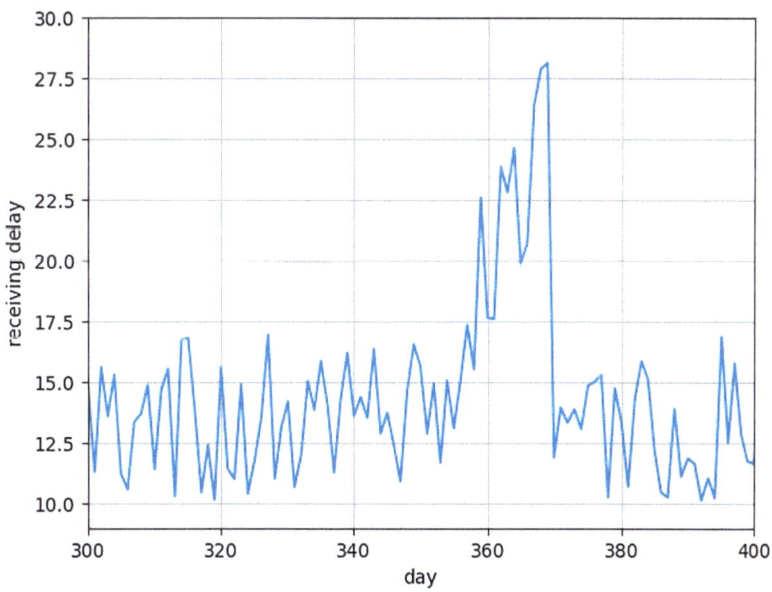

Fig. 7 Simulation data, a disruption is observed that begins on day 350 with an approximate duration of 20 days

The proposed algorithm for disruption detection was compared with two other methods: one based on support vector machines (SVMs), commonly used in detection literature [15] and another using autoencoders based on LSTM (LSTM AC) [19]. For the SVM, a regularization parameter of one and a radial basis function (RBF) kernel were used. No optimization was done to adjust further the hyperparameters. The LSTM-based network used two LSTM layers with 104 units each; the first was in sequence-to-sequence mode, and the second returned only the last output. The decoder had the same configuration, but at the end, a dense network with four layers consisting of 150, 50, 10, and 1 neurons was used. All layers had ReLU activation, except for the last one, which used a sigmoid activation function. To build the training and test sets, the data from the systems dynamics model were divided into subsequences of 14 samples. If a subsequence contained days within the disruption, it was labeled as a disruption; if the subsequence contained no days from the disruption, it was labeled as non-disruption. Note that, in this way, there may be more subsequences labeled as disruptions than the number of simulated disruptions. For training all methods, 70% of the simulated data were used, while the remaining 30% were reserved for testing.

Figure 8 shows the confusion matrices obtained by the different methods. The test data contained two disruptions of 15 days each, while the rest of the days contain no disruptions. The proposed method is the one that correctly detected the most days of disruption, while the SVM method is the one that correctly detects the least number of days of disruption, only 13 of the 30 days with disruptions.

Table 2 presents metrics summarizing the performance of each method. Our proposed method achieved the highest sensitivity at 0.8, indicating its superior ability to identify disruption days. The LSTM AC method followed with a sensitivity of 0.6, which is lower than our proposed method. For the specificity metric, all methods performed well, with LSTM AC achieving the highest score of 0.99, closely followed by SVM and our proposed method at 0.98. The proposed method also showed the lowest miss rate at 0.2, meaning it detected most disruption days correctly. In comparison, LSTM and SVM had higher miss rates of 0.4 and 0.56, respectively. All methods exhibited low fall-out rates (probability of false alarm), with LSTM achieving the lowest.

5 Conclusion

In this chapter was presented an approach to anomaly detection in smart manufacturing using an Intelligent Decision Support System. The proposed IDSS uses a prediction system to feed a system dynamics model, predicting future states of the industrial plant and the subsequent detection of disruptions through artificial neural networks. The integration of system dynamics modeling and deep learning algorithms allows for a more comprehensive understanding of the complex interactions within the supply chain, facilitating better decision for planning. The proposed deep learning-based network that incorporated an attention mechanism

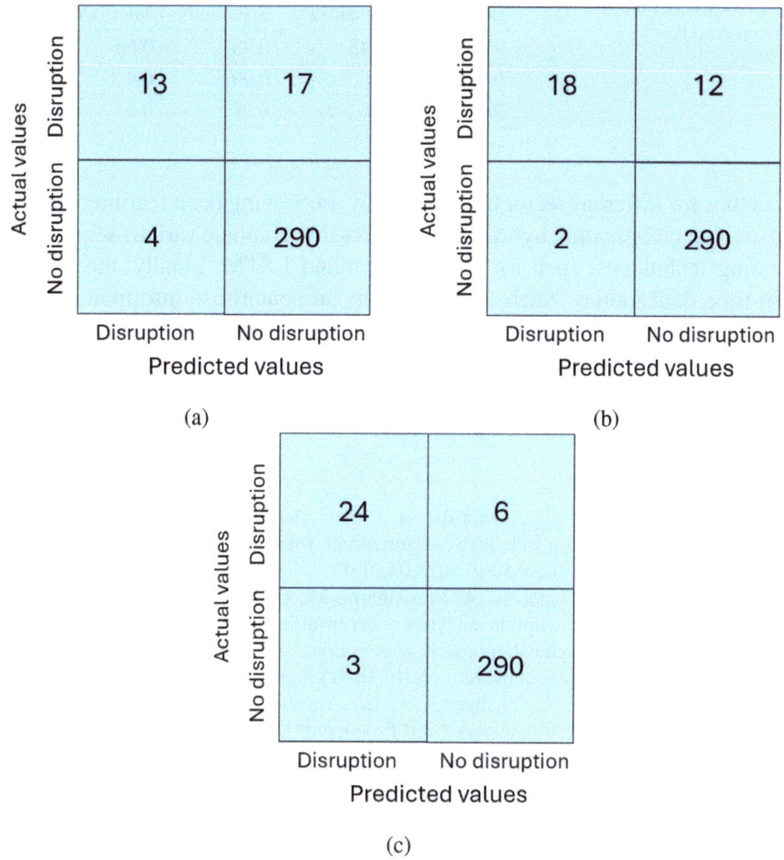

Fig. 8 Confusion matrices of the different methods: (**a**) SVM, (**b**) LSTM AC, and (**c**) proposed

has proven to be effective in identifying disruptions within the supply chain. In comparison it was demonstrated that the proposed approach outperforms the other methods for the task of disruption detection. The designed model using system dynamics was tailored for the metalworking industry, specifically for a supplier manufacturing membranes for automobiles, and, however, underscores the practical applicability of this research, and the model can be adapted and applied to other industries or manufacturing contexts. The work presented in this chapter could contribute to the advancement of smart manufacturing by providing a new tool that enhances the efficiency, reliability, and responsiveness of supply chains, by proactively detecting and addressing anomalies.

Based on the results obtained in this research on anomaly detection in smart manufacturing using an IDSS, several directions for future work can be explored. One possibility is the expansion of the system to other industries, focusing on the adaptation and validation of the system dynamics model and deep learning

Table 2 Metrics of the different methods

Detector	Sensitivity	Specificity	Miss rate	Fall-out
SVM	0.43	0.98	0.56	0.013
LSTM AC	0.6	0.99	0.4	0.006
Proposed	0.8	0.98	0.2	0.010

architecture for different sectors. Additionally, improving deep learning algorithms could involve investigating hybrid architectures that combine various sequential data processing techniques, such as Transformers and LSTM. Finally, the integration of real-time data sources, such as IoT sensors and enterprise information systems, could enhance the IDSS's ability to detect and respond to anomalies.

References

1. Achetoui, Z., Mabrouki, C., & Mousrij, A. (2019). Performance measurement system for automotive spare parts supply chain: a categorization. *Journal of Transportation and Logistics, 4*, 31–50. https://doi.org/10.26650/jtl.2018.04.01.03
2. Ali, S. M., Bari, A. M., Rifat, A. A. M., Alharbi, M., Choudhary, S., & Luthra, S. (2022). Modelling supply chain disruption analytics under insufficient data: A decision support system based on Bayesian hierarchical approach. *International Journal of Information Management Data Insights, 2*(2), 100121. https://doi.org/10.1016/j.jjimei.2022.100121
3. Andry, J. F., Nurprihatin, F., & Liliana, L. (2023). Developing a decision support system for supply chain component. *Management and Production Engineering Review, 14*(2), 124–133. https://doi.org/10.24425/mper.2023.146029
4. Azzi, R., Kilany, R., & Sokhn, M. (2019). The power of a blockchain-based supply chain. *Computers Amp; Industrial Engineering, 135*, 582–592. https://doi.org/10.1016/j.cie.2019.06.042
5. Badakhshan E., Humphreys, P., Maguire, L., & Mcivor, R. (2020). Using simulation-based system dynamics and genetic algorithms to reduce the cash flow bullwhip in the supply chain. *International Journal of Production Research, 58*(17), 5253–5279. https://doi.org/10.1080/00207543.2020.1715505
6. Brusset, X., Ivanov, D., Jebali, A., La Torre, D., & Repetto, M. (2023). A dynamic approach to supply chain reconfiguration and ripple effect analysis in an epidemic. *International Journal of Production Economics, 263*, 108935.
7. Bueno-solano A., & Cedillo-campos, M. G. (2014). Dynamic impact on global supply chains performance of disruptions propagation produced by terrorist acts. *Transportation Research Part E, 61*, 1–12. https://doi.org/10.1016/j.tre.2013.09.005
8. Cho, K., Van Merriënboer, B., Gulcehre, C., Bahdanau, D., Bougares, F., Schwenk, H., & Bengio, Y. (2014). Learning phrase representations using RNN encoder-decoder for statistical machine translation. arXiv preprint arXiv:1406.1078. https://doi.org/10.48550/arXiv.1406.1078
9. Cruz, L., & Pravia, M. (2022). Gestión integrada de riesgos de la seguridad de las cadenas de suministros con enfoque al servicio al cliente. *Ingeniería Y Competitividad, 24*(02), 1–14. https://doi.org/10.25100/iyc.v0i00.11197
10. Duan W., Ma, H., & Xu, D. S. (2021). Analysis of the impact of COVID-19 on the coupling of the material flow and capital flow in a closed-loop supply chain. *Advances in Production Engineering & Management, 16*(1), 5–22. https://doi.org/10.14743/apem2021.1.381

11. Fischer, M., Shin, H. S., & Hanssens, D. M. (2016). Brand performance volatility from marketing spending. *Management Science, 62*(1), 197–215. https://doi.org/10.1287/mnsc.2014.2102
12. Frankó, A., Vida, G., & Varga, P. (2020). Reliable identification schemes for asset and production tracking in industry 4.0. *Sensors, 20*(13), 3709. https://doi.org/10.3390/s20133709
13. Ghadge, A., Er, M., Ivanov, D., & Chaudhuri, A. (2022). Visualisation of ripple effect in supply chains under long-term, simultaneous disruptions: A system dynamics approach. *International Journal of Production Research, 60*(20), 6173–6186.
14. Gupta, S., Drave, V. A., Bag, S., & Luo, Z. (2019). Leveraging smart supply chain and information system agility for supply chain flexibility. *Information Systems Frontiers, 21*(3), 547–564. https://doi.org/10.1007/s10796-019-09901-5
15. Hosseinzadeh, M., Rahmani, A. M., Vo, B., Bidaki, M., Masdari, M., Zangakani, M. (2021). Improving security using SVM-based anomaly detection: Issues and challenges. *Soft Computing, 25*(4), 3195–3223. https://doi.org/10.1007/s00500-020-05373-x
16. Hu, K. H., Chen, F. H., Hsu, M. F., Yao, S., & Hung, M. C. (2021). Identification of the critical factors for global supply chain management under the COVID-19 outbreak via a fusion intelligent decision support system. *Axioms, 10*(2), 61. https://doi.org/10.3390/axioms10020061
17. Ivanov, D. (2020). Predicting the impacts of epidemic outbreaks on global supply chains: A simulation-based analysis on the coronavirus outbreak (COVID-19/SARS-CoV-2) case. *Transportation Research Part E: Logistics and Transportation Review, 136*, 101922. https://doi.org/10.1016/j.tre.2020.101922
18. Jonsson, P., Öhlin, J., Shurrab, H., Bystedt, J., Sheikh Muhammad, A., & Verendel, V. (2024). What are the root causes of material delivery schedule inaccuracy in supply chains?. *International Journal of Operations and Production Management, 44*(13), 34–68. https://doi.org/10.1108/IJOPM-12-2022-0806
19. Jung, G., Hyunsoo, H., Lee, S. (2021). Anomaly detection of facilities and non-disruptive operation of smart factory using kubernetes. *Journal of Information Processing Systems, 17*(6). https://doi.org/10.3745/JIPS.01.0083
20. Karadgi, S. (2022). A framework towards realization of smart manufacturing systems. *IOP Conference Series: Materials Science and Engineering, 1258*(1), 012018. https://doi.org/10.1088/1757-899x/1258/1/012018
21. Lai X., Chen, Z., Wang, X., & Chiu, C. H. (2023). Risk propagation and mitigation mechanisms of disruption and trade risks for a global production network. *Transportation Research Part E Logistics and Transportation Review, 170*(December 2022), 103013. https://doi.org/10.1016/j.tre.2022.103013
22. Lam, J. S. L., & Yip, T. L. (2012). Impact of port disruption on supply chains: a Petri net approach. In *Computational Logistics: Third International Conference, ICCL 2012, Shanghai, China, September 24–26, 2012. Proceedings 3* (pp. 72–85). Springer. https://doi.org/10.1007/978-3-642-33587-7_5
23. Lee, C. K. M., Huo, Y., Zhang, S. Z., & Ng, K. M. (2020). Diseño de un sistema de fabricación inteligente con la aplicación de edge computing de acceso múltiple y tecnología blockchain. *Access IEEE, 8*, 28659–28667. https://doi.org/10.1109/access.2020.2972284
24. Llaguno, A., Mula, J., & Campuzano-Bolarin, F. (2021). State of the art, conceptual framework and simulation analysis of the ripple effect on supply chains. *International Journal of Production Research, 60*(6), 2044–2066.
25. Marx, E., Stierle, M., Weinzierl, S., & Matzner, M. (2020). Closing the gap between smart manufacturing applications and data management. *WI2020 Community Tracks* (pp. 120–136). https://doi.org/10.30844/wi_2020_u1-marx
26. Menon, S., & Shah, S. (2019). An overview of digitalisation in conventional supply chain management. *MATEC Web of Conferences, 292*, 01013. https://doi.org/10.1051/matecconf/201929201013

27. Montiel H. N., Neme Castillo, J. A., & Mota Pantoja, S. I. (2021). Las anomalías: ¿qué son?,¿dónde surgen?,¿cómo detectarlas?. https://doi.org/10.22201/dgtic.26832968e.2021.4.3. https://www.ties.unam.mx/num04/index.html
28. Munim, Z. H., Fiskin, C. S., Nepal, B., & Chowdhury, M. M. H. (2023). Forecasting container throughput of major Asian ports using the Prophet and hybrid time series models. *The Asian Journal of Shipping and Logistics, 39*(2), 67–77. https://doi.org/10.1016/j.procs.2021.01.036
29. Nguyen, H., Tran, K., Thomassey, S., & Hamad, M. (2021). Forecasting and anomaly detection approaches using LSTM and LSTM autoencoder techniques with the applications in supply chain management. *International Journal of Information Management, 57*, 102282. https://doi.org/10.1016/j.ijinfomgt.2020.102282
30. Olivares-Aguila J., & ElMaraghy, W. (2021). System dynamics modelling for supply chain disruptions. *International Journal of Production Research, 59*(6), 1757–1775. https://doi.org/10.1080/00207543.2020.1725171
31. Oloke, O. C., Sholanke, A. B., Akindele, N. A., & Akinola, A. O. (2022). A short review of the concept and principles of supply chain management in building construction industry. *IOP Conference Series: Earth and Environmental Science, 1054*(1), 012050. https://doi.org/10.1088/1755-1315/1054/1/012050
32. Orlando, B., Tortora, D., Pezzi, A., & Bitbol-Saba, N. (2022). The disruption of the international supply chain: Firm resilience and knowledge preparedness to tackle the COVID-19 outbreak. *Journal of International Management, 28*(1), 100876. https://doi.org/10.1016/j.intman.2021.100876
33. Osorio, A. F., Brailsford, S. C., Smith, H. K., Forero-Matiz, S. P., & Camacho-Rodríguez, B. A. (2017). Simulation-optimization model for production planning in the blood supply chain. *Health Care Management Science, 20*(4), 548–564. https://doi.org/10.1007/s10729-016-9370-6
34. Park, Y. B., Yoon, S. J., & Yoo, J. S. (2018). Development of a knowledge-based intelligent decision support system for operational risk management of global supply chains. *European Journal of Industrial Engineering, 12*(1), 93-115. https://doi.org/10.1504/EJIE.2018.089878
35. Prasetia, F. T., & Imaroh, T. S. (2020). Contractor selection assessment strategy in the upstream oil and gas industry towards green supply chain management. *Dinasti International Journal of Economics, Finance Amp; Accounting, 1*(3), 373–383. https://doi.org/10.38035/dijefa.v1i3.314
36. Sánchez-ramírez, C., Ramos-hernández, R., Fong, J. R. M., Alor-Hernández, G., & García-alcaraz, J. L. (2020). A system dynamics model to evaluate the impact of production process disruption on order shipping. *Applied Sciences, 10*(1), 208. https://doi.org/10.3390/app10010208
37. Singh, S., Ghosh, S., Jayaram, J., & Tiwari, M. K. (2019). Enhancing supply chain resilience using ontology-based decision support system. *International Journal of Computer Integrated Manufacturing, 32*(7), 642–657. https://doi.org/10.1080/0951192X.2019.1599443
38. Sawik, T. (2018). Selection of a dynamic supply portfolio under delay and disruption risks. *International Journal of Production Research, 56*(1–2), 760–782. https://doi.org/10.1080/00207543.2017.1401238
39. Shawon, M. M. H., Akter, S., Islam, M. K., Ahmed, S., & Rahman, M. M. (2020). Forecasting PV panel output using Prophet time series machine learning model. In 2020 IEEE region 10 conference (TENCON) (pp. 1141–1144). IEEE. https://doi.org/10.1109/TENCON50793.2020.9293751
40. Schöneich, S., Saulich, C., & Müller, M. (2023). Traceability and foreign corporate accountability in mineral supply chains. *Regulation Amp; Governance, 17*(4), 954–969. https://doi.org/10.1111/rego.12527
41. Tang, Q., Wu, B., Chen, W., & Jinyue, Y. (2023). A digital twin-assisted collaborative capability optimization model for smart manufacturing system based on Elman-IVIF-TOPSIS. *IEEE Access, 11*, 40540–40564. https://doi.org/10.1109/access.2023.3269577

42. Tarei, P. K., Thakkar, J. J., & Nag, B. (2018). A hybrid approach for quantifying supply chain risk and prioritizing the risk drivers: A case of Indian petroleum supply chain. *Journal of Manufacturing Technology Management, 29*(3), 533–569. https://doi.org/10.1108/JMTM-10-2017-0218
43. Tarei, P. K., Thakkar, J. J., & Nag, B. (2020). Benchmarking the relationship between supply chain risk mitigation strategies and practices: an integrated approach. *Benchmarking: An International Journal, 27*(5), 1683–1715. https://doi.org/10.1108/BIJ-12-2019-0523
44. Tarei, P. K., Thakkar, J. J., & Nag, B. (2021). Development of a decision support system for assessing the supply chain risk mitigation strategies: An application in Indian petroleum supply chain. *Journal of Manufacturing Technology Management, 32*(2), 506–535. https://doi.org/10.1108/JMTM-02-2020-0035
45. Teniwut, W., & Hasyim, C. (2020). Decision support system in supply chain: A systematic literature review. *Uncertain Supply Chain Management, 8*(1), 131–148. https://doi.org/10.5267/j.uscm.2019.7.009
46. Thomas V. A., & Mahanty, B. (2020). Assessment of emergency sourcing strategy of a supply chain through dynamic simulation approach. *Journal of Industrial and Production Engineering, 37*(1), 56–69. https://doi.org/10.1080/21681015.2020.1729877
47. Tsiamas, K., & Rahimifard, S. (2021). A simulation-based decision support system to improve the resilience of the food supply chain. *International Journal of Computer Integrated Manufacturing, 34*(9). https://doi.org/10.1080/0951192X.2021.1946859
48. Wang, H., Tan, J., Guo, S., & Wang, S. (2018). High-value transportation disruption risk management: Shipment insurance with declared value. *Transportation Research Part E: Logistics and Transportation Review, 109*, 293–310. https://doi.org/10.1016/j.tre.2017.11.013
49. Wang, Y., Yu, Z., Shen, L., Yan, G., & Li, J. (2018). Different dominant models and fair-ness concern of e-supply chain. *Complexity, 2018*, 1–13. https://doi.org/10.1155/2018/8616595
50. Wei, S., Liu, W., Lin, Y., Wang, J., & Liu, T. (2021). Smart supply chain innovation model selection: exploitative or exploratory innovation?. *International Journal of Logistics Research and Applications, 26*(4), 478–497. https://doi.org/10.1080/13675567.2021.1965104
51. Wilson M. C. (2007). The impact of transportation disruptions on supply chain performance. *Transportation Research Part E Logistics and Transportation Review, 43*(4), 295–320. https://doi.org/10.1016/j.tre.2005.09.008
52. Yin, K., Wu, J., Wang, W., Lee, D. H., & Wei, Y. (2023). An integrated resilience assessment model of urban transportation network: A case study of 40 cities in China. *Transportation Research Part A: Policy and Practice, 173*, 103687. https://doi.org/10.1016/j.tra.2023.103687

Securing Financial Systems: The Role of DSS in Cybersecurity

Rabiu Abubakar and Fezile Ozdamli

1 Introduction

Financial cybersecurity has become popular in recent years due to increasing cyber-attacks through phishing, social engineering, and other ways employed by cybercriminals to steal sensitive financial data for companies, employees, and customers. Information has increasingly become a valuable asset for business institutions [17]. However, it is observed that the value of information differs among institutions based on the importance of the information assets [13]. Cyber-attacks are motivated for financial and personal benefits, costing the industry an unprecedented financial loss [21]. The aggregation of financial losses leads to a fall in financial institutions' total profits, which has a direct correlation with a country's economic growth [5]. Additionally, the inequalities resulting from negative economic growth will continue to widen among the affected countries, which is inconsistent with the sustainable development agenda of the United Nations, SDGs.

According to Rohmeyer and Bayuk [20], the motives of cybercriminals include theft for funds, theft for information, and threats for disruption. The authors stated that fund theft motivates cybercrimes in financial institutions, even though the goal might be for personal gain which is simple theft motive, or fund theft aiming to build powerful architecture to enable the scammers develop more robust attack capabilities. The increasing security features of the financial institutions face corresponding level of sophisticated cybercrimes. There are many potential motivations for the theft of information, which include facilitating a subsequent theft of funds, disruption to the company or employees workflow or setting up more opportunities

R. Abubakar (✉) · F. Ozdamli
Department of Innovation and Knowledge Management, Near East University, Nicosia, Turkey
e-mail: rabiu.abubakar@neu.edu.tr; fezile.ozdamli@neu.edu.tr

N. Gupta, J. D. Diaz Roman (eds.), *Decision Support Systems for Industrial and Scientific Applications*, Signals and Communication Technology,
https://doi.org/10.1007/978-3-031-89700-9_16

for further information theft. Nevertheless, of the motive of a scammer, the threat indicates potential attacks on an organization's valuable information, which may result in individual or competitive gains including disrupting the victim. The last motive for cybercrimes is the threat of disruption. In this motive, the attackers may aim to disrupt systems or data to deter the resources unavailable to system users. For instance, in a financial services setting, a competitor may sponsor an attack to disrupt a trading algorithm, resulting in the victim company's inability to effect transactions. This event could have short and long-run effects. In the short run, the customer may not transact with the company and may choose to transact with the other enterprise that might sponsor the disruption. Frequent occurrences of disruption might lead to reputational risk to the victim company in the long run and loss of customers to competitors by a rival company to another. All the motives mentioned above directly impact a company's decision support system.

The popularity of the Internet and financial services has increasingly exposed the financial sector to frequent attacks like prey to a hungry carnivore. This requires relentless efforts from various stakeholders such as financial sector regulators, financial institutions, and academics to launch innovative policy addressing cyber threats. The International Monetary Fund [14] revealed that the cost of cybercrimes in the financial sector stood at USD 11.50 trillion in 2023, and it is expected to rise to USD 14.57 trillion by the end of 2024. The implications are enormous; while many financial institutions may not sufficiently honor their obligations, leading to plummeting in their revenue streams and loss of customers, the government may also lose taxes due to cybercrimes. Therefore, making the Internet safer is critical for management and clients [15]. The study further outlined motivational factors that caused cyber terrorism, including website supportive nature, anonymity nature of the Internet, hacking, computer viruses, and password sniffing.

Pande [19] categorized cybercrime affecting an institution's decision support systems into insider, external, unstructured and structured attacks. Insider attack cybercrimes occur when an individual or group with authorized access to a company's system launches an attack against their institution. Unhappy or greedy employees or vendors mainly wage this attack. A company can prevent insider attacks by installing internal intrusion detection systems (IDS). An external attack is launched by either a recruit of the insider or an external entity to the company. The victim organization always suffers financial and reputational loss from stealing sensitive business or employee information. IDS can also be used to prevent the external attack on an organization. An unstructured attack is carried out by amateurs who may not have established motives and randomly try testing a tool over the Internet. A structured attack is conducted by expert(s), highly trained and skilled criminals whose goals of the attack are to perpetrate crimes. The attackers can use advanced tools to penetrate any network without being deterred by IDS. These attackers have sufficient knowledge to develop or modify the existing tool to achieve their objective. Structural attacks are mainly performed by white-color criminals, by a nation to another rival nation, or by politicians to harm the reputation of a rival politician's company.

Therefore, this chapter highlights cybersecurity decision support systems related to financial institutions. Various techniques cybercriminals use in attacking financial institutions and their protective countermeasures will be discussed. Similarly, decisions support systems aiding the detection, managing, and mitigating cyber-attack's impact are discussed.

2 Types of Cybercrimes in Financial Institutions

Cybercrimes in banks and other financial institutions have resulted in launching various illegal activities aimed at institutions offering financial services, to steal money, data or disrupt an institution's business operations.

There are various kinds of cybercrimes launched against financial institutions put forward by Pande [19], which are briefly discussed as follows:

2.1 Phishing and Spear Phishing

Phishing: Cybercriminals send unsolicited emails claiming to be from a reliable personality or institution to lure employees into divulging sensitive information like passwords, account details, the company's business strategy, credit card details, and business agreements.

Spear Phishing: A thin line difference with phishing is that attackers in spear phishing personalized a message targeting a particular person or institution to induce the victim to believe that such a message is genuine in order to access the victim's network easily.

2.2 Malware Attacks

Trojans: This is a vicious software camouflaged as genuine software that could grant attackers access if executed. Installing the software can enable attackers to monitor users and steal valuable information, including money.

Ransomware: Financial institutions are major targets of this attack (Fig. 1). One type of cybercrime targeting financial institutions is the ransomware attack. These attacks typically start with the infection of systems via email attachments, malicious websites, or compromised software. Ransomware is then installed and executed on a user's system. The ransomware encrypts the user's data. Once the files are encrypted, access to them is denied. The person then receives a notification to pay the ransom within a specified time. A decryption key is sent to decrypt the files once the payment is made. However, in some cases, the decryption key is not sent.

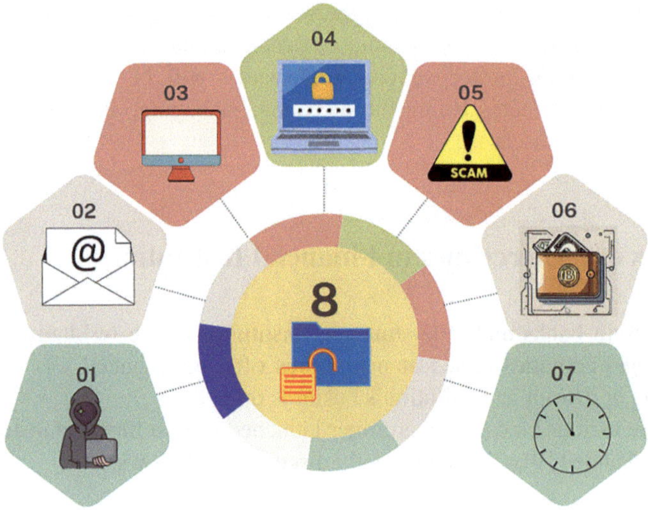

Fig. 1 Lifecycle of a ransomware attack in financial institutions

The result is permanent data loss. Such attacks result in financial losses, and loss of reputation.

Keyloggers: In this kind of attack, an institution suffers a software or hardware intrusion that records keystrokes to steal confidential information, such as access information.

2.3 *Man-in-the-Middle (MitM) Attacks*

In this attack, cybercriminals seize and change communications among or between employees or institutions. MitM can occur in the following forms:

Eavesdropping: Through hearing the interaction between or among users.
Data Manipulation: Changing the communication being sent.
Session Hijacking: Attackers take control of a current session between a customer and a financial institution.

2.4 *SQL Injection*

Cybercriminals often take advantage of weak and vulnerable financial institution applications and plant malicious and dangerous SQL code into input fields to gain illegitimate access to the bank's database and steal, change, or erase data.

2.5 Denial of Service (DoS) and Distributed Denial of Service (DDoS) Attacks

DoS Attacks: Attackers use these tactics to overwhelm an institution's system through excessively irrelevant requests, eventually collapsing the server.

DDoS Attacks: DDoS is similar to DoS. It occurs when cybercriminals attack an institution's system, but it is launched from multiple compromised systems and is difficult to defend against.

2.6 Insider Threats

Cybercriminality could occur by existing or former staff, vendors, or business partners who can access the company's data. This threat could be devastating due to the access and knowledge of the company's confidence.

2.7 Credential Stuffing

Another tactic cybercriminals use is employing enormous combinations of stolen usernames and passwords using automated tools on different financial institutions' websites to access users' sensitive information.

2.8 Card Skimming

Attackers often use some devices to capture a user's card details during a legitimate transaction at ATMs or PoS to access sensitive information for fraud purposes.

2.9 Social Engineering

Scammers employ various strategies, including influencing customers to reveal sensitive information through pre-texting, baiting, or quid-pro-quo deals.

2.10 Advanced Persistent Threats (APTs)

Cyber-criminals employ APTs, where advanced and sustained cybercrimes are carried out on a financial institution's network without being detected for a long time while stealing the company's money and data.

2.11 Business Email Compromise (BEC)

BEC allows cybercriminals to compromise a genuine official email account through scamming or hacking to carry out illegitimate transactions, such as social engineering or hacking to conduct unauthorized transactions, such as rerouting payment transactions to scam accounts.

2.12 Crypto-Jacking

Attackers employ crypto-jacking to use unauthorized computing resources of a financial institution for cryptocurrency mining.

2.13 Brute Force Attacks

Cybercriminals use sophisticated tools to systematically predict passwords, testing potential combinations through trial and error until they find the correct credentials.

3 Protective Strategies from Cybercrimes in Financial Institutions

Many defensive techniques aid financial institutions in mimicking the risks related to cybercrimes. These strategies are explained by [7, 10], as follows:

Multi-factor authentication (MFA): This implies putting additional defensive layers above using usernames and passwords. It involves using a one-time password (OTP) with a username and Password.

Encryption: Encryption enables financial institutions to encrypt their data to protect against unauthorized access.

Regular Security Audits: Financial institutions should conduct regular cyber threat audits to identify and fix flaws.

Employee Capacity Building: Capacity building programs enlighten employees about new tactics attackers use to avoid being victims.

Plan for incident response: Financial institutions should have a well-articulated policy, guideline, or plan to guide employees in responding promptly to cyber-attacks.

Segmentation of Network: Network segmentation allows institutions to segment their main network into units to mimic the permeation of malware and illegitimate access.

4 Cybersecurity in the Financial Sector

Cybercriminals are always at work to penetrate financial companies' data, including employees and vendors. Therefore, financial institutions must provide adequate protection against cybercriminals by strengthening their cybersecurity protocols. The US Federal Trade Commission (FTC) [10] has provided information about cybercrimes in businesses and highlighted measures that businesses can take in a bid to protect their organizations, employees, and vendors' data against cybercrimes;

4.1 Protecting Company's Files and Devices in Cybersecurity Environment

To protect its files and devices, a company can focus on the following measures:

Software Update: Regular software updates, including applications, web browsers, and operating systems, protect companies against cybercrimes.

Secure Files Offline: Files should be securely backed up offline to an external drive or the cloud to protect an organization's data from ransomware.

Use Password across the devices: All devices such as laptops, smartphones, tablets, and any devices authorized to work in the company's network environment should have a strong password and not be left unattended in a public place. Strong password requires a minimum of 12 characters with a combination of numbers, symbols, capital and lowercase letters.

Encrypt Devices: Devices and other media containing sensitive information should be encrypted. These include laptops, tablets, smartphones, removable drives, backup tape, and cloud storage solutions.

Multi-factor authentication: Multi-factor authentication requires extra procedures beyond password logging. Companies are encouraged to adopt the use of multi-factor authentication in order to access their network with sensitive information.

Staff training: Companies should develop a culture of training and re-training staff on cybersecurity and update staff on new risks and vulnerabilities. The companies should penalize employees who do not participate in the cybersecurity training.

Enforce physical security: Cybersecurity begins with strong physical security; negligence in physical security can uncover sensitive data to identity theft, potentially damaging consequences against companies.

Organizations can create a checklist similar to the example provided (Fig. 2) to ensure the necessary digital security measures against cyber threats.

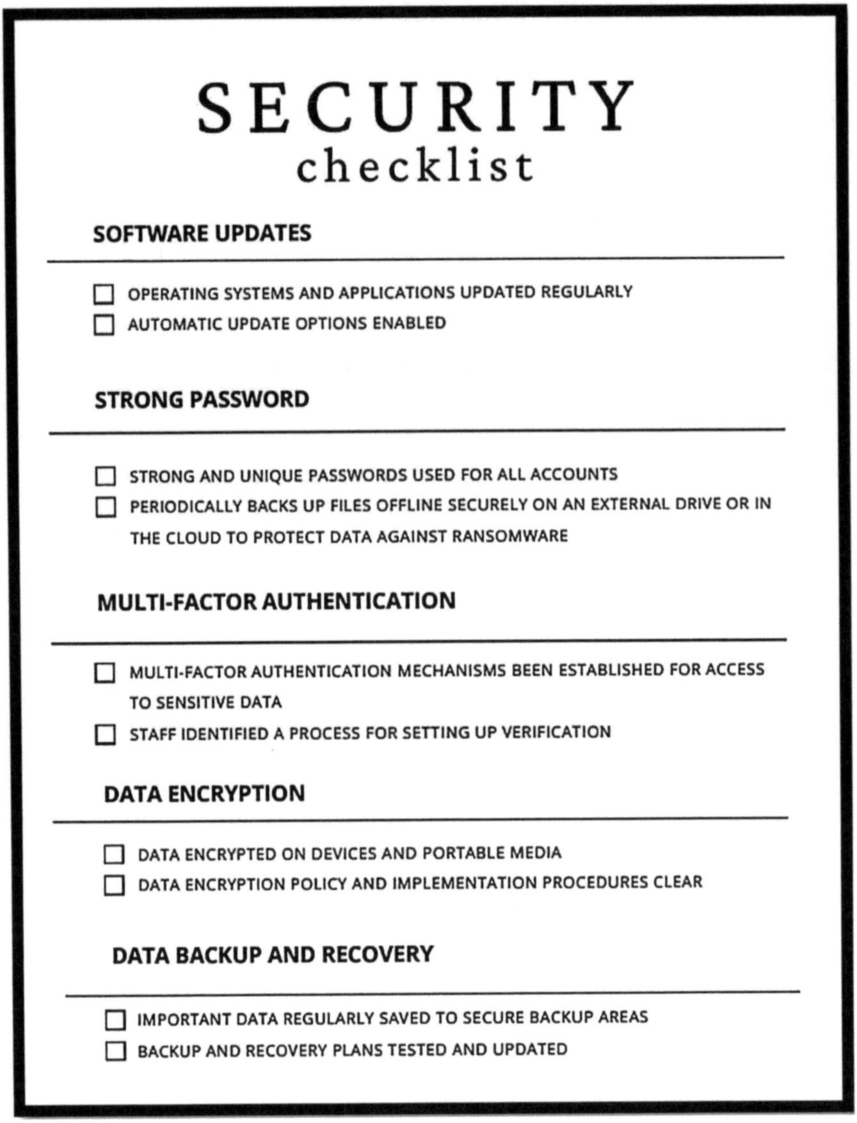

Fig. 2 Security checklist: digital measures

4.2 *Protecting a Company's Cybersecurity Environment*

Ransomware attacks on companies occur when an unsolicited email containing a link or attachment is opened, which will download software that locks companies or personal data holding the data hostage. The attackers request money or cryptocurrency as a ransom before allowing the company to use its data. However, even if the ransom was paid, there was no certainty that the cybercriminals would keep the data uncompromised. In the meantime, the information a company requires to run its business and sensitive details about the company, customers, and employees are now in criminal hands. Ransomware can negatively affect companies' business. However, it is not recommended for a company to pay the ransom. Though it depends on the company to decide whether the risks and costs of paying the ransom are worth the possibility of recovering the data, it may not guarantee the recovery of the company data. Companies can protect their business against ransomware through the following measures:

Develop a business continuity plan: This plan will help a company stay in business even if ransomware attacks occur. The plan should enable the company to access its business, customers, and employees' data via its cloud-backed repository. This plan should be made available to employees.

Regular Backup: Companies should create a culture of regular Backup of their data on a server not connected to its network. The Backup should be made during routine operations.

Regular security update: The company should always install the latest patches and updates. It should also provide further protection, such as email authentication and intrusion prevention software, and configure them to update automatically on the companies' computers.

Regular sensitization of staff: Organizations should train their staff to avoid phishing scams, and various possible ways computers and devices are infected.

4.3 *What Companies Should Do If Already Attacked by Cybercriminals*

In a situation where the attack has already occurred, companies can immediately take the following measures:

Reduce the damage: Companies can reduce the magnitude of the attack by instantly disconnecting the affected computers or devices from their network connection. If the company's data has been stolen, the IT unit should immediately take steps to protect other IT infrastructure and notify those affected units or customers.

Inform the appropriate authorities: The company should immediately notify the information technology regulator in the country where it does business. Doing so will prompt the regulator to monitor the attack against other industry participants.

Implement business continuity plan: The company should keep its business running by implementing its business continuity plan.

Notify the affected parties: The company should notify the affected parties, especially if the data or personal information was compromised, this is because the parties could be at risk of identity theft.

4.4 Tactics Use for Phishing by Cybercriminals

Scammers employ different strategies to trap an employee to access companies' or personal data. Below are some of these tactics:

Send unsolicited email or text: Scammers induce an employee of a company by sending an unsolicited email that seems to be official from someone you know. The email requires the target to click a link, open the folder, or provide a password, business bank account, or other sensitive information.

The email or text looks authentic: With current technology, it is easy for scammers to spoof logos (spoofing is creating a false email or website address) and produce fake email addresses. Cybercriminals use familiar company names or pretend to be someone the company's employees know.

Urgency: The scammers use messages that show urgency and pressure the target to act promptly. They use phrases such as "act now", "something bad will happen", "it is urgent", and similar to these.

Consequences of compliance: Scammers' main aim when sending a message is to get an employee to comply by clicking the link, opening the folder, or providing the information requested. If an employee complies promptly, the scammers can install ransomware or programs that lock the user out and spread to the entire company network. If you share passwords, scammers now have access to all those accounts.

4.5 What to Do If Companies Receive a Scammer's Phishing Request

To protect the company's valuable assets, such as data, devices, computers, and software, an employee should cautiously perform the following before acting on any unsolicited email containing links, folders, or requesting sensitive information;

Check the website or email address: It is very important to scrutinize the website or email address behind the message and ensure that it is from a real company, not a scammer. Employees can hover on the website address to ensure it has genuine protocol, https://not http://. Employees should ensure the sender's domain name is clear for the email address.

Escalate to a colleague: An employee can talk to a colleague to ascertain whether the request is real or phishing.

Make confirmation: An employee can call the vendor, colleague, or client who sent the message to confirm if the email is authentic using the phone number you know, not the phone number provided in the requested email.

4.6 *What an Employee Can Do If Complied with a Scammers Request*

Immediately an employee realizes that a scammer made them click a link, open a folder, or share sensitive information, it is recommended to perform the following action:

Alert colleagues: Inform your co-workers about the incident and share your experience. Phishing attacks might affect more than one person, as they often affect many employees.

Reduce the damage: Change any compromised password and disconnect any computer or devices from the company's main network infected with the malware.

Notify customers: The company should inform the customers who is their data was compromised by the attack.

Report the incident: The employee should notify his company's information technology unit immediately. This will prompt the unit to monitor the attack against the entire organization.

5 The Role of Cybersecurity in the Financial Sector

Cybersecurity is essential to the development of the financial sector. Other sectors relied heavily on the financial industry to operate efficiently. Several reasons emphasize cybersecurity's importance in the financial sector [1, 2, 4]. These reasons are highlighted below:

It Protects Sensitive Information: Banks, finance companies, and other financial institutions interact with large amounts of sensitive data, such as trade agreements, customer transaction records, and employee information. Cybersecurity controls protect this vast amount of data from being attacked by unauthorized personnel, ascertaining privacy and compliance with regulations.

It Prevents Financial Losses for the institutions: Cybercrimes such as phishing, ransomware, and malware may result in colossal loss of funds. However, stronger and functional cybersecurity helps to deter these cybercrimes, safeguarding institutions, employees, and customers from potential economic loss.

Sustaining Customer Trust: The financial sector is built on trust. Customers and suppliers expect secured financial information that cannot be breached. After all, a breach of financial information can significantly impact on institution's reputation and result in the loss of market share and customers. Therefore, effective cybersecurity controls help in maintaining and building customer trust.

Compliance with Standards & Regulations: Standards and Regulations such as the Payment Card Industry Data Security Standard (PCI DSS) and General Data Protection Regulation (GDPR) mandate financial institutions to comply with certain standards and regulations for rigorous cybersecurity and data protection practices. Lack of compliance with this standard and regulation can lead to heavy penalties and litigation.

Operational or Business Continuity: Cybercrimes can interrupt services in the financial sector, triggering downtime and negatively affecting critical services. However, cybersecurity enables operational and business continuity by protecting systems and networks from disruptions.

Safeguarding the Financial Sector Against Emerging Threats: The financial industry is a main target for advanced cyber-attacks. Cybersecurity approaches must be developed continuously to secure against emerging dangers such as advanced persistent threats (APTs), zero-day vulnerabilities, state-sponsored attacks, etc.

6 Decision Support Systems for Cybersecurity in Financial Institutions

Decision Support Systems (DSS) aim to address problems of organizations deemed to be semi-structured or unstructured. In the cybersecurity ecosystem, DSS are specialized systems developed to help in decision-making processes relating to safeguarding financial institutions' information systems against cyber-attacks. These information systems, comprised data analysis, threat intelligence, and machine learning, proffer adequate implementable insights, guidance, and automated feedback to improve the company's cybersecurity position.

6.1 Components of Cybersecurity DSS in Financial Institutions

Chen et al. [8] identified various components of cybersecurity DSS in financial institutions, which are briefly discussed as follows:

Collection and Integration of Data: Financial institution's data is obtained from different sources such as network logs, employee or customer activity, threat intelligence feeds, and security incident-based reports. The system consolidates and regularizes this information into a comprehensive view of the security landscape.

Analysis and Processing of Data: Sophisticated algorithms and ML models are employed to analyze the data gathered, identify patterns, and discover inconsistencies in the pattern for possible action. Similarly, data is also being tallied from various sources to determine if there is a potential threat based on the other available data points.

Threat Intelligence: Sophisticated real-time threat intelligence feeds are built into the system to keep updated on emerging cyber threats and vulnerabilities. The intelligence feeds utilize past data and patterns to predict imminent threats and suggest cautious measures.

Visualization and Reporting: This DSS cybersecurity component furnishes reliable visual dashboards that show essential security metrics, alerts, and reports of cyber-attack incidents. It also generates detailed reports for regulatory or standards compliance, auditing, and management review.

Decision-Making Support: This provides recommendations to mimic threats and enhance security controls while allowing users to test various threat scenarios and assess the efficacy of various responses.

Automation and Feedback: This component focuses on causing an automated response to identified threats, like blockage of IP addresses and installing patches. It also facilitates harmonizing incident response by providing important information and suggesting actions accordingly.

6.2 *Functionalities/Features of Cybersecurity DSS in Financial Institutions*

NIST [18] provided insight on functionalities regarding cybersecurity DSS as enumerated as follows:

Intrusion Detection and Prevention: This feature is aimed at helping the Cybersecurity DSS to monitor network traffic for suspicious activities and provide automatic responses to prevent unauthorized access. Tools used to deter intrusion against cyber threats by financial companies comprised Snort, Cisco IDS/IPS, and Suricata.

Managing Systems Vulnerability: This helps ascertain the existing vulnerabilities in systems and applications, prioritizes them based on their specific risk exposure, and suggests remedial measures. OpenVas, Qualys, and Nessus are tools for managing systems or application vulnerabilities.

Risk Assessment: This appraises the quantum of risk of different assets and processes in an institution to classify security efforts. Financial institutions employ risk assessment tools like RiskWatch, RSA Archer, and factor analysis of information risk (FAIR) in analyzing and assessing the level of risks of different processes and assets.

Compliance Management: This functionality certifies conformity with regulatory requirements to deter cyber threats by ensuring monitoring status and providing necessary documentation. Tools used in managing regulatory compliance by financial institutions against cyber-attacks are Compliance 360 and LogicGate, among others.

Incident Management: This cybersecurity DSS functionality assists the overall incident lifecycle, from detection, analysis, response, and recovery of affected data. Splunk and IBM Resilient are examples of tools in incident management.

7 Benefits of Cybersecurity DSS

Kotenko and Victor Skormin [16] enumerated the benefits of cybersecurity DSS in improving cybersecurity controls and provided some challenges hindering implementation of cybersecurity. The benefits are explained as follows:

Improving Cyber Threat Detection: The Cybersecurity Decision Support System enhances automatic threat detection and response capability, reducing the risk of possible cybercrimes.

Informed Decision-Making: Cybersecurity DSS proffers the necessary insights and information for an organization to deliver prompt and effective decisions.

Optimization of Resource: Another benefit of the cybersecurity DSS is to allocate scarce resources effectively by prioritizing cyber threats and vulnerabilities vis-a-vis their impact.

Improved Compliance: Cybersecurity DSS facilitates compliance with regulatory requirements and standards by automating monitoring and reporting processes.

Proactive Security: Cybersecurity DSS facilitates financial institutions to take a proactive measure to cybersecurity by reducing threats before causing harm to the institution.

8 Challenges of Implementing Cybersecurity DSS

Quality and Integration of Data: Ascertaining the integrity and quality of data drawn from different sources posed a serious challenge to cybersecurity DSS.

Complexity: The complexity of putting together and maintaining a decision support system tends to cause problems for a company with inadequate IT resources such as infrastructure and workforce.

Cost: Setting up and maintaining a comprehensive DSS is exorbitant and requires huge capital investment.

Skilled Manpower Requirements: Cybersecurity DSS requires a workforce with the requisite knowledge to process and interpret data and manage the system professionally.

False Positives and False Negatives: Another challenge of cybersecurity DSS is the ability to balance the sensitivity of DSS to mitigate false positives (friendly events identified as cyber threats) and false negatives (real threats not discovered).

9 Cyber Threats in Capital Markets

Due to specific operations and regulatory environment, capital markets encounter unique cyber threats different from other financial institutions. These unique threats are discussed as follows:

Market Manipulation Attacks: Attackers can access and manipulate a stockbroker's order management systems, resulting in fake feeds incorrect orders, or disrupting trade surveillance systems. This kind of attack leads to unlawful and unethical trading practices, causing automatic illegal trading activities resulting in flash crashes endangering confusion in the capital markets system. Attackers take advantage of trading systems of the capital market if adequate measures had not been put in place [3, 9].

Insider Threats: Staff, contractors, and vendors of a firm with authorized access to important data constitute significant threats. Insider threats can result in the initiation of unscrupulous trades, reveal of confidential customer data, and modify with electronic or digital assets of the company. These risks are complicated because they seldom entail trusted parties within the company, making them hard to discover and fix the threats. This threat's financial, reputational regulatory consequences can be enormous [3, 9].

Third-Party Risk: Another unique cyber threat to the capital market is increasing dependency on the service providers and technology platforms, which presents another vulnerability. Security breaches due to vendors (service providers) and technology platforms used in the organization may lead to serious implications. The growing adoption of cloud services brings more cybersecurity risks [6].

Data Breaches: Capital market firms process large sensitive datasets such as financial information and proprietary trading algorithms. Cyber threats attacking sensitive data can lead to weighty financial losses and litigations [3].

Distributed Denial of Service (DDoS) Attacks: These threats are meant to cause interruptions on a firm's trading platforms and other important infrastructure. By derailing systems with unnecessary traffic, cybercriminals can cause serious downtime, affecting market operations and eroding investor confidence. DDoS threats affect capital markets because of transactions' high volume and speed [9].

Regulatory Compliance and Penalties: Non-conformity with growing cybersecurity policies, guidelines, and regulations can result in heavy penalties and possible operational restrictions. Therefore, capital markets firms must adhere to extant policies, regulations, and guidelines like the General Data Protection Regulation (GDPR), which enforces prompt reporting of data breaches and imposes substantial fines for non-compliance with such regulations [3].

10 Measures to Counter Cyber Threats in Capital Markets

Organizations must adopt a comprehensive and multi-layered approach to mitigate the cyber threats encountered by capital markets. These approaches are outlined as follows:

Robust Algorithm Testing and Monitoring: Firms should diligently examine trading algorithms to detect and reduce vulnerabilities before installation and utilization. Regular monitoring of the algorithm's performance is crucial in detecting and preventing manipulative activities by attackers [3, 9].

Trade Surveillance Systems: Firms should install sophisticated trade surveillance systems that efficiently monitor and analyze trading patterns automatically to identify suspicious activities. These trade surveillance systems should use ML and AI to improve detection abilities [3, 9].

Regulatory Compliance: Strict adherence to existing regulatory requirements enforces stringent measures and ensures monitoring of trading activities to avert abuse of the capital market [3].

Employee Training and Awareness: Training and re-training of employees on the mechanics of cybersecurity is recommended. Firms should emphasize the risks of cyber-attacks, such as phishing, social engineering, and employee negligent behaviors [3].

Behavioral Analytics: Firms should also deploy User and Entity Behavior Analytics (UEBA) to observe and examine user behaviors. This assists in identifying suspicious activities that may endanger the company through insider behavior [3].

Access Controls: Access controls should be enforced and monitored squarely. Tailgating and other forms of unauthorized entering into a sensitive area should be watched and dealt with accordingly [3].

Vendor Risk Management: Firms should conduct a thorough vendor risk management program to examine and observe third-party security practices. This comprised carrying out regular security checks and the need for stakeholder vendors to comply with the security standard of the capital markets industry [3].

Data Encryption: Firms should ensure a strong policy of sensitive data encryption in which both data at rest and in use are encrypted using strong algorithms to protect it from cyber attackers. Also, the encryption protocols should be regularly updated [3].

Data Loss Prevention (DLP): Utilizing DLP solutions to observe and supervise data movement within and outside the company. This assists in preventing illegitimate data exfiltration and ascertains compliance with the regulations of data protection [3].

Incident Response Plan: An effective incident response plan should be developed, clearly describing various steps to be followed when a data breach occurs. Simulations and drills should be conducted regularly for preparedness [3].

DDoS Mitigation Services: Firms should use specific DDoS mitigation services which can engage and avert malicious traffic. These remedial services seldom leverage global server networks to provide protective measures against large-scale attacks [3].

Redundant Infrastructure: Firms to design a strong infrastructure with redundant systems and failover capacities to assure business continuity at the time of an attack. This involves load balancing and distributed network architecture for effective business continuity [9].

Compliance Frameworks: Develop a thorough compliance framework that conforms to applicable regulations like GDPR, MiFID II, etc. Constantly review and update guidelines and policies to ensure effective compliance [3, 9].

Regular Audits and Assessments: Firms should also conduct regular security audits and evaluations to determine compliance gaps and areas that require further improvement. Similarly, secure the services of third-party auditors to give an objective appraisal of the security posture [3].

Incident Reporting Procedures: Establish a clear process for reporting cybersecurity incidents within required timeframes to appropriate regulatory authorities. Firms should ensure that their staff know their roles and responsibilities in the event of an incident [3].

11 Conclusion and Recommendations

It is important to note that the financial sector faces an unprecedented cyber-attack than any other sector, and this may be attributed due to the immediate access to funds and sensitive data that could potentially be turned into money [22]. Sophisticated cybersecurity tools are continuously invented and used in financial institutions to check the corresponding efforts by advanced tactics employed by attackers [13, 20]. The chapter has addressed the major and minor techniques cybercriminals use to attack businesses and financial institutions, such as phishing, malware attacks, DoS, DDoS, APTs, etc. Cybersecurity DSS tools have shown to be productive against cyber-attacks. Finally, various tactics employed by attackers and preventive strategies were highlighted.

This investigation has proposed the following recommendations. Firstly, sophisticated technological tools expose financial institutions to cybersecurity risks [11, 12]. Therefore, financial institutions should think "outside the box" to counter cybercriminals. The attackers continued to invent sophisticated means to access sensitive data, so the institutions should invent more advanced cybersecurity DSS tools to deter attackers. Secondly, financial institutions should intensify cyber threat awareness within their organizations through training and re-training and penalize those who refuse to partake [10]. No matter how much money an organization spends or how sophisticated the tools it deploys, they cannot impact without blocking employee-initiated phishing and social engineering intrusions. Finally, IT policies and guidelines should be frequently updated to capture the emerging cyber threat concerns. The policies and guidelines should also be made available to employees and contractors for compliance.

References

1. Accenture Security. (2019). The COST of cybercrime. https://iapp.org/resources/article/the-cost-of-cybercrime-annual-study-by-accenture/
2. Adelmann, F., Elliott, J., Ergen, I., Gaidosch, T., Jenkinson, N., Khiaonarong, T., Morozova, A., Schwarz, N., & Wilson, C. (2020). *Cyber risk and financial stability: It's a small world after all.* https://www.imf.org/en/Publications/Staff-Discussion-Notes/Issues/2020/12/04/Cyber-Risk-and-Financial-Stability-Its-a-Small-World-After-All-48622
3. Agarwal, S. (2024). *Cybersecurity essentials for capital markets firms in the digital age.* Wipro.
4. Bank for International Settlements (BIS), & International Organization of Securities Commissions (IOSCO). (2016). *Guidance on cyber resilience for financial market infrastructures consultative report.* Bank for International Settlements.
5. Bilych, G. (2012). Profit and economic growth. *Business and Economic Research, 2*(2). https://doi.org/10.5296/ber.v2i2.2285
6. Brauneis, M. (2024). The economy, cyber and third parties among issues portending riskier times in financial services. Protiviti.chrome-extension://efaidnbmnnnibpcajpcglclefindmkaj/https://www.protiviti.com/sites/default/files/2023-12/research-brief-fsi-top-risks-2024-2034-protiviti.pdf
7. Chavez, N., & Bichler, G. (2019). Guarding against cyber-trespass and theft: Routine precautions from the hacking community. *International Journal of Cyber Criminology, 13*(1), 106–110. https://doi.org/10.5281/zenodo.3551489
8. Chen, H., Chiang, R. H. L., Storey, V. C., & Robinson, J. M. (2012). Business intelligence and analytics: From big data to big impact. *MIS Quarterly, 36*(4) www.freakonomics.com/2008/02/25/hal-varian-answers-your-questions/
9. de Koning, N. (2022, December 30). AFM report on capital markets sector's vulnerability to cyber threats. *Global Regulation Tomorrow*
10. Federal Trade Commission (FTC). (2024). *Small Business*. https://www.ftc.gov/business-guidance/small-businesses
11. Fink, L., Shao, J., Lichtenstein, Y., & Haefliger, S. (2020). The ownership of digital infrastructure: Exploring the deployment of software libraries in a digital innovation cluster. *Journal of Information Technology, 35*(3), 251–269. https://doi.org/10.1177/0268396220936705
12. Gaudenzi, B., & Siciliano, G. (2018). *Managing IT and cyber risks in supply chains*. Springer.
13. Iijima, T., & Curtis, J. (2004). Need to justify IT security? Measure your risk! *Journal of Corporate Accounting and Finance, 15*(5), 47–51. https://doi.org/10.1002/jcaf.20037
14. International Monetary Fund. (2024, January 10). *Cybercrime expected to skyrocket in the coming years.*
15. Kalakuntla, R., Vanamala, A. B., & Kolipyaka, R. R. (2019). Cyber Security. *HOLISTICA—Journal of Business and Public Administration, 10*(2), 115–128. https://doi.org/10.2478/hjbpa-2019-0020
16. Kotenko, I., & Skormin, V. (2013). *Computer network security*. Springer.
17. Misra, S. C., Kumar, V., & Kumar, U. (2007). A strategic modeling technique for information security risk assessment. *Information Management and Computer Security, 15*(1), 64–77. https://doi.org/10.1108/09685220710738787
18. National Institute of Standards and Technology (NIST). (2020). *Cybersecurity framework.* chrome-extension://efaidnbmnnnibpcajpcglclefindmkaj/ https://isacala.org/wp-content/uploads/2020/04/NIST-CSF-2020.pdf
19. Pande, J. (2017). *Introduction to cyber security*. http://uou.ac.in.
20. Rohmeyer, P., & Bayuk, J. L. (2019). *Financial cybersecurity risk management* (Apress Media LLC: Welmoed Spahr, Susan McDermont, Laura Berendson, & Rita Fernando, Eds.). Springer.

21. Verizon. (2012). *2012 Data breach investigations report*. www.verizonbusiness.com/resources/reports/rp_data-breach-investigations-report-2011_en_xg.pdf
22. Wang, J., Ho, C. Y. (Chloe), & Shan, Y. G. (2024). Does cybersecurity risk stifle corporate innovation activities? *International Review of Financial Analysis, 91*, 103028. https://doi.org/10.1016/j.irfa.2023.103028

Application of the Sugeno-Takagi-Kang Fuzzy Logic Model as a Decision Support System with a Mathematical Approach for Managing Uncertainty in Supply Chain for the Demand Forecasting: Proposal of a Fuzzy Time Series Forecasting Model (FTS)

Francisco Trejo, Rafael Torres-Escobar, and Alberto Ochoa

1 Introduction

At the beginning of the COVID-19 pandemic, manufacturing and services companies, and practically any industry in the world, had to adapt to changes that nobody during our generation had experienced. Markets were witnesses of how fragile and unreliable short and mid-term goals could be, how vulnerable supply chains are in the whole world, and maybe never in the last 50 years for those of us who are alive, we had the chance to see the challenges a globalized and interconnected to unprecedented levels supply chain had to face. Demand with high volatility, but above all, a high degree of uncertainty in the supply. All these changes obligated us to rethink the ways of doing business and how to face day to day challenges we were encountering as the pandemic progressed. Supply chains not only needed to be more agile but more flexible; however, almost none was really prepared. This demonstrates that there is little control and information that, in moments of crisis, may prepare us to face the unexpected.

In this sense, inventory is a collective term to show the supply of inputs or outputs as the anticipation to satisfy demand. In this context, not only are tangible goods subject to forecasts but also services are even though they cannot be stored, they also require resources that are subject to estimates and that require planning to maintain

F. Trejo (✉) · R. Torres-Escobar
Universidad Anáhuac Mexico, Facultad de Ingeniería, Mexico City, Mexico

A. Ochoa
Universidad Anáhuac Mexico, Facultad de Ingeniería, Mexico City, Mexico

Doctorado en Tecnología, UACJ, Ciudad Juárez, Mexico
e-mail: alberto.ochoa@uacj.mx

N. Gupta, J. D. Diaz Roman (eds.), *Decision Support Systems for Industrial and Scientific Applications*, Signals and Communication Technology,
https://doi.org/10.1007/978-3-031-89700-9_17

availability and supply. In this work, we focus on tangible goods, specifically those that are stored in third-party logistics (3PL). Whether as the final product is stored, which will later be utilized to be sold in a determined period, or as those necessary inputs for production of the same in a future period. Demand anticipation is key here, acquisition of supplies for a future sale. The model proposed is an alternative to the current different and widely utilized models, with the difference that the proposed model uses fuzzy logic (FL) to perform demand forecasting with a relatively small data amount (<20) in a high uncertainty level environment.

The proposed model makes use of fuzzy logic (FL), specifically through the development of a Sugeno-Takagi-Kang controller, whose purpose is to develop a linear combination of smooth transition functions. Part of what a diffuse controller does: (1) fuzzifies, (2) cuts, (3) unifies, and (4) defuzzifies; this is, it makes the following: an amount of entries from the controller comes in, in this case, the actual demand value, as data, with this data, diffuse sets of each rule are evaluated "IF-THEN", this is, they are evaluated in the entering value based on membership function μ_{A1}, is fuzzified; afterward, the consequent sets of each rule are cut at the height of the corresponding fuzzification value μ_{B1}; subsequently they are united to calculate the resulting set from inference $\mu_{B'}$, defuzzification is applied to this in order to obtain a unique scale value resulting from the controller, and this is repeated every certain time interval, except that in this paper, it is employed instead of a defuzzifier, so many resulting linear functions as means of inference to obtain the forecast value. This process simplifies calculations and rules of inference and streamlines system sensitivity analysis; the model employed is the one described by Sugeno and Takagi [1].

This paper is divided into five sections, section one provides an introduction by making a review of different methods used traditionally in the supply chain, and since there has not been an ample application of the diffuse methods on this, section two shows the fundamentals of fuzzy logic; in section three the used procedure is shown; section four deals with application to a practical example developed in a storing company (3PL) utilizing MATLAB Fuzzy Logic Designer (R2022a) which offers quick prototypes of finished products, and provides a great amount of precision and control over the codification process; section five presents the conclusions.

2 Fuzzy Logic

Fuzzy logic provides an effective way to treat problems involving imprecision and phenomena, and vague or blurred concepts. It allows the examiners to utilize the linguistic terms to evaluate indicators in expressions from natural language, and each linguistic term may be associated with a membership function. Also, fuzzy logic has found significant applications in decision processes [2, 3]. Based on this methodology, a system has been built based on fuzzy rules based on experiences from experts to estimate supply chain agility. This has allowed us to apply it in a model proposal with a real case study in a 3PL company that stores raw materials for maquila companies in the city of Tijuana, BC, Mexico (3PL).

The nature of the data generated over time is classified in general terms as stationary, seasonal, tendency, and strictly non-stationary. However, most of the historical data generated over time in the real world are not so statistically consistent as it is considered for data and forecasting purposes. Precise results may be obtained if data in a stationary environment is handled with some dedicated techniques in place of the traditional algorithms. Generally, traditional algorithms based on mathematical and statistical forecast models cannot address the problem in the real world, in which the historical data is imprecise [4, 5] and vague [6]. The fuzzy time series forecast is an emerging investigation area. These are problems associated with uncertainty, vagueness, and imprecision. There are certain characteristics of the temporary fuzzy series that make them more adequate than conventional fuzzy systems. First, it can process concrete (crispy) and fuzzy values. Besides that, they do not require large data sets as the statistical forecast models. If data or phenomena have multiple seasonal patterns, then those, too, require large data sets in the FTS models. Fuzzy temporary series forecast models (FTS) have the following steps: (1) discourse universe definition, (2) partitioning of the discourse universe into intervals, (3) fuzzification of data sets, (4) establishment of the fuzzy logic relation, and (5) defuzzification.

This article presents a fuzzy model based on the inference from Sugeno-Takagi-Kang, which is a proposed inference, a simplification of the Mandami inference method, where the premise of an implication is the description of a subspace of fuzzy entries, and its consequence is a linear output. In this paper, a series of linear functions is introduced corresponding to demand behavior to evaluate the ability to forecast the required space required in a storage company (3PL). However, the proposed model may be utilized to forecast the demand of a supply chain, considering different factors or criteria. Thus, a fuzzy inference system (FIS) is a system based on rules with concepts and operations associated with fuzzy sets and fuzzy logic [7]. These systems start from an entry space function with an exit function; they allow to build structures or linguistic rules based on experience or knowledge from the experts to generate answers (sure outputs or crisp outputs). The model also allows us to make certain simulations based on this stored knowledge as how the system shall respond and allows to obtain outputs that represent what an expert would say or value in function of the input information. This knowledge is stored in the form of a rule base, which is a set of rules expressing the relation between the inputs of a system and expected outputs. The first step to building a fuzzy model is to select a membership function for each variable. A "membership function" is a curve that defines the behavior of the value of a fuzzy variable; this variable is assigned with a value or degree of membership between 0–1; in the case of the Sugeno-Takagi-Kang inference, this is represented through a linear function. The following step consists in capturing the experience from the experts, in which the most common way of representing this knowledge is through expressions in natural language in the form of an "IF" premise, which is known as antecedent, and "THEN", which is the conclusion or consequent. This form is commonly referred to as "IF-THEN" to make the different rules of decision, this also known as the *modus ponendo ponens*, which is a valid argument form in deductive reasoning, and one of the propositional inference rules; this could be defined as if $\boldsymbol{P}$ implies $\boldsymbol{Q}$ $(\boldsymbol{P} \rightarrow \boldsymbol{Q})$, and if $\boldsymbol{P}$ is true, then $\boldsymbol{Q}$ is also true. Normally expresses an inference that says that if we know the fact (premise, hypothesis, antecedent), then we can infer or derive another fact (conclusion). This form of reasoning exists entirely in the context of linguistics since it expresses empirical and heuristic knowledge in our own communication language [8].

The fuzzy implication presented in this work is simple enough. It is based on a fuzzy partition of the entry space. In each fuzzy subspace, a linear entry-exit relation is formed. Most fuzzy controllers have been designed based on human experience; however, it often occurs that a user may not linguistically tell which type of action takes in a particular situation.

3 Forecast Fuzzy Model

In this work, we establish the membership function of a fuzzy set $\boldsymbol{A}$ as $\boldsymbol{\mu_{An}}$ $\boldsymbol{(x)}$, $\boldsymbol{x} \in \boldsymbol{X}$. All fuzzy sets associated with membership functions are linear. Thus, a membership function is characterized by two parameters giving a greater membership degree 1 and a minor degree 0; respectively, this is $\boldsymbol{\mu_n}$: $\boldsymbol{U} \rightarrow$ **[0,1]**. Additionally, the true value of a proposition "x is A and y is B" is expressed by:

$$\mid \boldsymbol{x} \text{ is } \boldsymbol{A} \text{ and } \boldsymbol{y} \text{ is } \boldsymbol{B} \mid = \boldsymbol{A}(\boldsymbol{x}) \wedge \boldsymbol{B}(\boldsymbol{y})$$

This considers the following implications C:

$$\boldsymbol{C} : \textbf{If } \boldsymbol{f}(\boldsymbol{x}_1 \textbf{ is } \boldsymbol{A}_1, \ldots, \boldsymbol{x}_n \textbf{ is } \boldsymbol{A}_n), \textbf{ then } \boldsymbol{y} = \boldsymbol{g}(\boldsymbol{x}_1, \ldots, \boldsymbol{x}_n)$$

where:

$\boldsymbol{y}$, is the consequent variable for the inferred values.
$\boldsymbol{x_1} - \boldsymbol{x_n}$, are the premise variables that also appear in the consequence part.
$\boldsymbol{\mu_n}$, is the membership function of each n fuzzy model A_n.
$\boldsymbol{A_1} - \boldsymbol{A_n}$, are fuzzy sets with linear membership functions that represent a fuzzy subspace in which the implication of C may be applied to reasoning, this is, they may establish decision rules.
$\boldsymbol{f_n}$, is a function of logic connecting the propositions in the premises.
$\boldsymbol{g}$, is a function implicating the value of y and when $x_1 - x_n$ premises are satisfied.

Where premise $\boldsymbol{A_i}$ equals $\boldsymbol{X_i}$, for a determined $\boldsymbol{i}$, where $\boldsymbol{X_i}$ is the discourse universe for $\boldsymbol{x_i}$. Thus, for example, if $\boldsymbol{x_1}$ is high, and $\boldsymbol{x_2}$ is low, then $\boldsymbol{y = 2x_1 - 3x_2 + c}$, this is, the value of $\boldsymbol{y}$ would be, two times $\boldsymbol{x_1}$ minus 3 times $\boldsymbol{x_2}$ plus a $\boldsymbol{c}$ value. This may be defined by practically any means, whether by the experience of the users, historic data, an econometric model, temporality data, or simple analytic geometry. In such way that an implication would be written as:

$$C : \textbf{If } x_1 \textbf{ is } A_1 \textbf{ and } x_2 \textbf{ is } A_2 \textbf{ and } x_n \textbf{ is } A_n, \textbf{ then } y = p0 + p_1x_1 + \cdots + p_nx_n$$

The model establishes the following elements:

The input condition for the forecast is given by:

$$x \textbf{ is } A' \textbf{and } y \textbf{ is } B'$$

then come the rules and their antecedents:

$$\textbf{if } x \textbf{ is } A_1 \textbf{ and } y \textbf{ is } B_1, \textbf{ then } z = f_1(x, y)$$
$$\textbf{if } x \textbf{ is } A_2 \textbf{ and } y \textbf{ is } B_2, \textbf{ then } z = f_2(x, y)$$

.

.

n

$$z = z_0$$

This model does not have a consequent fuzzy set, but the function $f_1, f_2, \ldots, f_n$, which depend on the system entries, which are x and y, and a z conclusion, which is not a fuzzy set, but a scale amount represented by z_0, this is due to the consequences of the rules which are functions.

Where z_0 is calculated:

$$z_0 = \frac{\omega_1 f_1(x_0, y_0) + \omega_2 f_2(x_0, y_0) + \ldots}{\omega_1 + \omega_2 + \ldots}$$

This is, each of the functions in the entries is evaluated for the fact x_0, y_0, which arc the value of the Singleton sets for A' and B', in such a manner that they are weighed, and a linear combination is made **from** $f_1, f_2, \ldots, f_n$. The values ω_n, are calculated as minimum resulting values from the fuzzification of each rule. Thus:

$$\omega_n = \min\left(\mu_{An}(x_0), \mu_{Bn}(y_0)\right)$$

The output is a linear combination of the functions that are in each rule.

Utilizing the Sugeno-Takagi-Kang controller [1], it is considered that the function or control surface is known, in this case, the demand behavior in each period emerges from the function of control surface, and from this, the control rules are inferred, this premise is fundamental for the application of this model, also, it considers that all functions f_n are convex.

Determination of the optimal interval number of linear functions may be established in different forms: (a) from the simple analysis of graphic data, where

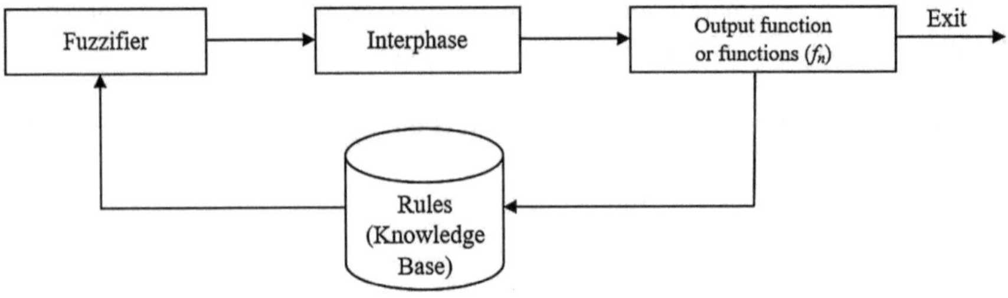

Fig. 1 Representation of the inference system proposed for demand forecast based on the Sugeno-Takagi-Kang model

Table 1 Demand data from a client in a storage company 3PL

Month	Demand	Month	Demand
Jan-22	3806	Jul-22	2330
Feb-22	4449	Aug-22	128
Mar-22	4746	Sep-22	224
Apr-22	4259	Oct-22	404
May-22	4089	Nov-22	410
Jun-22	2859	Dec-22	413

group data is evident, (b) through the model established by Shah [9], where the discourse universe is divided in equal intervals and uniform longitude, and the total number of intervals is defined as that integer $\leq (N/2 + 1)$, where N is the total number of data. Obtaining the variation percentage Y of a series of time X between two consecutive time periods [4]:

$$Y = Y_t, Y_{t+1}, Y_{t+2}, \ldots \text{ where } Y_t = \frac{X_{t+1} - X_t}{X_t} * 100, t = 1, 2, \ldots, n-1 \tag{1}$$

Where data series t will belong to the same group if: (a) Y_t is descendent or ascendent, but it does not change sign and will not belong in a contrary case, or well if Y_t keeps constant.

This gives as a result a system of inference as the one shown in Fig. 1.

4 Example of Application of the Sugeno-Takagi-Kang Control Fuzzy Model, for Demand forecast in a Storage Company 3PL

Step 1 We start from the known demand data d and the behavior of these over time.

Whether monthly demand data during the last 12 months, shown in Table 1, and the resulting graphic from these same data in Fig. 2.

Fig. 2 Demand graphic from the last 12 months of a client from a storage company 3PL

Step 2 Let these be now the resulting demand functions $\boldsymbol{f_1}, \boldsymbol{f_2}, \boldsymbol{f_3}, \boldsymbol{f_4}, \boldsymbol{f_5}$, and $\boldsymbol{f_6}$, as shown in Fig. 3**,** the functions representing the demand intervals, also need to determine the transition points, in this case: $\boldsymbol{p_1}, \boldsymbol{p_2}, \boldsymbol{p_3}, \boldsymbol{p_4}$, and $\boldsymbol{p_5}$. The coordinates for these points are:

$$p_1 = (3, 4750)\ ,\ p_2 = (5, 4100)\ ,\ p_3 = (7, 2350)\ ,\ p_4 = (8, 130)\ ,\ \text{and}\ p_5 = (10, 400)$$

Step 3 Let these be now the control rules which represent these linear functions, starting from the form:

If $\boldsymbol{x}$ belongs to $\boldsymbol{EM}$, then $\boldsymbol{y} = \boldsymbol{a_1}\ \boldsymbol{x} + \boldsymbol{b_1}$, and like this for each one of the rules, applying analytic geometry, they are the following:

$$\text{If } d \text{ is } EM, \text{ then } d = f_1(EM) = 475d + 3325 \tag{2}$$

$$\text{If } d \text{ is } MM, \text{ then } d = f_2\ (\text{MM}) = -325d + 5725 \tag{3}$$

$$\text{If } d \text{ is } MJ, \text{ then } d = f_3\ (\text{MJ}) = -875d + 8475 \tag{4}$$

$$\text{If } d \text{ is } JA, \text{ then } d = f_4(JA) = -2220d + 17890 \tag{5}$$

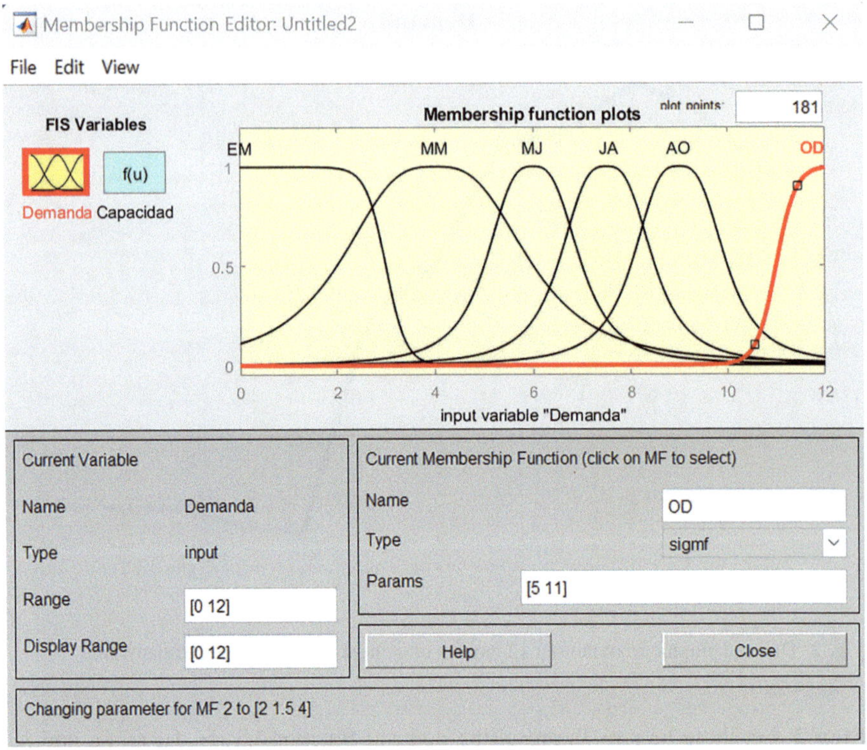

Fig. 3 Membership functions $f_1 \ldots f_6$ representation in Matlab

$$\text{If } d \text{ is } AO\text{, then } d = f_5(AO) = 135d - 950 \tag{6}$$

$$\text{If } d \text{ is } OD\text{, then } d = f_6(OD) = 500 \tag{7}$$

In Appendix, Fig. 9, the application of these equations over the demand curve can be seen.

Step 4 Define the semantics of the linguistic terms; for this, we chose the membership function that better adapts to data behavior. In this case, we observe that f_1 belongs to a membership function open on the left, a Sigmoidal function is selected, centered where the transition occurs, in this case, 3 or where p_1 is, a similar case occurs for f_6, only that this a membership function, but open on the right, also centered on the transition, this is, in p_5 or 10. In the case for other functions (f_2, f_3, f_4, **and** f_5), general bells are selected, centered in 4, 6, 7.5, and 9, respectively. Finally, the widths of these functions are proposed as the width of the corresponding intervals. It is only left to define the slopes or borders for the membership functions, this is, the value that allows to transition from one function to another, how smooth is the transition from one interval to another, are parameters ***a*** and ***b***, these parameters will be adjusted later to smoothen the transitions between f_1 and f_2, and so forth, the smaller these parameters are, the smoother the control curve will be. Semantic rules are expressed by the following membership functions and their parameters:

$$\mu(EM) = sigmf_1\,(d; -a, 3) \tag{8}$$

$$\mu(MM) = gbellmf_2\,(d; 2, b, 4) \tag{9}$$

$$\mu(MJ) = gbellmf_3\,(d; 1, b, 6) \tag{10}$$

$$\mu(JA) = gbellmf_4\,(d; 1, b, 7.5) \tag{11}$$

$$\mu(AO) = gbellmf_5\,(d; 1, b, 9) \tag{12}$$

$$\mu(OD) = gbellmf_6\,(d; a, 11) \tag{13}$$

Step 5 Upon having the Sugeno-Takagi-Kang, having the control rules defined by Eqs. (2, 3, 4, 5, 6, and 7), semantic with linguistic variables defined by membership functions defined in Eqs. (8, 9, 10, 11, 12, and 13), through the Fuzzy Logic Designer tool from Matlab, to apply this model. For this model values $\boldsymbol{a} = \boldsymbol{5}$ and $\boldsymbol{b} = \boldsymbol{1.5}$ were selected. This gives the curves shown in Fig. 3.

Step 6 In the same Fuzzy Logic Designer from Matlab, now the linear functions are introduced previously designed by Eqs. (2, 3, 4, 5 and 6), and only f_6 as a constant defined by Eq. (7), as previously defined. In Fig. 4 such functions are shown.

Step 7 Rules are defined. Thus, for example, if demand has an EM behavior, then the membership function corresponds to f_1, and so forth. This is shown in Fig. 5.

Step 8 To observe the control surface, which in this case, by being only one entry, is a control curve, representing the model, we will have as a result, what is shown on Fig. 6. As the obtained control curve is observed, it is very similar to the original; defined by the data table and shown in Fig. 1. Also, as observed in Fig. 7, the transitions are smooth, which is the purpose of the Sugeno-Takagi-Kang control model. If the intention was to smooth the transitions even more, what needs to be done is to reduce the values for ***a*** and ***b***. Finally, and for simulation purposes for the inference system, the rules can be visualized, where we can select the entry value for ***d***, and we could obtain the exit value for the forecasted capacity for that demand, as shown in Fig. 7. Thus, for example, before an entry of 5.73 of demand period, the output will be approximately 3690 from the required capacity, the red line represents the fuzzification values, and the result from the evaluation of functions $f_1 \ldots f_6$, and weigh is shown on the lower part of the graphic.

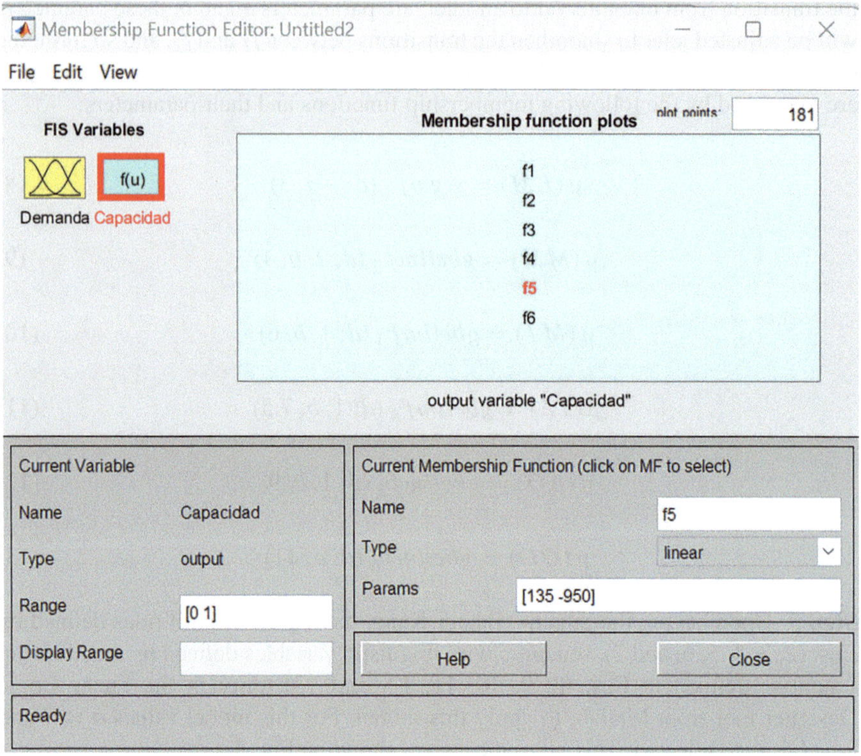

Fig. 4 Linear functions $f_1 \ldots f_6$ representation in Matlab

4.1 Performance Evaluation of the Model

Model performance may be evaluated directly from rule visualization and selecting the corresponding input. These results are set forth in Table 2. Also, the performance of the same may be observed directly in Fig. 8.

Also, there are many ways to evaluate the performance of a forecast [10]. Among the most utilized in the industry, MAPE was selected [11].

The mean absolute percentage error (MAPE) is defined by Eq. (14).

$$\text{MAPE} = \frac{100\%}{n} \sum_{j=1}^{n} \left| \frac{x_j - \hat{x}_j}{x_j} \right| \tag{14}$$

The result obtained from the forecasted data in Table 2. Was: MAPE = 30.42%, (9.95%, if we rule out the resulting value from August), which could be smoothen out even more. Significantly acceptable if we compare it to the results obtained by

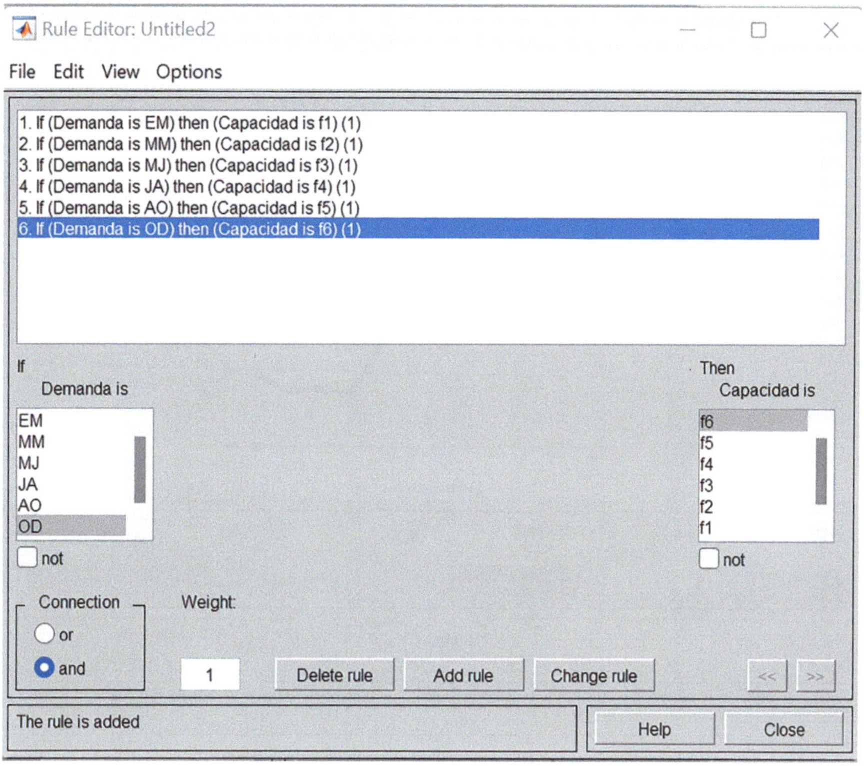

Fig. 5 Representation of model rules defined in step 3

more sophisticated techniques, such as data mining, which requires a very superior amount of information, obtaining barely better marginally results (5.1%) in [12].

4.2 Correct Implementation of the FTS Model Associated with Correct Supply Chain Demand Planning Decision-Making

The implementation of forecasting models in the supply chain is a critical task that significantly impacts strategic and operational decision-making. In this context, the use of fuzzy logic and the Sugeno-Takagi-Kang (STK) model offers an innovative approach to deal with uncertainty and improve accuracy in demand forecasting. This chapter proposes a fuzzy time series forecasting (FTS) model that incorporates the advantages of fuzzy logic, offering a robust methodology for supply chain resource planning. Demand forecasting is essential for optimizing inventory levels, reducing warehousing costs, and improving customer satisfaction. The ability to accurately

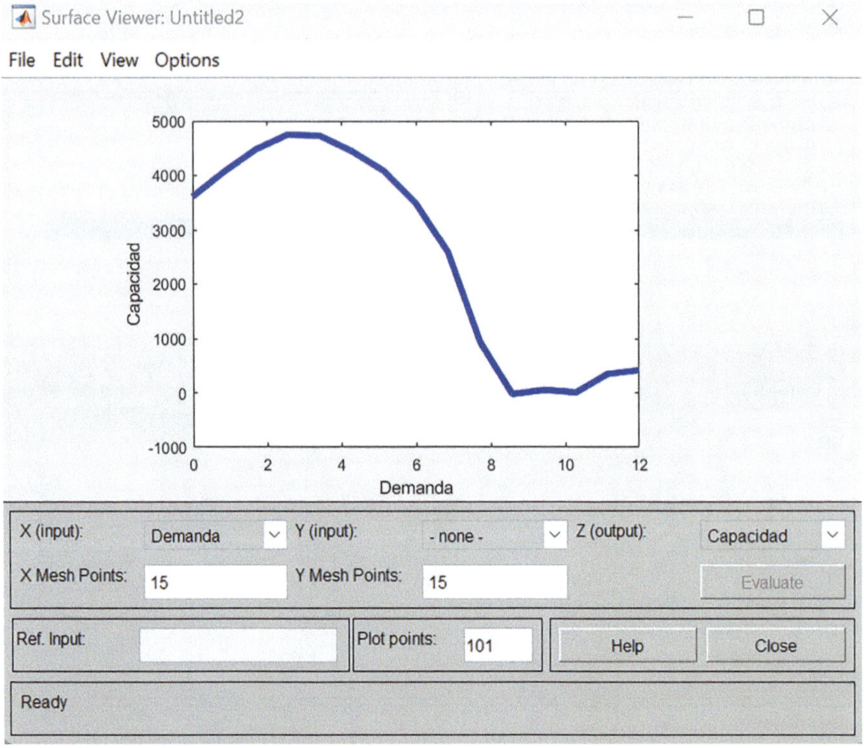

Fig. 6 Resulting control curve from membership functions from Eqs. (8, 9, 10, 11, 12, and 13) and rules from Eqs. (2, 3, 4, 5, 6, and 7)

forecast future demand enables companies to efficiently plan production, raw material procurement, and logistics, avoiding both inventory excess and shortages. One of the biggest challenges in demand forecasting is the availability of accurate and complete data. In many cases, the available data is limited or incomplete, making it difficult to apply traditional forecasting methods. Fuzzy logic, by allowing the handling of uncertainty and imprecision in the data, presents itself as a viable solution for these scenarios [13]. The STK model is based on fuzzy logic to transform imprecise data into useful information for decision making. This model uses fuzzy rules to combine different variables and generate accurate predictions. By applying the STK model to demand forecasting in the supply chain, it is possible to improve the accuracy of predictions even with limited data sets.

One of the strengths of the STK model is its ability to incorporate the knowledge and experience of experts in the field. This is achieved through the definition of fuzzy rules based on heuristics and previous experience, which allows the model to be adapted to different contexts and operating conditions. Results from the application of the FTS model show a significant improvement in prediction accuracy compared to traditional approaches as innovative and novel metaheuristics [17]. By

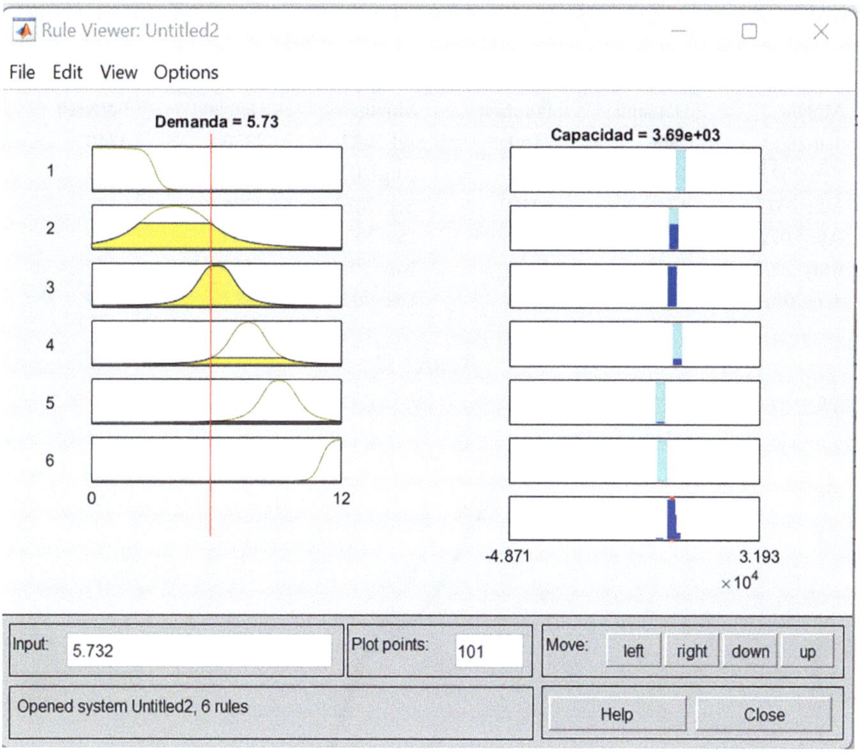

Fig. 7 Visualization of input analysis to the system, and the inference or forecast model

effectively managing uncertainty and leveraging expert knowledge, the FTS model offers a powerful tool for supply chain demand planning. The metric used to evaluate model performance, mean absolute percentage error (MAPE), demonstrates that the FTS model outperforms other methodologies in terms of accuracy. Therefore, the Decision-Making Implications, associated with the ability to accurately forecast demand, has direct implications for strategic and operational decision-making. Implementing the FTS model enables companies to make informed decisions based on accurate data and expert knowledge, resulting in improved resource planning, optimized inventory levels, and increased supply chain efficiency. In addition, the fuzzy logic-based approach allows companies to quickly adapt to changes in the environment and fluctuations in demand, ensuring continuous and efficient operation [14]. Therefore, the application of the Sugeno-Takagi-Kang model considering fuzzy logic for demand forecasting including time series as is proposed in [16] focuses on the supply chain and represents a significant advance in forecasting methodology. The combination of limited data with expert knowledge allows for more accurate forecasts and more informed decisions, improving the efficiency and effectiveness of the supply chain [15].

Table 2 Results for the Sugeno-Takagi-Kang forecast model, result from linear functions defined in Eqs. (2, 3, 4, 5, 6, and 7) and semantic rules defined in Eqs. (8, 9, 10, 11, 12, and 13)

Month	Demand	Forecast	Month	Demand	Forecast
Jan-2022	3806	4150	Jul-2022	2330	2340
Feb-2022	4449	4590	Aug-2022	128	429
Mar-2022	4746	4810	Sep-2022	224	172
Apr-2022	4259	4540	Oct-2022	404	457
May-2022	4089	4160	Nov-2022	410	302
Jun-2022	2859	3490	Dec-2022	413	419

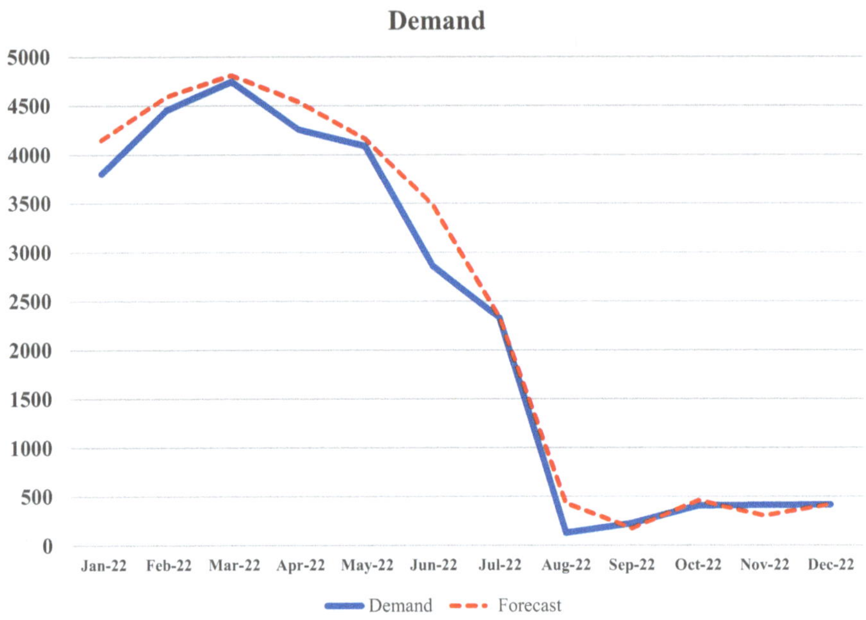

Fig. 8 Comparison in entry data and the inference or forecast model

5 Conclusions and Future Research

Since the Sugeno-Takagi-Kang model has been used, it is not necessary to make a data defuzzification, this saves time in calculation, and allows to establish several linear or constant functions to form the inference system. The mathematical requirement of these is minimum. According to Bose and Mali [10], there are two important issues related to the design of fuzzy temporary series which are necessary to perform more job: (1) number orders (past values) to be used, (2) number of intervals to be considered.

This research provides an answer to the second issue, establishing that so many intervals have to be created as linear functions are identified, to create an inference system to smoothen the transitions among them. Upon doing this, it allows to identify group or cluster time series to attenuate such effects which are present in data, and therefore avoid distortions in modeling. To determine the number of intervals representing these linear functions, one can approach through the eq. (1), and the subsequent criteria. The quality of the results may be bettered even more, depending on the smoothing between the overlaps of the linear functions, and if we consider the results obtained, these are equivalent to those obtained by other methods requiring a greater amount of information, as data mining is.

The nature of data generated during time is classified in general terms as stationary, seasonal, trend, and random. The type of data considered in this paper proposes a forecast model, considering that these may be stationary, with temporality or seasonal and tendency series, yet it makes especial emphasis in vagueness or uncertainty of these, this is, a great number of factors exists which are not known, or from which there is no certainty or more information, to make an inferred statistic. It is considered that most of the historical data generated over time in the real world is, despite not being statistically consistent, what is available for analysis and forecast purposes.

Appendix (Fig. 9)

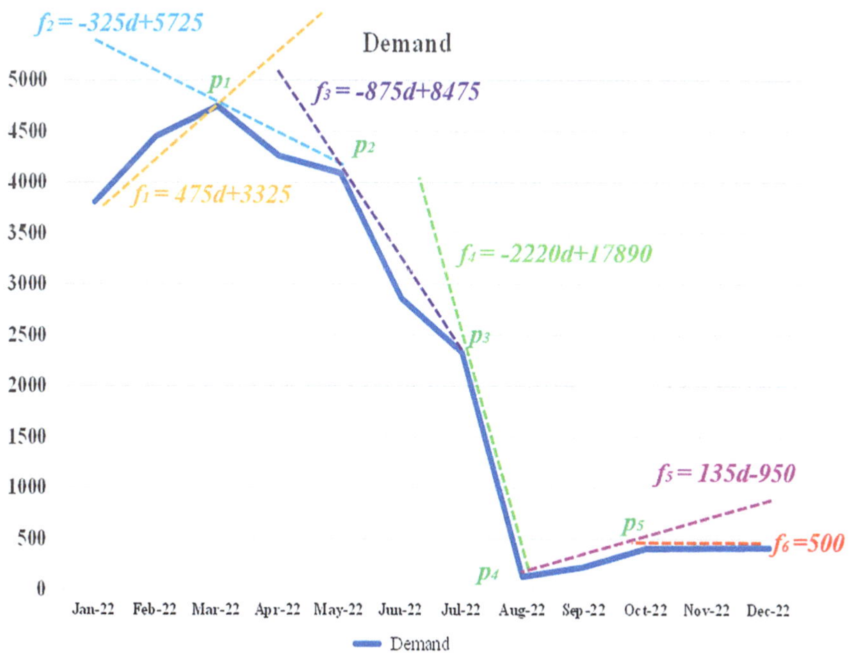

Fig. 9 Representation of control rules through linear functions (f_n) over the demand curve d

References

1. Takagi, T., & Sugeno, M. (1985). Fuzzy identification of systems and its applications to modeling and control. *IEEE Transactions on Systems, Man, and Cybernetics, 15*(1), 116–132.
2. Pilevari, N., SeyedHosseini, S., &. Jassbi, J. (2008). *Fuzzy logic supply chain agility assessment methodology*. In IEEE International conference on industrial engineering and engineering management (pp. 1113–1117).
3. Lin, C., & Chen, C. (2004). A fuzzy logic-based approach for new product. *IEEE Transactions on Systems and Humans, 34*, 134–142.
4. Dixit, A., & Jain, S. (2022). Intuitionistic fuzzy time series forecasting method for non-stationary time series data with suitable number of clusters and different window size for fuzzy rule generation. *Information Sciences, 623*, 132–145.
5. Aliev, R., Pedrycz, W., Guirimov, B., & Huseynov, O. (2020). Clustering method for production of Z-number based if-then rules. *Information Sciences, 520*, 155–176.
6. Zhou, Y., Ren, H., Zhao, D., Li, Z., & Pedrycz, W. (2022). A novel multi-level framework for anomaly detection in time series data. *Applied Intelligence*, 1–18.
7. Cortés Ríos, J. C., Ortiz de la Vega, H. A., Melin Olmeda, E. P., & Castillo López, O. (2017). *El algoritmo de sintonización simple de controladores difusos (ASSCD)*. Parmenia.
8. Ross, T. J. (2017). *Fuzzy logic with engineering applications*. Wiley.
9. Shah, M. (2012). Fuzzy based trend mapping and forecasting for time series data. *Expert Systems with Applications, 39*(7), 6351–6358.
10. Bose, M., & Mali, K. (2019). Designing fuzzy time series forecasting models: A survey. *International Journal of Approximate Reasoning, 111*, 78–99.
11. Merkuryeva, G., Valberga, A., & Smirnov, A. (2019). Demand forecasting in pharmaceutical supply chains: A case study. *Procedia Computer Science, 149*, 3–10.
12. Zhang, B., Tseng, M.-L., Qi, L., Guo, Y., & Wang, C.-H. (2022). A comparative online sales forecasting analysis: Data mining techniques. *Computer and Industrial Engineering, 173*.
13. Gooya, H. R., Katibeh, H., & Maleki, A. (2024). Forecasting groundwater fluctuations caused by earthquakes using fuzzy logic and AHP Method: A case study from Iran. *Earth Science Informatics, 17*(3), 2143–2158.
14. Peláez-Rodríguez, C., Pérez-Aracil, J., Marina, C. M., Prieto-Godino, L., Casanova-Mateo, C., Gutiérrez, P. A., & Salcedo-Sanz, S. (2024). A general explicable forecasting framework for weather events based on ordinal classification and inductive rules combined with fuzzy logic. *Knowledge-Based Systems, 291*, 111556.
15. Castillo, O., Castro, J. R., & Melin, P. (2023). Forecasting the COVID-19 with interval type-3 fuzzy logic and the fractal dimension. *International Journal of Fuzzy Systems, 25*(1), 182–197.
16. Gaxiola, F., Melin, P., Valdez, F., Castro, J. R. (2016). *Optimization with genetic algorithm and particle swarm optimization of type-2 fuzzy integrator for ensemble neural network in time series*. In FUZZ-IEEE 2016 (pp. 2067–2074).
17. Guajardo, H. M., & Valdez, F. (2024). Dragonfly algorithm for benchmark mathematical functions optimization. *Computación y Sistemas (CyS), 28*(2).

Index

A
Anomaly detection, 25, 27, 183–185, 187, 244, 245, 247, 248, 252–253, 256, 257
Artificial Intelligence (AI), 4, 23, 39, 58, 75, 92, 109, 133, 147, 160, 179, 190, 195, 211, 225, 278

B
Benign tumor, 111, 112, 114, 116, 118, 120, 147
Big data, 71–88
Bioinformatics, 160, 161
Blockchain, 6, 24, 216, 225, 246
Brain tumor detection, 148, 157

C
Case studies, 25, 34, 48–49, 64, 80–81, 84–85, 186–188, 225, 226, 230–234, 285
Computational tools, 161, 163, 168–171
Convolution neural network, 151, 152, 157, 163
Cryptography, 201–203, 217, 220
Cyber-physical systems, 39
Cybersecurity, 51, 52, 196, 197, 208, 211–222, 263–279

D
Data analysis, 26, 30–32, 52, 53, 59, 86, 92, 93, 167, 182, 184, 198, 203, 219, 229, 253, 274
Data analytics techniques, 4, 5, 23, 30, 34, 59, 63, 80, 83, 92–93, 160, 169, 178
Data integration, 25, 30, 34, 35, 82, 161
Decision support system (DSS), 40, 57, 110, 127, 162, 190, 195, 211, 225, 244, 264
Deep learning, 4, 6, 8, 10–17, 20, 26, 27, 66, 76, 91–106, 109, 110, 117, 121, 126, 128, 131, 132, 134, 147–157, 160, 163–165, 169–171, 175–191, 243–258
Diagnosis, 4, 62, 91–106, 109, 110, 121, 126, 127, 138, 147, 160, 161, 163, 170, 215
Digital transformation, 28, 39
Digital twins, 39–53, 61, 62, 245
Driven decision support systems, 49, 57–67, 110

E
Environmental management, 57–67
Ethical considerations, 66, 168

F
Finance, 197, 226, 229, 239, 273
Forecasting, 64, 87, 163, 227, 232, 233, 237, 239, 246, 283–297
Fuzzy logic, 283–297

G
Generative AI, 199

N. Gupta, J. D. Diaz Roman (eds.), *Decision Support Systems for Industrial and Scientific Applications*, Signals and Communication Technology, https://doi.org/10.1007/978-3-031-89700-9

H
Hardware security, 195–208
Health 4.0, 211–214, 217, 218, 221
Healthcare, 24, 25, 29–31, 33–35, 41, 80, 91–106, 110, 119–121, 127, 142, 159–171, 195, 198, 211, 212, 229
Healthcare engineering, 159–171

I
Industry 4.0, 41, 211–214, 216, 218, 221, 244–246
Intelligent decision support system (IDSS), 177–186, 188, 189, 197, 243, 244–253, 256–258
Intelligent traffic management (ITMS), 71–73, 75, 77, 79, 81, 82, 71–88
Intelligent transportation system (ITS), 3–20, 83, 86
Internet of things (IoT), 12, 23, 28, 39, 82, 190, 211, 221, 225, 226, 236, 245

L
Lung adenocarcinoma, 111, 112, 117, 118

M
Machine learning (ML), 4, 23, 40, 58, 67, 71, 92, 119, 127, 148, 160, 178, 195, 211, 248, 274
Medical decision support, 161
Medical imaging, 92, 93, 121, 127, 147, 157, 159, 163, 164, 171, 183
Melanoma, 91–93, 95, 96, 105
MRI Images, 147–157, 164

P
Peripheral arterial occlusive disease (PAOD), 125–144
Personalized medicine, 169
Predictive modeling, 30, 58, 163
Public transport system, 79–85

Q
Quantum computers, 212, 213, 218–222

R
Radiographic inspection, 178, 179, 185–191

S
Sensor fusion, 9, 16, 19, 27, 31, 32, 65
Sensor technologies, 5–8, 15, 20, 51, 163
Skin cancer, 91–95, 106
Smart manufacturing, 25, 30, 39–53, 181, 245, 246, 257
SMOTE, 92, 94, 96–97, 101–104, 106
Squamous cell lung carcinoma, 109, 118, 120
Sugeno-Takagi-Kang, 283–297
Super-resolution imaging, 125–144
Supply chain, 26, 33, 225–239, 243–258
Supply chain management (SCM), 225–239, 243, 245–247
System dynamics, 243–258

T
Transparency and security, 225–228, 234, 238

V
Vehicle classification system, 4–6, 10–12, 14, 15, 20

W
Weather data, 73, 74

MIX
Papier aus verantwortungsvollen Quellen
Paper from responsible sources
FSC® C105338

If you have any concerns about our products,
you can contact us on
ProductSafety@springernature.com

In case Publisher is established outside the EU,
the EU authorized representative is:
Springer Nature Customer Service Center GmbH
Europaplatz 3, 69115 Heidelberg, Germany

Printed by Libri Plureos GmbH
in Hamburg, Germany